ETHICS IN INFORMATION TECHNOLOGY

Third Edition

ETHICS IN INFORMATION TECHNOLOGY

Third Edition

George W. Reynolds

COURSE TECHNOLOGY
CENGAGE Learning

Australia • Brazil • Japan • Korea • Mexico • Singapore • Spain • United Kingdom • United States

COURSE TECHNOLOGY
CENGAGE Learning™

**Ethics in Information Technology,
Third Edition**
by George W. Reynolds

VP/Editorial Director: Jack Calhoun

Publisher: Joe Sabatino

Senior Acquisitions Editor:
  Charles McCormick Jr.

Senior Product Manager:
  Kate Hennessy Mason

Development Editor: Mary Pat Shaffer

Editorial Assistant: Nora Heink

Marketing Manager: Bryant Chrzan

Marketing Coordinator: Suellen Ruttkay

Content Product Manager: Jennifer Feltri

Senior Art Director: Stacy Jenkins Shirley

Cover Designer: Itzhack Shelomi

Cover Image: iStock Images

Technology Project Manager: Chris Valentine

Manufacturing Coordinator: Julio Esperas

Copyeditor: Green Pen Quality Assurance

Proofreader: Suzanne Huizenga

Indexer: Alexandra Nickerson

Composition: Pre-Press PMG

> For product information and technology assistance, contact us at
> **Cengage Learning Customer & Sales Support, 1-800-354-9706.**
>
> For permission to use material from this text or product,
> submit all requests online at **cengage.com/permissions**
> Further permissions questions can be emailed to
> **permissionrequest@cengage.com**

Microsoft, Windows 95, Windows 98, Windows 2000, Windows XP, and Windows Vista are registered trademarks of Microsoft® Corporation. Some of the product names and company names used in this book have been used for identification purposes only and may be trademarks or registered trademarks of their manufacturers and sellers.

Library of Congress Control Number: 2009935605

ISBN-13: 978-0-538-74622-9

ISBN-10: 0-538-74622-X

Course Technology
20 Channel Center Street
Boston, MA 02210
USA

Cengage Learning is a leading provider of customized learning solutions with office locations around the globe, including Singapore, the United Kingdom, Australia, Mexico, Brazil, and Japan. Locate your local office at: **international.cengage.com/region**

Cengage Learning products are represented in Canada by Nelson Education, Ltd.

For your lifelong learning solutions, visit **www.cengage.com/coursetechnology**

Visit our corporate Web site at **www.cengage.com**

Printed in the United States of America
2 3 4 5 6 7 13 12 11 10

BRIEF CONTENTS

TABLE OF CONTENTS

We are excited to publish the third edition of *Ethics in Information Technology*. This new edition builds on the success of the previous editions and meets the need for a resource that helps readers understand many of the legal, ethical, and societal issues associated with IT. We have responded to the feedback from our first and second edition adopters, students, and other reviewers to create an improved text. We think you will be pleased with the results.

Ethics in Information Technology, Third Edition, fills a void of practical business information for business managers and IT professionals. The typical introductory information systems book devotes one chapter to ethics and IT, which cannot possibly cover the full scope of ethical issues related to IT. Such limited coverage does not meet the needs of business managers and IT professionals—the people who are primarily responsible for addressing ethical issues in the workplace. What is missing is an examination of the different ethical situations that arise in IT as well as practical advice for addressing these issues.

Ethics in Information Technology, Third Edition, has enough substance for an instructor to use it in a full-semester course in computer ethics. You can also use the book as a reading supplement for such courses as Introduction to Management Information Systems, Principles of Information Technology, Managerial Perspective of Information Technology, Computer Security, E-Commerce, and so on.

WHAT'S NEW

Ethics in Information Technology, Third Edition, has been updated and revised to incorporate the many new developments and ethical issues that have arisen since the last edition. Two new chapters have been added: Chapter 9, "Social Networking," and Chapter 10, "Ethics of IT Organizations." All opening vignettes, as well as over 80 percent of the end-of-chapter cases, are new. Most of the real-world examples in each chapter have also been updated. Based on reviewer feedback, we have substantially increased the number of "Self-Assessment Questions" as well as the number of "What Would You Do?" exercises. We think you will like these changes and additions.

ORGANIZATION

Each of the 10 chapters in this book addresses a different aspect of ethics in information technology:

- Chapter 1, "An Overview of Ethics," provides an introduction to ethics, ethics in business, and the relevance of discussing ethics in IT. The chapter also discusses philosophical approaches to ethical decision making, and suggests a model for ethical decision making.

- Chapter 2, "Ethics for IT Workers and IT Users," explains the importance of ethics in the business relationships of IT professionals, and it discusses the roles that certification and licensing can play in legitimizing professional standards. The chapter also emphasizes the significance of IT professional organizations and their codes of ethics. The chapter touches on some ethical issues faced by IT users—including software piracy, inappropriate use of computing resources, and inappropriate sharing of information—and offers advice on how to support the ethical practices of IT users.

- Chapter 3, "Computer and Internet Crime," describes the types of ethical decisions that IT professionals must make, as well as the business needs they must balance when dealing with security issues. In addition to providing a useful classification of computer crimes and their perpetrators, the chapter explains both how to implement trustworthy computing to manage security vulnerabilities and how to respond to specific security incidents to fix problems quickly and improve ongoing security measures.

- Chapter 4, "Privacy," covers the issue of privacy, explains how the use of IT affects privacy rights, and discusses several key pieces of legislation that have addressed privacy rights over the years. The chapter explains how the personal information that businesses gather using IT can be used to obtain or keep customers (or to monitor employees). It also discusses the concerns of privacy advocates regarding how much information can be gathered, with whom it can be shared, how the information is gathered in the first place, and how it is used. These concerns also extend to the information-gathering practices of law enforcement and government.

- Chapter 5, "Freedom of Expression," addresses issues raised by the growing use of the Internet as a means for freedom of expression, while examining the types of speech that are protected by the First Amendment of the U.S. Constitution. The chapter also covers the ways in which the ease and anonymity with which Internet users communicate can pose problems for people who might be adversely affected by such communication, and it discusses attempts (using legislation and technology) to control access to Internet content that is unsuitable for children or unnecessary in a business environment.

- Chapter 6, "Intellectual Property," defines intellectual property and explains the varying degrees of ownership protection offered by copyright, patent, and trade secret laws. The chapter discusses several key issues that are relevant to ethics in IT, including plagiarism, reverse engineering of software, open source code, competitive intelligence gathering, and cybersquatting.

- Chapter 7, "Software Development," provides a thorough discussion of the software development process and the importance of software quality. The chapter covers issues that software manufacturers must consider when deciding "how good is good enough?" with regard to their software products—particularly when the software is safety-critical and its failure can cause loss of human life. Topics include product liability, risk analysis, different approaches to quality assurance testing, and processes and standards that may be adopted to achieve quality goals.

- Chapter 8, "The Impact of Information Technology on Productivity and the Quality of Life," examines the effect that IT has had on the standard of living and worker productivity in developed countries. The chapter also discusses the digital divide, and profiles some programs that are designed to close that gap. The chapter closes with a look at IT's impact on health care and healthcare costs.

- Chapter 9, "Social Networking," discusses how people use social networks, identifies common business uses of social networks, and examines many of the ethical issues associated with the use of social networks. The chapter also touches on virtual life communities and the ethical issues associated with virtual worlds.

- Chapter 10, "Ethics of IT Organizations," covers a range of ethical issues facing IT organizations, including those associated with the use of nontraditional workers, such as temporary workers, contractors, consulting firms, H-1B workers, and offshore-outsourced workers. The chapter discusses the risks, protections, and ethical decisions related to whistle-blowing, and presents a process for safely and effectively handling a whistle-blowing situation. In addition to introducing the concept of green computing, the chapter discusses the ethical issues that both IT manufacturers and IT users face when a company is considering how to transition to green computing—and at what cost. Finally, the chapter examines a code of conduct for the electronics and information and communications technology (ICT) industries designed to address ethical issues in the areas of worker safety and fairness, environmental responsibility, and business efficiency. The chapter explains what has been done so far and what still needs to be done to continue this work.

- Appendix A provides an in-depth discussion of how ethics and moral codes developed through time. Appendices B through F consist of the codes of ethics for several important IT professional organizations. Appendix G provides answers to the end-of-chapter Self-Assessment Questions.

PEDAGOGY

Ethics in Information Technology, Third Edition, employs a variety of pedagogical features to enrich the learning experience and provide interest for the instructor and student:

- **Opening Quotation.** Each chapter begins with a quotation to stimulate interest in the chapter material.
- **Vignette.** At the beginning of each chapter, a brief real-world example illustrates the issues to be discussed and piques the reader's interest.
- **Questions to Consider.** Carefully crafted focus questions follow the vignette to further highlight topics that are covered in the chapter.
- **Learning Objectives.** Learning objectives appear at the start of each chapter. They are presented in the form of questions for students to consider while reading the chapter.
- **Key Terms.** Key terms appear in bold in the text and are defined in the glossary at the end of the book.
- **Manager's Checklist.** Each checklist provides a practical and useful list of questions to consider when making a business decision.

End-of-Chapter Material

To help students retain key concepts and expand their understanding of important IT concepts and relationships, the following sections are included at the end of every chapter:

- **Summary.** Each chapter includes a summary of the key issues raised. These items relate to the Learning Objectives for each chapter.
- **Self-Assessment Questions.** These questions help students review and test their understanding of key chapter concepts. The answers to the Self-Assessment Questions are included in Appendix G.
- **Discussion Questions.** These more open-ended questions help instructors generate class discussion to move students deeper into the concepts and help them explore the numerous aspects of ethics in IT.
- **What Would You Do?** These exercises present realistic dilemmas that encourage students to think critically about the ethical principles presented in the text.
- **Cases.** In each chapter, three real-world cases reinforce important ethical principles and IT concepts, and show how real companies have addressed ethical issues associated with IT. Questions after each case focus students on its key issues and ask them to apply the concepts presented in the chapter.

ABOUT THE AUTHOR

George W. Reynolds brings a wealth of computer and industrial experience to this project, with more than 30 years of experience in government, institutional, and commercial IS organizations. He has authored over 20 texts and has taught at the University of Cincinnati, Xavier University (Ohio), Miami University (Ohio), and the College of Mount St. Joseph.

TEACHING TOOLS

The following supplemental materials are available when this book is used in a classroom setting. All of these tools are provided to the instructor on a single CD-ROM. You can also find some of these materials on the Cengage Learning Web site at *www.cengage.com/mis*.

- **Electronic Instructor's Manual.** The Instructor's Manual that accompanies this textbook includes additional instructional material to assist in class preparation, including suggestions for lecture topics. It also includes solutions to all end-of-chapter exercises.
- **ExamView®.** This textbook is accompanied by ExamView, a powerful testing software package that allows instructors to create and administer printed, computer (LAN-based), and Internet exams. ExamView includes hundreds of questions that correspond to the topics covered in this text, enabling students to generate detailed study guides that include page references for further review. The computer-based and Internet testing components allow students to take exams at their computers, and save the instructor time by grading each exam automatically.

- **PowerPoint Presentations.** This book comes with Microsoft PowerPoint slides for each chapter. The slides can be included as a teaching aid for classroom presentation, made available to students on the network for chapter review, or printed for classroom distribution. The slides are fully customizable. Instructors can either add their own slides for additional topics they introduce to the class or delete slides they won't be covering.
- **Figure Files.** Figure files allow instructors to create their own presentations using figures taken directly from the text.
- **Blackboard® and WebCT™ Level 1 Online Content.** If you use Blackboard or WebCT, the test bank for this textbook is available at no cost in a simple, ready-to-use format. Go to *www.cengage.com/coursetechnology* and search for this textbook to download the test bank.

ACKNOWLEDGMENTS

I want to thank a number of people who helped greatly in the creation of this book: Charles McCormick Jr., Senior Acquisitions Editor, for his belief in and encouragement of this project; Jennifer Feltri, Content Product Manager, for guiding the book through the production process; Kate Hennessy Mason, Senior Product Manager, for overseeing and directing this effort; Mary Pat Shaffer, Development Editor, for her tremendous support, many useful suggestions, and helpful edits; Naomi Friedman, for writing many of the vignettes and cases; and my many students at Miami University and the University of Cincinnati, who provided excellent ideas and constructive feedback on the text. I also wish to thank Clancy Martin for writing Appendix A. In addition, I want to thank a marvelous set of reviewers who offered many useful suggestions:

Valerie Akuna, Estrella Mountain Community College
Pat Artz, Bellevue University
Kathryn Baalman, Webster University–Saint Louis
Scott Burrell, Collins College
Kuan-Chou Chen, Purdue University Calumet
Elaine Crable, Xavier University
Becky Cunningham, Arkansas Tech University
Lee Falta, Bellingham Technical College
Maria Raquel Garcia, Franklin Pierce University
Jonathan Jelen, Polytechnic Institute of NYU
Cynthia Jensen, Jacksonville State University
Diana Johnson, The College of St. Scholastica
Carol Kalen, Rockford College
Chuck Lund, Central Lakes College
Elizabeth McCarthy, Kirkwood Community College
Michael Nelson, Hodges University
Nancy O'Geary-Smith, Kirkwood Community College
Richard Smith, The University of Findlay
Sandy Vitale, McHenry County College

I'd also like to thank the reviewers who helped shape the second edition:

Karen Williams, University of Texas at San Antonio
Catharine Kuchar, Southern Institute for Business and Professional Ethics, Decatur, GA
Cherie Ann Sherman, Ramapo College of New Jersey
Jeff Stewart, PhD, Macon State College

Last of all, thanks to my family for all their support, and for giving me the time to write this text.

—George W. Reynolds

ETHICS IN INFORMATION TECHNOLOGY

Third Edition

CHAPTER **1**

AN OVERVIEW OF ETHICS

VIGNETTE

Dubious Methods Used to Investigate Leaks by Hewlett-Packard Board Members

On September 5, 2006, *Newsweek* published a story revealing that Hewlett-Packard (HP) chairman Patricia Dunn authorized an internal investigation of HP board members beginning sometime in 2005. Dunn suspected one or more board members of leaking information about HP's long-term strategy to the news media. One such article appeared on the technology Web site CNET in January 2006.[2] Making such confidential information public could have a significant impact on the competitiveness of the company and impact its share price.

Three private detectives involved in the investigation allegedly engaged in **pretexting**—the use of false pretenses—to gain access to the telephone records of HP directors, certain employees, and nine journalists. The detectives allegedly obtained and used the targeted individuals' Social Security numbers to impersonate those individuals in calls to the phone company, with the goal of obtaining private phone records.

The California state attorney general filed criminal charges against Dunn, Kevin Hunsaker—senior legal counsel and director of ethics and standards of business conduct—and the three outside investigators.[3] Dunn admitted that she oversaw the investigation but said she never had any knowledge that illegal methods were used. All charges were eventually dropped against Dunn. Hunsaker and two of the detectives involved pleaded no contest to fraudulent wire communications, a misdemeanor. The judge agreed to drop the charges if they completed 96 hours of community service. The third detective charged in the case agreed to act as a witness for the prosecution and thus escaped criminal proceedings.[4]

Eventually, the state settled a separate civil complaint against the company. HP agreed to pay $14.5 million to cover fines and legal costs, and also agreed to strengthen its corporate governance practices. The settlement did not involve any admission or conclusion of guilt on the part of HP.[5]

When the scandal broke, one HP board member, Tom Perkins, resigned to protest the methods Dunn used in the investigation. In the aftermath, Dunn and Hunsaker also resigned from the board. George Keyworth, a 21-year member of the HP board of directors, was identified by Dunn as being the one who leaked the information, and he also resigned.

Throughout the scandal, investors continued to show faith in HP, and the price of the stock rose steadily from a level of $32 per share in July 2006 to over $43 per share in early January 2007.

Questions to Consider

1. Which issue is more disconcerting—the fact that a board member leaked confidential information about the firm or the tactics used to investigate the leak? Defend your position.
2. Can the use of pretexting to gain information ever be justified? Is it considered legal under any circumstances?

WHAT IS ETHICS?

Each society forms a set of rules that establishes the boundaries of generally accepted behavior. These rules are often expressed in statements about how people should behave, and they fit together to form the **moral code** by which a society lives. Unfortunately, the different rules often have contradictions, and people are sometimes uncertain about which rule to follow. For instance, if you witness a friend copy someone else's answers while taking an exam, you might be caught in a conflict between loyalty to your friend and the value of telling the truth. Sometimes the rules do not seem to cover new situations, and an individual must determine how to apply existing rules or develop new ones. You may strongly support personal privacy, but what rules do you think are acceptable for governing the appropriate use of company resources, such as e-mail and Internet access?

The term **morality** refers to social conventions about right and wrong that are so widely shared that they become the basis for an established consensus. However, individual views of what is moral may vary by age, cultural group, ethnic background, religion, life experiences, education, and gender. There is widespread agreement on the immorality of murder, theft, and arson, but other behaviors that are accepted in one culture might be unacceptable in another. Even within the same society, people can have strong disagreements over important moral issues. In the United States, for example, issues such as abortion, the death penalty, and gun control are continuously debated, and both sides feel that their arguments are on solid moral ground.

Definition of Ethics

Ethics is a set of beliefs about right and wrong behavior within a society. Ethical behavior conforms to generally accepted norms—many of which are almost universal. However, although nearly everyone would agree that lying and cheating are unethical, opinions about what constitutes ethical behavior often vary dramatically. For example, attitudes toward **software piracy**—that is, the practice of illegally making copies of software or enabling others to access software to which they are not entitled—range from strong opposition to acceptance of the practice as a standard approach to conducting business. In 2007, 38 percent of all software in circulation worldwide was pirated—at a cost of nearly $48 billion (USD). The highest piracy rate—93 percent—was in Armenia; Bangladesh, Azerbaijan, and

Moldova all had piracy rates of 92 percent. The lowest piracy rates were in the United States (20%), Luxembourg (21%), and New Zealand (22%).[6]

As children grow, they learn complicated tasks—walking, talking, swimming, riding a bike, writing the alphabet—that they perform out of habit for the rest of their lives. People also develop habits that make it easier to choose between what society considers good or bad. **Virtues** are habits that incline people to do what is acceptable, and **vices** are habits of unacceptable behavior. Fairness, generosity, and loyalty are examples of virtues, while vanity, greed, envy, and anger are considered vices. People's virtues and vices help define their personal value system—the complex scheme of moral values by which they live.

The Importance of Integrity

Your moral principles are statements of what you believe to be rules of right conduct. As a child, you may have been taught not to lie, cheat, or steal. As an adult facing more complex decisions, you often reflect on your principles when you consider what to do in different situations: Is it okay to lie to protect someone's feelings? Should you intervene with a coworker who seems to have a chemical dependency problem? Is it acceptable to exaggerate your work experience on a résumé? Can you cut corners on a project to meet a tight deadline?

A person who acts with **integrity** acts in accordance with a personal code of principles. One approach to acting with integrity—one of the cornerstones of ethical behavior—is to extend to all people the same respect and consideration that you expect to receive from others. Unfortunately, consistency can be difficult to achieve, particularly when you are in a situation that conflicts with your moral standards. For example, you might believe it is important to do as your employer requests while also believing that you should be fairly compensated for your work. Thus, if your employer insists that you do not report the overtime hours that you have worked due to budget constraints, a moral conflict arises. You can do as your employer requests or you can insist on being fairly compensated, but you cannot do both. In this situation, you may be forced to compromise one of your principles and act with an apparent lack of integrity.

Another form of inconsistency emerges if you apply moral standards differently according to the situation or people involved. To be consistent and act with integrity, you must apply the same moral standards in all situations. For example, you might consider it morally acceptable to tell a little white lie to spare a friend some pain or embarrassment, but would you lie to a work colleague or customer about a business issue to avoid unpleasantness? Clearly, many ethical dilemmas are not as simple as right versus wrong but involve choices between right versus right. As an example, for some people it is "right" to protect the Alaskan wildlife from being spoiled and also "right" to find new sources of oil to maintain U.S. reserves, but how do they balance these two concerns?

The Difference Between Morals, Ethics, and Laws

Morals are one's personal beliefs about right and wrong, while the term ethics describes standards or codes of behavior expected of an individual by a group (nation, organization, profession) to which an individual belongs. For example, the ethics of the law profession demand that defense attorneys defend an accused client to the best of their ability, even if they know that the client is guilty of the most heinous and morally objectionable crime one could imagine.

Law is a system of rules that tells us what we can and cannot do. Laws are enforced by a set of institutions (the police, courts, law-making bodies). Legal acts are acts that conform to the law. Moral acts conform with what an individual believes to be the right thing to do. Laws can proclaim an act as legal, although many people may consider the act immoral—for example, abortion.

The remainder of this chapter provides an introduction to ethics in the business world. It discusses the importance of ethics in business, outlines what businesses can do to improve their ethics, provides advice on creating an ethical work environment, and suggests a model for ethical decision making. The chapter concludes with a discussion of ethics as it relates to information technology (IT).

ETHICS IN THE BUSINESS WORLD

Ethics has risen to the top of the business agenda because the risks associated with inappropriate behavior have increased, both in their likelihood and in their potential negative impact. In the past decade, we have seen the failure of major corporations such as Enron and WorldCom due to accounting scandals. We have watched the collapse of financial institutions due to unwise and unethical decision making over the approval of mortgages and lines of credit to unqualified individuals and organizations. We have also witnessed numerous corporate officers and senior managers sentenced to prison terms for their unethical behavior. Clearly, unethical behavior has led to serious negative consequences that have had a major global impact.

Several trends have increased the likelihood of unethical behavior. First, for many organizations, greater globalization has created a much more complex work environment that spans diverse cultures and societies, making it much more difficult to apply principles and codes of ethics consistently. For example, numerous U.S. companies have garnered negative publicity for moving operations to third-world countries, where employees work in conditions that would not be acceptable in most developed parts of the world.

Second, in today's recessionary economic climate, organizations are extremely challenged to maintain revenue and profits. Some organizations are sorely tempted to resort to unethical behavior to maintain profits. For example, the Peanut Corporation of America allegedly shipped tainted products from its plant in Georgia, which led to a salmonella outbreak in 2008 that killed at least eight people and sickened over 550 people in 43 states.[7]

Employees, shareholders, and regulatory agencies are increasingly sensitive to violations of accounting standards, failures to disclose substantial changes in business conditions, nonconformance with required health and safety practices, and production of unsafe or substandard products. Such heightened vigilance raises the risk of financial loss for businesses that do not foster ethical practices or that run afoul of required standards. There is also a risk of criminal and civil lawsuits resulting in fines and/or incarceration for individuals.

A classic example of the many risks of unethical decision making can be found in the Enron accounting scandal. In 2000, Enron employed over 22,000 people and had annual revenue of $101 billion. During 2001, it was revealed that much of Enron's revenue was the result of deals with limited partnerships, which it controlled. In addition, as a result of faulty accounting, many of Enron's debts and losses were not reported in its financial statements. As the accounting scandal unfolded, Enron shares dropped from $90 per share to less than $1 per share, and the company was forced to file for bankruptcy in December 2001. The Enron case was notorious, but many other corporate scandals have recently occurred in spite of safeguards enacted as a result of the Enron debacle. Here are just a few examples from around the world:

- As discussed in the opening vignette, Hewlett-Packard hired investigators to identify members of its board of directors who were responsible for leaking confidential company information to the press.
- Comverse Technology develops, designs, manufactures, and supports computer and telecommunications systems for multimedia communications. It provides its products to more than 500 service providers in over 130 countries. In 2008, the company's founder and former CEO Jacob Alexander admitted he backdated Comverse stock options for personal gain and is being sued by the firm for $70 million.[8]
- The chairman of the India-based outsourcing firm Satyam Computer Services admitted he falsified his firm's profits for years—by as much as $1 billion. The revelation represents India's largest ever corporate scandal.[9]
- Two separate scandals, just weeks apart in 2006, caused China to lose ground in its effort to catch up with advanced countries in building its own computer chips. Work on the Haxin chip—a digital signal processing chip—was halted and its lead designer disgraced when it was revealed to be a copy of a U.S. design. Shortly after, the state-supported R & D effort to build the Arca-3 CPU chip had to be delayed when it was uncovered that the research funds had been embezzled.[10]

This is just a small sample of the scandals that have led to an increased focus on business ethics within many IT organizations.

Why Fostering Good Business Ethics Is Important

Organizations have at least five good reasons for promoting a work environment in which employees are encouraged to act ethically when making business decisions:

1. Gaining the good will of the community
2. Creating an organization that operates consistently
3. Fostering good business practices
4. Protecting the organization and its employees from legal action
5. Avoiding unfavorable publicity

Gaining the Good Will of the Community

Although organizations exist primarily to earn profits or provide services to customers, they also have some fundamental responsibilities to society. Often they declare these responsibilities in a formal statement of their company's principles or beliefs (see Figure 1-1 for an example).

> **Our Values**
>
> As a company, and as individuals, we value integrity, honesty, openness, personal excellence, constructive self-criticism, continual self-improvement, and mutual respect. We are committed to our customers and partners and have a passion for technology. We take on big challenges, and pride ourselves on seeing them through. We hold ourselves accountable to our customers, shareholders, partners, and employees by honoring our commitments, providing results, and striving for the highest quality.

Source: Accessed at www.microsoft.com/about/default.mspx.

FIGURE 1-1   Microsoft's statement of values

"Technology companies are waking up to the fact that they have to attract and maintain loyalty with their customers," says Carol Cone, head of Boston marketing consulting firm Cone, Inc., which helps develop corporate giving programs. "Philanthropy allows a company to demonstrate its values in action and present a human face to its stakeholders."[11] (A **stakeholder** is someone who stands to gain or lose, depending on how a situation is resolved.) As a result, many organizations initiate or support socially responsible activities, which include making contributions to charitable organizations and nonprofit institutions, providing benefits for employees in excess of any legal requirements, and devoting organizational resources to initiatives that are more socially desirable than profitable. Table 1-1 provides a few examples of some of the many socially responsible activities supported by major IT organizations.

TABLE 1-1   Examples of IT organizations' socially responsible activities

Organization	Example of socially responsible activity
Google, Inc.	Donated $33 million in free ads to nonprofit organizations in low-income areas[12]
Hewlett-Packard	Employees work to implement technology solutions to benefit residents of central city communities[13]
IBM	Awards millions of dollars of grants each year to support the arts
Intel	Supplied 100,000 computers to schools in low-income neighborhoods[14]
Microsoft	Matches its employees' direct contributions to thousands of nonprofit organizations[15]
SAP America	Awards up to nine undergraduate students a $10,000 scholarship each academic year[16]
Yahoo!	Allows employees to take time off to develop technology applications to aid charitable organizations[17]

The good will that socially responsible activities create can make it easier for corporations to conduct their business. For example, a company known for treating its employees well will find it easier to compete for the best job candidates. On the other hand, companies viewed as harmful to their community may suffer a disadvantage. For example, a corporation that pollutes the air and water (see Figure 1-2) may find that adverse publicity reduces sales, impedes relationships with some business partners, and attracts unwanted government attention.

FIGURE 1-2   Corporations that harm a community place themselves at a disadvantage

Creating an Organization That Operates Consistently

Organizations develop and abide by values to create an organizational culture and to define a consistent approach for dealing with the needs of their stakeholders—shareholders, employees, customers, suppliers, and the community. Such consistency means that employees know what is expected of them and can employ the organization's values to help them in their decision making. Consistency also means that shareholders, customers, suppliers, and the community know what they can expect of the organization—that it will behave in the future much as it has in the past. It is especially important for multinational or global organizations to present a consistent face to their shareholders, customers, and suppliers no matter where those stakeholders live or operate their business. Although each company's value system is different, many share the following values:

- Operate with honesty and integrity, staying true to organizational principles
- Operate according to standards of ethical conduct, in words and action
- Treat colleagues, customers, and consumers with respect
- Strive to be the best at what matters most to the organization
- Value diversity
- Make decisions based on facts and principles

Fostering Good Business Practices

In many cases, good ethics can mean good business and improved profits. Companies that produce safe and effective products avoid costly recalls and lawsuits. Companies that provide excellent service retain their customers instead of losing them to competitors. Companies that develop and maintain strong employee relations suffer lower turnover rates and enjoy better employee morale. Suppliers and other business partners often place a priority on working with companies that operate in a fair and ethical manner.

On the other hand, bad ethics can lead to bad business results. For example, according to the American Customer Satisfaction Index—an annual quality survey conducted by the University of Michigan—consumers rated their satisfaction with their personal computer manufacturer 3 percent lower in 2007 than in 2006.[18] Dell was especially hard hit. Its rating dropped 5 percent due to customers' perception of service and product reliability, a drop that likely cost it millions of dollars in lost sales. While this drop in customer satisfaction was generally linked to service and reliability on the survey, a lawsuit filed by New York Attorney General Andrew Cuomo proved that some of those issues were actually the result of deliberate consumer fraud on the part of Dell. In May 2007, a judge found that Dell failed to provide the timely on-site technical support that customers were entitled to, frustrated customers seeking telephone tech support through long wait times and frequent transfers, failed to provide promised rebates, and failed to deliver on promises of no-interest-rate financing. At one time Dell was the world's largest personal computer manufacturer, but over a period of a few years it lost that spot to Hewlett-Packard, with many industry experts citing poor customer service as a major factor.[19]

Likewise, bad ethics can have a negative impact on employees, many of whom can develop negative attitudes if they perceive a difference between their own values and those stated or implied by an organization's actions. In such an environment, employees may suppress their tendency to act in a manner that seems ethical to them and instead act in a manner that will protect them against anticipated punishment. When such a discrepancy between employee and organizational ethics occurs, it destroys employee commitment to organizational goals and objectives, creates low morale, fosters poor performance, erodes employee involvement in organizational improvement initiatives, and builds indifference to the organization's needs.

Protecting the Organization and Its Employees from Legal Action

In a 1909 ruling (*United States v. New York Central & Hudson River Railroad Co.*), the U.S. Supreme Court established that an employer can be held responsible for the acts of its employees even if the employees act in a manner contrary to corporate policy and their employer's directions.[20] The principle established is called *respondeat superior*, or "let the master answer." An example of the application of this principle can be found in the collapse in 2002 of Arthur Andersen, one of the "Big Five" international accounting firms. Andersen was indicted by the Department of Justice for obstruction of justice for the shredding of documents associated with the auditing work that a few of its partners performed for Enron.[21] Andersen was forced to relinquish its auditing license. It closed its U.S. offices due to lack of clients, and some 26,000 employees lost their jobs.

A coalition of several legal organizations, including the Association of Corporate Counsel, the U.S. Chamber of Commerce, the National Association of Manufacturers, the National Association of Criminal Defense Lawyers, and the New York State Association of

Criminal Defense Lawyers, argues that organizations should "be able to escape criminal liability if they have acted as responsible corporate citizens, making strong efforts to prevent and detect misconduct in the workplace."[22] One way to do this is to establish effective ethics and compliance programs.

Indeed, in 1991, the Department of Justice established sentencing guidelines that suggest more lenient treatment for convicted executives if their companies have ethics programs. Fines for criminal violations can be lowered by up to 80 percent if the organization has implemented an ethics management program and cooperates with authorities.[23]

Avoiding Unfavorable Publicity

The public reputation of a company strongly influences the value of its stock, how consumers regard its products and services, the degree of oversight it receives from government agencies, and the amount of support and cooperation it receives from its business partners. Thus, many organizations are motivated to build a strong ethics program to avoid negative publicity. If an organization is perceived as operating ethically, customers, business partners, shareholders, consumer advocates, financial institutions, and regulatory bodies will usually regard it more favorably.

Improving Corporate Ethics

Research by the Ethics Resource Center found that only one in four organizations has a well-implemented ethics and compliance program. The Ethics Resource Center has defined the following characteristics of a successful ethics program:

- Employees are willing to seek advice about ethics issues.
- Employees feel prepared to handle situations that could lead to misconduct.
- Employees are rewarded for ethical behavior.
- The organization does not reward success obtained through questionable means.
- Employees feel positively about their company.[24]

The risk of unethical behavior is increasing, so the improvement of business ethics is becoming more important. The following sections explain some of the actions corporations can take to improve business ethics.

Appointing a Corporate Ethics Officer

A **corporate ethics officer** (also called a **corporate compliance officer**) provides an organization with vision and leadership in the area of business conduct. Organizations send a clear message to employees about the importance of ethics and compliance in their decision about who will be in charge of the effort and to whom that individual will report. Ideally, the corporate ethics officer should be a well-respected, senior-level manager who reports directly to the CEO. Ethics officers come from diverse backgrounds, such as legal staff, human resources, finance, auditing, security, or line operations.

The presence of a corporate ethics officer has become increasingly common. Not surprisingly, a rapid increase in the appointment of corporate ethics officers typically follows the revelation of a major business scandal. The first flurry of appointments began following a series of defense-contracting scandals during the administration of Ronald Reagan—when firms used bribes to gain inside information that they could use to improve their contract bids. A second spike in appointments came in the early 1990s, following the new U.S. sentencing guidelines that promised decreased fines for firms that adopted ethics programs. A third surge followed the myriad accounting scandals of the early 2000s. Another increase in appointments can be expected in the aftermath of the mortgage loan scandals uncovered in 2008.

Typically the ethics officer tries to establish an environment that encourages ethical decision making through the actions described in this chapter. Specific responsibilities include:

- Responsibility for compliance—that is, ensuring that ethical procedures are put into place and consistently adhered to throughout the organization
- Responsibility for creating and maintaining the ethics culture that the highest level of corporate authority wishes to have
- Responsibility for being a key knowledge and contact person on issues relating to corporate ethics and principles[25]

Unfortunately, simply naming a corporate ethics officer does not automatically improve an organization's ethics; hard work and effort are required to establish and provide ongoing support for an organizational ethics program.

Ethical Standards Set by Board of Directors

The board of directors is responsible for the careful and responsible management of an organization. In a for-profit organization, the board's primary objective is to oversee the organization's business activities and management for the benefit of all stakeholders, including shareholders, employees, customers, suppliers, and the community. In a nonprofit organization, the board reports to a different set of stakeholders, particularly the local community that the nonprofit serves.

The board fulfills some of its responsibilities directly and assigns others to various committees. The board is not normally responsible for day-to-day management and operations; these responsibilities are delegated to the organization's management team. However, the board is responsible for supervising the management team.

Board members are expected to conduct themselves according to the highest standards for personal and professional integrity, while setting the standard for company-wide ethical conduct and ensuring compliance with laws and regulations. Employees will "get the message" if board members set an example of high-level ethical behavior. If they don't set a good example, employees will get that message as well. Importantly, board members must create an environment in which employees feel they can seek advice about appropriate business conduct, raise issues, and report misconduct through appropriate channels. Unfortunately, while nearly half of all employees surveyed saw some form of ethical misconduct in 2007, less than 60 percent of those employees reported the misconduct to management, primarily because they feared retaliation of some kind or felt that no action would be taken even if they did report it. Regrettably, one in eight employees did experience some form of retaliation.[26]

Establishing a Corporate Code of Ethics

A **code of ethics** is a statement that highlights an organization's key ethical issues and identifies the overarching values and principles that are important to the organization and its decision making. The code frequently includes a set of formal, written statements about the purpose of the organization, its values, and the principles that should guide its employees' actions. An organization's code of ethics applies to its directors, officers, and employees. The code of ethics focuses employees on areas of ethical risk relating to their role in the organization, offers guidance to help them recognize and deal with ethical issues, and provides mechanisms for reporting unethical conduct and fostering a culture of honesty and accountability within the organization. The code of ethics helps ensure that employees abide by the law, follow necessary regulations, and behave in an ethical manner.

The Sarbanes-Oxley Act of 2002 was passed in response to public outrage over several major accounting scandals, including those at Enron, WorldCom, Tyco, Adelphia, Global Crossing, and Qwest—plus numerous restatements of financial reports by other companies, which clearly demonstrated a lack of oversight within corporate America. The goal of the bill was to renew investors' trust in corporate executives and their firm's financial reports. The act led to significant reforms in the content and preparation of disclosure documents by public companies.

For example, Section 404 of the act states that annual reports must contain a statement signed by the CEO and CFO attesting that the information contained in all of the firm's Securities and Exchange Commission filings is accurate. The company must also submit to an audit to prove that it has controls in place to ensure accurate information. The penalties for false attestation can include up to 20 years in prison and significant monetary fines for senior executives. Section 406 of the act also requires public companies to disclose whether they have a code of ethics and to disclose any waiver of the code for certain members of senior management. The Securities and Exchange Commission also approved significant reforms by the NYSE and NASDAQ that, among other things, require companies listed with them to have codes of ethics that apply to all employees, senior management, and directors.

A code of ethics cannot gain company-wide acceptance unless it is developed with employee participation and fully endorsed by the organization's leadership. It must also be easily accessible by employees, shareholders, business partners, and the public. The code of ethics must continually be applied to a company's decision making and emphasized as an important part of its culture. Breaches in the code of ethics must be identified and dealt with appropriately so that the code's relevance is not undermined.

In March 2007, *Business Ethics* magazine rated U.S.-based, publicly held companies based on a statistical analysis of corporate service to seven stakeholder groups—employees, customers, community, minorities and women, shareholders, the environment, and non-U.S. stakeholders. The top IT company, based on performance between 2000 and 2007, was Intel Corporation, the world's largest computer chip maker. A summary of Intel's code of ethics is shown in Figure 1-3. A more detailed version is spelled out in a 22-page document (Intel Code of Conduct, May 2007, found at *www.intel.com/intel/finance/docs/code-of-conduct.pdf*), which offers employees guidelines designed to deter wrongdoing, promote honest and ethical conduct, and comply with applicable laws and regulations. Intel's Code of Conduct also expresses its policies regarding the environment, health and safety, intellectual property, diversity, nondiscrimination, supplier expectations, privacy, and business continuity.

1. Intel conducts business with honesty and integrity
2. Intel follows the letter and spirit of the law
3. Intel employees treat each other fairly
4. Intel employees act in the best interests of Intel and avoid conflicts of interest
5. Intel employees protect the company's assets and reputation

Source: Intel's public conduct code for employees, published in Intel's Social Media Guidelines, www.intel.com/sites/
sitewide/en_US/social-media.htm (accessed January 26, 2009).

FIGURE 1-3   Intel's five principles of conduct

Conducting Social Audits

An increasing number of organizations conduct social audits of their policies and practices. In a **social audit**, an organization reviews how well it is meeting its ethical and social responsibility goals, and communicates its new goals for the upcoming year. This information is shared with employees, shareholders, investors, market analysts, customers, suppliers, government agencies, and the communities in which the organization operates. For example, each year Intel prepares its Corporate Responsibility Report, which summarizes the firm's progress toward meeting its ethical and social responsibility goals. A partial summary of this report is presented in Table 1-2.

TABLE 1-2   Partial Intel 2007 Corporate Responsibility Report

2007 Goals	2007 Performance	Results
Audit 20% of our suppliers who may be at high risk for nonconformance to the EICC	We did not reach our 20% goal; challenges included industry-wide supplier classification and auditor training	Did not meet goal
Reduce greenhouse gas emissions per production unit by 30% from 2004 levels by 2010	Goal remains on track; absolute greenhouse gas emissions were down 6%	Met our goal
Reduce water usage per production unit below 2005 levels by 2010	Absolute water use was down 2%; usage was up 4% per chip	Did not meet goal
Recycle more than 70% of both chemical and solid waste generated from our worldwide facilities	In 2007, Intel recycled 89% of the solid waste and 87% of the chemical waste generated at our facilities worldwide	Met our goal
Empower students and teachers by donating 20,000 computers to schools in developing nations	Donated 27,000 full-featured PCs with Internet connectivity to more than 500 schools in 22 countries as part of our education donation program	Met our goal

Source: Intel 2007 Corporate Responsibility Report, www.intel.com/go/responsibility.

Requiring Employees to Take Ethics Training

The ancient Greek philosophers believed that personal convictions about right and wrong behavior could be improved through education. Today, most psychologists agree with them. Lawrence Kohlberg, the late Harvard psychologist, found that many factors stimulate a person's moral development, but one of the most crucial is education. Other researchers have repeatedly supported the idea that people can continue their moral development through further education, such as working case studies and examining contemporary issues.

Thus, an organization's code of ethics must be promoted and continually communicated within the organization, from top to bottom. Organizations can do this by showing employees examples of how to apply the code of ethics in real life. One approach is through a comprehensive ethics education program that encourages employees to act responsibly and ethically. Such programs are often presented in small workshop formats in which employees apply the organization's code of ethics to hypothetical but realistic case studies. Employees may also be given examples of recent company decisions based on principles from the code of ethics. It is critical that such training increase the percentage of employees who report incidents of misconduct; thus, employees must be shown effective ways of reporting such incidents. In addition, they must be reassured that such feedback will be acted on and that they will not be subjected to retaliation.

Ethics training not only makes employees more aware of a company's code of ethics and how to apply it, but also demonstrates that the company intends to operate in an ethical manner. The existence of formal training programs can also reduce a company's liability in the event of legal action.

Including Ethical Criteria in Employee Appraisals

Managers can ensure that employees are meeting performance expectations if they monitor employee behavior and provide feedback; however, a recent survey of HR professionals revealed that only 43 percent of organizations include ethical conduct as part of an employee's performance appraisal.[27] Those that do so base a portion of their employees' performance evaluations on treating others fairly and with respect; operating effectively in a multicultural environment; accepting personal accountability for meeting business needs; continually developing others and themselves; and operating openly and honestly with suppliers, customers, and other employees. These factors are considered along with the more traditional criteria used in performance appraisals, such as an employee's overall contribution to moving the business ahead, successful completion of projects and tasks, and maintenance of good customer relations.

Creating an Ethical Work Environment

Most employees want to perform their jobs successfully and ethically, but good employees sometimes make bad ethical choices. Employees in highly competitive workplaces often feel pressure from aggressive competitors, cutthroat suppliers, unrealistic budgets, unforgiving quotas, tight deadlines, and bonus incentives. Employees may also be encouraged to do "whatever it takes" to get the job done. Such environments can make some employees feel pressure to engage in unethical conduct to meet management's expectations, especially if the organization has no corporate code of ethics and no strong examples of senior management practicing ethical behavior. Table 1-3 shows how management's behavior can result in unethical employee behavior; Table 1-4 provides a manager's checklist for establishing an ethical workplace (the preferred answer to each question in this table is *yes*).

TABLE 1-3 How management can affect employees' ethical behavior

Managerial behavior that can encourage unethical behavior	Possible employee reaction
Set and hold people accountable to meet "stretch" goals, quotas, and budgets	"My boss wants results, not excuses, so I have to cut corners to meet the goals my boss has set."
Fail to provide a corporate code of ethics and operating principles to guide decisions	"Because the company has not established any guidelines, I don't think my conduct is really wrong or illegal."
Fail to act in an ethical manner and instead set a poor example for others to follow	"I have seen other successful people take unethical actions and not suffer negative repercussions."
Fail to hold people accountable for unethical actions	"No one will ever know the difference, and if they do, so what?"
When employees are hired, put a 3-inch thick binder entitled "Corporate Business Ethics, Policies, and Procedures" on their desks. Tell them to "read it when you have time and sign the attached form that says you read and understand the corporate policy."	"This is overwhelming. Can't they just give me the essentials? I can never absorb all this."

TABLE 1-4 A manager's checklist for establishing an ethical work environment

Question	Yes	No
Does your organization have a code of ethics?		
Do employees know how and to whom to report any infractions of the code of ethics?		
Do employees feel that they can report violations of the code of ethics safely and without fear of retaliation?		
Do employees feel that action will be taken against those who violate the code of ethics?		
Do senior managers set an example by communicating the code of ethics and using it in their own decision making?		
Do managers evaluate and provide feedback to employees on how they operate with respect to the values and principles in the code of ethics?		
Are employees aware of sanctions for breaching the code of ethics?		
Do employees use the code of ethics in their decision making?		

Employees must have a knowledgeable resource with whom they can discuss perceived unethical practices. For example, Intel expects employees to report suspected violations of its code of conduct to a manager, the Legal or Internal Audit Departments, or a business unit's legal counsel. Employees can also report violations anonymously through an internal

Web site dedicated to ethics. Senior management at Intel has made it clear that any employee can report suspected violations of corporate business principles without fear of reprisal or retaliation.

Including Ethical Considerations in Decision Making

We are all faced with difficult decisions in our work and in our personal life. Most of us have developed a decision-making process that we execute automatically, without thinking about the steps we go through. For many of us, the process generally follows the steps outlined in Figure 1-4.

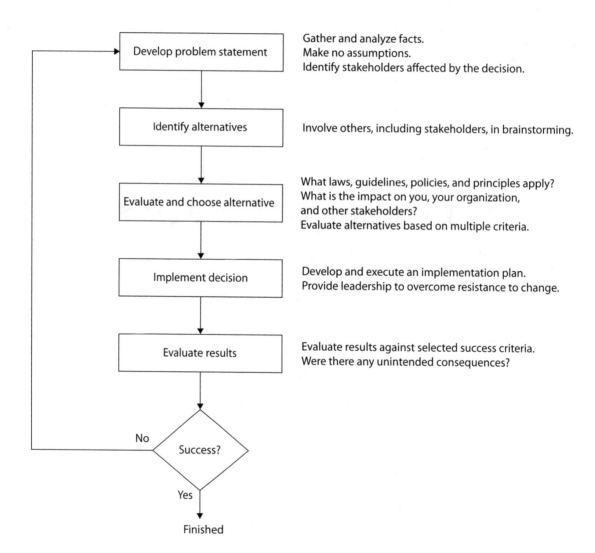

FIGURE 1-4   Decision-making process

The following sections discuss this decision-making process further and point out where and how ethical considerations need to be brought into the process.

Develop a Problem Statement

A **problem statement** is a clear, concise description of the issue that needs to be addressed. A good problem statement answers the following questions: What do people observe that causes them to think there is a problem? Who is directly affected by the problem? Is there anyone else affected? How often does it occur? What is the impact of the problem? How serious is the problem? Development of a problem statement is the most critical step in the decision-making process. Without a clear statement of the problem or the decision to be made, it is useless to proceed. Obviously, if the problem is stated incorrectly, the decision will not solve the problem.

One must gather and analyze facts to develop a good problem statement. Seek information and opinions from a variety of people to broaden your frame of reference. During this process, you must be extremely careful not to make assumptions about the situation. Simple situations can sometimes turn into complex controversies because no one takes the time to gather the facts. For example, you might see your boss receive what appears to be an employment application from a job applicant and then throw the application into the trash after the applicant leaves. This would violate your organization's policy to treat each applicant with respect and to maintain a record of all applications for one year. You could report your boss for failure to follow the policy, or you could take a moment to speak directly to your boss. You might be pleasantly surprised to find out that the situation was not as it appeared. Perhaps the "applicant" was actually a salesperson promoting a product for which your company had no use, and the "application" was marketing literature.

Part of developing a good problem statement involves identifying the stakeholders and their positions on the issue. Stakeholders often include others beyond those directly involved in an issue. Identifying the stakeholders helps you understand the impact of your decision and could help you make a better decision. Unfortunately, it may also cause you to lose sleep from wondering how you might affect the lives of others. By involving stakeholders in the decision, you gain their support for the recommended course of action. What is at stake for each stakeholder? What does each stakeholder value, and what outcome does each stakeholder want? Do some stakeholders have a greater stake because they have special needs or because the organization has special obligations to them? To what degree should they be involved in the decision?

The following list includes one example of a good problem statement as well as two examples of poor problem statements:

- Good problem statement: Our product supply organization is continually running out of stock of finished products, creating an out-of-stock situation on over 15 percent of our customer orders, resulting in over $300,000 in lost sales per month.
- Poor problem statement: We need to implement a new inventory control system. (This is a possible solution, not a problem statement.)
- Poor problem statement: We have a problem with finished product inventory. (This is not specific enough.)

Identify Alternatives

During this stage of decision making, it is ideal to enlist the help of others, including stakeholders, to identify several alternative solutions to the problem. Brainstorming with just one other person will reduce your chances of identifying a broad range of alternatives and determining the best solution. On the other hand, there are times when it is inappropriate to involve others in solving a problem that you are not at liberty to discuss. In providing participants information about the problem to be solved, offer just the facts, without your opinion, so you don't influence others to accept your solution.

During any brainstorming process, try not to be critical of ideas, as any negative criticism will tend to "shut down" the group, and the flow of ideas will dry up. Simply write down the ideas as they are suggested.

Evaluate and Choose an Alternative

Once a set of alternatives has been identified, the group attempts to evaluate them based on numerous criteria, such as effectiveness at addressing the issue, the extent of risk associated with each alternative, cost, and time to implement. An alternative that sounds attractive but that is not feasible will not help solve the problem.

As part of the evaluation process, weigh various laws, guidelines, and principles that may apply. You certainly do not want to violate a law that can lead to a fine or imprisonment for yourself or others. Are there any corporate policies or guidelines that apply? Does the organizational code of ethics offer guidance? Do any of your own personal principles apply?

Also consider the likely consequences of each alternative from several perspectives—What is the impact on you, your organization, other stakeholders (including your suppliers and customers), and the environment?

The alternative selected should be ethically and legally defensible; be consistent with the organization's policies and code of ethics; take into account the impact on others; and, of course, provide a good solution to the problem.

Philosophers have developed many approaches to aid in ethical decision making. Four of the most common approaches, which are summarized in Table 1-5 and discussed below, provide a framework for decision makers to reflect on the acceptability of their actions and evaluate their moral judgments. People must find the appropriate balance between all applicable laws, corporate principles, and moral guidelines to help them make decisions. (For a more in-depth discussion of ethics and moral codes, see Appendix A.)

TABLE 1-5   Four common approaches to ethical decision making

Approach to dealing with moral issues	Principle
Virtue ethics approach	The ethical choice best reflects moral virtues in yourself and your community.
Utilitarian approach	The ethical choice produces the greatest excess of benefits over harm.
Fairness approach	The ethical choice treats everyone the same and shows no favoritism or discrimination.
Common good approach	The ethical choice advances the common good.

Virtue Ethics Approach

The **virtue ethics approach** to decision making focuses on how you should behave and think about relationships if you are concerned with your daily life in a community. It does not define a formula for ethical decision making, but suggests that when faced with a complex ethical dilemma, people do either what they are most comfortable doing or what they think a person they admire would do. The assumption is that people are guided by their virtues to reach the "right" decision. A proponent of virtue ethics believes that a disposition to do the right thing is more effective than following a set of principles and rules, and that people should perform moral acts out of habit, not introspection.

Virtue ethics can be applied to the business world by equating the virtues of a good businessperson with those of a good person. However, businesspeople face situations that are peculiar to business, so they may need to tailor their ethics accordingly. For example, honesty and openness when dealing with others are generally considered virtuous; however, a corporate purchasing manager who is negotiating a multimillion dollar deal might need to be vague in discussions with potential suppliers.

A problem with the virtue ethics approach is that it doesn't provide much of a guide for action. The definition of *virtue* cannot be worked out objectively; it depends on the circumstances—you work it out as you go. For example, bravery is a great virtue in many circumstances, but in others it may be foolish. The right thing to do in a situation depends on which culture you're in and what the cultural norm dictates.

Utilitarian Approach

The **utilitarian approach** to ethical decision making states that you should choose the action or policy that has the best overall consequences for all people who are directly or indirectly affected. The goal is to find the single greatest good by balancing the interests of all affected parties.

Utilitarianism fits easily with the concept of value in economics and the use of cost-benefit analysis in business. Business managers, legislators, and scientists weigh the benefits and harm of policies when deciding whether to invest resources in building a new plant in a foreign country, to enact a new law, or to approve a new prescription drug.

A complication of this approach is that measuring and comparing the values of certain benefits and costs is often difficult if not impossible. How do you assign a value to human life or to a pristine wildlife environment? It can also be difficult to predict the full benefits and harm that result from a decision.

Fairness Approach

The **fairness approach** focuses on how fairly actions and policies distribute benefits and burdens among people affected by the decision. The guiding principle of this approach is to treat all people the same. However, decisions made with this approach can be influenced by personal bias toward a particular group, and the decision makers may not even realize their bias. If the intended goal of an action or a policy is to provide benefits to a target group, other affected groups may consider the decision unfair.

Common Good Approach

The **common good approach** to decision making is based on a vision of society as a community whose members work together to achieve a common set of values and goals. Decisions and policies that use this approach attempt to implement social systems, institutions, and environments that everyone depends on and that benefit all people. Examples include an effective education system, a safe and efficient transportation system, and accessible and affordable health care.

As with the other approaches to ethical decision making, there are complications with the common good approach. People clearly have different ideas about what constitutes the common good, which makes consensus difficult. In addition, maintaining the common good often requires some groups to bear greater costs than others—for instance, homeowners pay property taxes to support public schools, but apartment dwellers do not.

Implement Decision

Once the alternative is selected, it should be implemented in an efficient, effective, and timely manner. This is much easier said than done, since people tend to resist change. In fact, the bigger the change, the greater is the resistance to it. Communication is the key to helping people accept a change. It is imperative that someone whom the stakeholders trust and respect answer the following questions: Why are we doing this? What is wrong with the current way we do things? What are the benefits of the new way for you? A transition plan must be defined to explain to people how they will move from the old way of doing things to the new way. It is essential that the transition be seen as relatively easy and pain free.

Evaluate the Results

After the solution to the problem has been implemented, monitor the results to see if the desired effect was achieved, and observe its impact on the organization and the various stakeholders. Were the success criteria fully met? Were there any unintended consequences? This evaluation may indicate that further refinements are needed. If so, return to the problem development step, refine the problem statement as necessary, and work through the process again.

ETHICS IN INFORMATION TECHNOLOGY

The growth of the Internet, the ability to capture and store vast amounts of personal data, and greater reliance on information systems in all aspects of life have increased the risk that information technology will be used unethically. In the midst of the many IT breakthroughs in recent years, the importance of ethics and human values has been underemphasized—with a range of consequences. Here are some examples that raise public concern about the ethical use of information technology:

- Many employees might have their e-mail and Internet access monitored while at work, as employers struggle to balance their need to manage important company assets and work time with employees' desire for privacy and self-direction.
- Millions of people have downloaded music and movies at no charge and in apparent violation of copyright laws at tremendous expense to the owners of those copyrights.
- Organizations contact millions of people worldwide through unsolicited e-mail (spam) as an extremely low-cost marketing approach.

- Hackers break into databases of financial and retail institutions to steal customer information, then use it to commit identity theft—opening new accounts and charging purchases to unsuspecting victims.
- Students around the world have been caught downloading material from the Web and plagiarizing content for their term papers.
- Web sites plant cookies or spyware on visitors' hard drives to track their online purchases and activities.

This book is based on two fundamental tenets. First, the general public does not understand the critical importance of ethics as it applies to IT, as too much emphasis has been placed on technical issues. Unlike most conventional tools, IT has a profound effect on society. IT professionals and users need to recognize this fact when they formulate policies that will have legal ramifications and affect the well-being of millions of consumers.

The second tenet on which this book is based is that in the business world, important decisions are too often left to the technical experts. General business managers must assume greater responsibility for these decisions, but to do so they must be able to make broad-minded, objective decisions based on technical savvy, business know-how, and a sense of ethics. They must also try to create a working environment in which ethical dilemmas can be discussed openly, objectively, and constructively.

Thus, the goals of this text are to educate people about the tremendous impact of ethical issues in the successful and secure use of information technology; to motivate people to recognize these issues when making business decisions; and to provide tools, approaches, and useful insights for making ethical decisions.

Summary

- Even within the same society, people can have strong disagreements over important moral issues.

- Ethics has risen to the top of the business agenda because the risks associated with inappropriate behavior have increased, both in their likelihood and in their potential negative impact.

- Organizations have at least five good reasons for promoting a work environment in which they encourage employees to act ethically: (1) to gain the good will of the community, (2) to create an organization that operates consistently, (3) to foster good business practices, (4) to protect the organization and its employees from legal action, and (5) to avoid unfavorable publicity.

- An organization with a successful ethics program is one in which employees are willing to seek advice about ethical issues that arise; employees feel prepared to handle situations that could lead to misconduct; employees are rewarded for ethical behavior; the organization does not reward success gained through questionable means; and employees feel positively about their company.

- The corporate ethics officer (or corporate compliance officer) ensures that ethical procedures are installed and consistently adhered to throughout the organization, creates and maintains the ethics culture, and serves as a key resource on issues relating to corporate principles and ethics.

- Managers' behavior and expectations can strongly influence employees' ethical behavior.

- Most of us have developed a simple decision-making model that includes these steps: (1) develop a problem statement, (2) identify alternatives, (3) evaluate and choose an alternative, (4) implement the decision, and (5) evaluate the results.

- One can incorporate ethical considerations into decision making by identifying and involving the stakeholders; weighing various laws, guidelines, and principles— including the organization's code of ethics—that may apply; and considering the impact of the decision on you, your organization, stakeholders, your customers and suppliers, and the environment.

- Philosophers have developed many approaches to ethical decision making. Four common philosophies are the virtue ethics approach, the utilitarian approach, the fairness approach, and the common good approach.

The answers to the Self-Assessment Questions can be found in Appendix G.

Choose the word(s) that best complete the following sentences.

1. Statements about how people should behave fit together to form the _____ by which a society lives.

2. A(n) _____ is a moral habit that inclines people to do what is considered acceptable.

3. The _____ that socially responsible activities create can make it easier for corporations to conduct their business.

4. _____ means that employees know what is expected of them, and they can employ the organizational values to help them in their decision making.

5. That an employer can be held responsible for the acts of its employee even if the employee acts in a manner contrary to corporate policy and the employer's direction is based on the principle called _____.

6. The public _____ of an organization strongly influences the value of its stock, how consumers regard its products and services, the degree of oversight it receives from government agencies, and the amount of support and cooperation it receives from its business partners.

7. The corporate ethics officer provides the organization with _____ and _____ in the area of business conduct.

8. The _____ is responsible for the careful and responsible management of an organization.

9. _____ requires public companies to disclose whether they have codes of ethics and disclose any waiver to their code of ethics for certain members of senior management.

10. The goal of the Sarbanes-Oxley Act was to _____.

11. _____ highlights an organization's key ethical issues and identifies the overarching values and principles that are important to the organization and its decision-making process.

12. A(n) _____ enables an organization to review how well it is meeting its ethical and social responsibility goals, and communicate new goals for the upcoming year.

13. _____ makes employees more aware of a company's code of ethics and how to apply it, as well as demonstrates that the company intends to operate in an ethical manner.

14. The most important part of the decision-making process is _____.

15. The _____ approach to ethical decision making states that you should choose the action or policy that has the best overall consequences for all people who are directly or indirectly affected.

16. _____ is a process for generating a number of alternative solutions to a problem.

Discussion Questions

1. There are many ethical issues about which people hold very strong opinions—abortion, gun control, and the death penalty, to name a few. If you were a team member on a project with someone whom you knew held an opinion different from yours on one of these issues, would it affect your ability to work effectively with this person? Why or why not?

2. What do you think are the most important factors that helped you define your own personal code of ethics?

3. Do you think that ethics in business is improving or getting worse? Defend your position.

4. Do you believe that an organization should be able to escape criminal liability for the acts of its employees if it has acted as a responsible corporate citizen, making strong efforts to prevent and detect misconduct in the workplace? Why or why not?

5. The Ethics Resource Center identified five characteristics of a successful ethics program. Suggest a sixth characteristic, and defend your choice.

6. Which incident has a greater negative impact on an organization—an unethical act performed by an hourly worker or the same act performed by a senior manager of the organization? Explain your answer fully. Should the hourly worker be treated differently than the senior manager who committed the unethical act? Why or why not?

7. It is a common and acceptable practice for managers to hold people accountable to meet "stretch" goals, quotas, and budgets. How can this be done in a way that does not encourage unethical behavior on the part of employees?

8. Is every action that is legal also ethical? Can you describe an action that is legal but ethically wrong? Is every ethical action also legal? Is the law, not ethics, the only guide that business managers need to consider? Explain.

9. Do you think it is easier to establish an ethical work environment in a nonprofit organization? Why or why not?

10. This chapter discusses four approaches to dealing with moral issues. Identify and briefly summarize each one. Do you believe one perspective is better than the others? If so, which one and why?

11. Is it possible for an employee to be successful in the workplace without acting ethically?

What Would You Do?

Use the five-step decision-making process to analyze the following situations and recommend a course of action.

1. Imagine that you are Hewlett-Packard's new chairman of the board, CEO, and president Mark Hurd, who succeeded Patricia Dunn. What actions would you take for HP to regain its reputation as a highly ethical organization?

2. You are the customer service manager for a small software manufacturer. The newest addition to your 10-person team is Aubrey, a recent college graduate. She is a little overwhelmed by the volume of calls, but is learning quickly and doing her best to keep up. Today, as you

performed your monthly review of employee e-mail, you were surprised to see that Aubrey is corresponding with employment agencies. One message says, "Aubrey, I'm sorry you don't like your new job. We have lots of opportunities that I think would much better match your interests. Please call me and let's talk further." You're shocked and alarmed. You had no idea she was unhappy, and your team desperately needs her help to handle the onslaught of calls generated by the newest release of software. If you're going to lose her, you'll need to find a replacement quickly. You know that Aubrey did not intend for you to see the e-mail, but you can't ignore what you saw. Should you confront Aubrey and demand to know her intentions? Should you avoid any confrontation and simply begin seeking her replacement? Could you be misinterpreting the e-mail? What should you do?

3. A coworker calls you at 9:00 a.m. at work and asks for a favor. He is having car trouble and will be an hour late for work. He explains that he has already been late for work twice this month and that a third time will cost him four hours of pay. He asks you to stop by his cubicle, turn his computer on, and place some papers on the desk so that it appears that he is in. You have worked on some small projects with this coworker and have gone to lunch together. He seems nice enough and does his share of the work, but you are not sure what to tell him. What would you do?

4. While mingling with friends at a party, you mention a recent promotion that has put you in charge of evaluating bids for a large computer hardware contract. A few days later, you receive a dinner invitation at the home of an acquaintance who also attended the party. Over cocktails, the conversation turns to the contract you are managing. Your host seems remarkably well-informed about the bidding process and likely bidders. You volunteer information about the potential value of the contract and briefly outline the criteria your firm will use to select the winner. At the end of the evening, the host surprises you by revealing that he is a consultant for several companies in the computer hardware market. Later that night your mind is racing. Did you reveal information that could provide a supplier with a competitive advantage? What are the potential business risks and ethical issues in this situation? Should you report the conversation to someone? If so, whom should you talk to, and what would you say?

5. You are a recent graduate of a well-respected business school, but you are having problems getting a job. You worked with a professional résumé service to develop a well-written résumé and placed it on several Web sites; you also sent it directly to contacts at a dozen companies. So far you have not even had an invitation for an interview. You know that one of your shortcomings is that you have no real job experience to speak of. You are considering beefing up your résumé by exaggerating the extent of the class project you worked on for a few weeks under the supervision of your brother-in-law. You could reword the résumé to make it sound as if you were actually employed and that your responsibilities were greater than they actually were. What should you do?

6. You have just completed interviewing three candidates for an entry-level position in your organization. One candidate is the friend of a coworker who has implored you to give his friend a chance. The candidate is the weakest of the three but has sufficient skills and knowledge to adequately fill the position. Would you hire this candidate?

Cases

1. Computer Associates: A Firm with a Scandal-Riddled Past

Computer Associates (CA) is a multinational computer software company founded in 1980 and headquartered in Islandia, New York. In 1989, it became the first software organization to generate $1 billion in sales.[28] Today CA employs nearly 14,000 people in 150 offices spread across more than 45 countries. Its 2008 annual revenue was $4.3 billion.

In 1997, the chairman of Computer Sciences Corporation (CSC), Van Honeycutt, filed a $50 million bribery and extortion suit against CA's founder, chairman, and CEO Charles Wang. Honeycutt claimed that Wang had offered him a $102.5 million bribe to sell the company to CA for $100 per share.[29] Van Honeycutt further alleged that when he did not accept the offer, CA executives "threatened to wrongfully harm CSC if it refused to agree to a transaction" at $98 per share. It was a few weeks after this alleged incident that CA launched a hostile takeover bid. Eventually the two companies dropped all lawsuits related to CA's takeover attempt and announced a major expansion of their global software licensing agreements.[30]

In 1999, a CA shareholder objecting to Wang's compensation package took Wang and two other executives to court and kept them from receiving a $1.1 billion payout. As a result, Wang had to settle for a total compensation of $675 million for 1999—making him the highest paid executive in the United States. This amounted to one of the largest executive compensation packages in history at that time, and came at a time when CA's earnings and stock price had fallen.[31]

The vice president of finance of CA pleaded guilty to conspiracy to commit securities fraud and obstruction of justice in April 2004. The fraud involved backdating contracts worth hundreds of millions of dollars to pump up the company's quarterly earnings both to meet analysts' expectations and to make the firm's stock look more attractive to investors.[32] The former CFO and a former senior vice president were also suspected of playing a role in the fraud.

In April 2004, Sanjay Kumar—chairman and CEO since Wang's retirement in 2002—resigned under pressure from the board, which feared he would become embroiled in the growing accounting scandal.[33] Also that month, the company restated its financial results from 2000 and 2001 to reflect $2.2 billion in revenue that was booked prematurely. Kumar was charged with securities fraud, conspiracy, and obstruction of justice in September 2004. Kumar was eventually found guilty and sentenced to 12 years in jail.

CA agreed to pay $225 million to shareholders in restitution in order to defer criminal prosecution. CA said it would cut its workforce by 800 to help pay the restitution. Also as part of the agreement, an outside monitor was assigned to track CA's financial reporting for one and a half years. CA also agreed to assist the government in retrieving any compensation and bonuses awarded based on the fraudulent financial results.[34]

In more bad news for CA's Kumar, the indictment against him was revised in July 2005 to include charges that he offered a $3.7 million bribe to discourage a business client from revealing CA's fraudulent accounting practices.[35]

As a result of these various scandals, many CA board members and executives were replaced between 2004 and 2006, including the CEO, chairman of the board, executive vice president of development, CFO, COO, CTO, chief marketing officer, chief administrative officer, and co-general counsel. Most of these executives were sentenced to jail, fined millions of dollars, or both.

Discussion Questions

1. CA executives involved in the accounting scandal were not accused of reporting bogus contracts or hiding major problems in the business. The contracts that were backdated were real sales agreements. Was this really a crime? Should the individuals have been punished so harshly?

2. In December 2004, CA appointed Patrick J. Gnazzo as senior chief compliance officer to demonstrate to the government and shareholders that the firm would take measures to operate ethically. Gnazzo served in this role at United Technologies for 10 years and had been a member of the board of directors of the Ethics Officers Association. Gnazzo reported to a new executive vice president and general counsel at CA as well as the board's Compliance Committee. Outline some of the actions Gnazzo might have taken in his first six months on the job.

3. John Swainson, a 26-year veteran of IBM, joined CA in November 2004 as CEO and president. His first few months with the firm were rough—major customers threatened to dump the firm; some products were behind schedule and were of poor quality; executives had to be fired for breaking company rules; accountants continued to find past mistakes; and many newly hired executives had to be brought on board. What sort of leadership could he have demonstrated to show that he was determined to avoid future scandals at CA?

4. CA has been hit with numerous scandals since the late 1990s. These scandals raise questions about how successful the firm might have been if not for the amount of time its executives had to spend on these distractions. Compare the revenue growth and stock price of CA to that of some of its competitors over the time period 2004–2008. (Be sure to use CA's corrected figures!) Can you detect any impact of these scandals on CA's performance? What else might explain the difference in performance?

2. Dell Inc. Merely Tough on Suppliers or a Bully?

Dell Computers designs, develops, manufactures, markets, sells, and supports a range of products, including desktop personal computers, servers and networking products, storage, mobile products, software and peripherals, and services. In many cases, these products are customized to individual customer requirements. Dell's recent annual sales were $64 billion, which generated $2.8 billion in net income. The company employs about 80,000 people and is headquartered in Round Rock, Texas.

A key component of Dell's business model is the elimination of resellers by selling directly to customers. Dell employs a build-to-order manufacturing process that enables customers to direct-order servers and workstations custom-made to their specifications. In addition, Dell can fulfill and ship orders within four days. This all requires an amazing supply-chain system that includes hundreds of suppliers around the globe functioning seamlessly to enable Dell to operate with a near-zero amount of finished goods inventory. Indeed, Dell has no warehouses to store finished goods but is still able to assemble 80,000 computers every day.[36] Two advantages of carrying a minimal amount of inventory are:

1. When a new computer component, say a faster CPU chip, comes out, Dell can more quickly sell its existing inventory of computers with the old chip and get to market with computers based on the new faster CPU weeks earlier than its competitors.

2. Because advances come so quickly in computer components, they depreciate at a rate of up to 1 percent per week. Carrying a minimal inventory means that Dell avoids getting stuck with obsolete parts and suffering a significant financial loss.

Dell collects payment from the customer once it receives an order. It then pulls the needed parts directly from its suppliers, builds the computer, and ships the order within four days. Dell makes payments to suppliers 36 days after it receives payment from the customer. The lag between Dell's receipt of payment from the customer and Dell's payment of its suppliers means that its suppliers are helping finance the cost of Dell's operations.[37] This gives Dell a major competitive advantage over non-direct-order computer manufacturers, who must typically pay their suppliers 30 days before the product is shipped to market, put into inventory, bought by a customer, and paid for.

Closer analysis shows that Dell holds inventory for roughly two days—eight hours as parts travel across its assembly line and another day while the completed computer is transported to one of its distribution centers, where the unit is combined with a monitor and shipped to the customer.

Each Dell supplier, however, is required to maintain a minimum of eight to 10 days' worth of buffer stock in multivendor warehouses located nearby each Dell factory. This buffer helps ensure that Dell always has on hand the parts needed to assemble a customer order. Many of the direct-to-Dell suppliers combine components from their own suppliers to build subassemblies, which in turn are shipped to Dell. For example, MMC Technology ships some 50 million disks to Maxtor (Dell's hard drive supplier) each year. MMC Technology performs one week of performance testing before it releases its drives. It takes another week for the disks to arrive at Maxtor's Singapore factory, and Maxtor insists that MMC carry one week of buffer stock, just as Dell requires of its direct suppliers. As a result, MMC Technology must carry at least three weeks of inventory. So suppliers who ship directly to Dell suppliers carry at least three weeks of inventory, suppliers who ship to those direct suppliers carry additional inventory, and raw material suppliers carry even more inventory. If one examines the entire supply chain closely, Dell's suppliers hold anywhere from three to 11 weeks' worth of inventory.[38] Meanwhile, Dell carries just three days of inventory.

Dell is a very measures-driven organization and does not hesitate to use measures to pressure its suppliers to improve. It meticulously gathers data to rate each of its suppliers on the basis of cost, quality, service, and on-time delivery. These scores are posted on a Web site so that individual suppliers can view their scores. Dell's procurement managers meet with all the suppliers' executive teams at the firm's headquarters for quarterly business reviews. The suppliers are given feedback and told how they rank against the competition. Based on their ranking, the suppliers are informed as to what percentage of Dell's purchases for the upcoming quarter they will be awarded.[39]

Says Marty Garvin, Dell's procurement chief, "You want competition—suppliers that are fighting very hard for Dell's business. You set a high bar and keep raising it, so suppliers have to continually redefine themselves in terms of their efficiency and reliability. Otherwise, they won't be around for very long."[40]

"If you aren't performing, Dell won't hesitate to take some of your business and give it to a competitor, so boo-hoo," says Jerry Gregoire, Dell's CEO from 1995 to 2000. "That's bullying? It's called holding the supplier's feet to the fire. If you want my business, you're going to have to meet my expectations."[41]

While firms are happy to receive the high volume of business that comes with being a Dell supplier, they recognize that their performance will be closely scrutinized. Suppliers also know that they will be required to finance a portion of Dell's operating costs, bear the cost of carrying Dell's inventory, and run the risk of getting stuck with obsolete parts. "Dell does business with suppliers who are willing to hold its inventory," says International Data Corporation analyst Roger Kay. "And if they're not willing, Dell will find suppliers who are."[42]

Discussion Questions

1. Identify at least two other companies that manage their suppliers in a manner similar to Dell Computers. Do these companies have anything in common with Dell?

2. Can you identify any disadvantages for Dell in taking this approach to managing suppliers?

3. Do you think that this approach to managing suppliers is unethical, or is it acceptable and ethical? Explain your reasoning.

4. Would you recommend any changes to the way Dell manages its suppliers? Defend your position.

3. Is There a Place for Ethics in IT?

On March 15, 2005, Michael Schrage published an article in *CIO* magazine entitled "Ethics, Schmethics," which stirred up a great deal of controversy in the IT community. In the article, Schrage proposed that CIOs (chief information officers) "should stop trying to do the 'right thing' when implementing IT and focus instead on getting their implementations right." Schrage argued that *ethics* had become a buzzword, just like *quality* in the 1980s; he asserted that the demand for ethical behavior interferes with business efficiency.

In the article, Schrage provided a few scenarios to back up his opinion. In one such example, a company is developing a customer relationship management (CRM) system, and the staff is working very hard to meet the deadline. The company plans to outsource the maintenance and support of the CRM system once it is developed, meaning that there is a good chance that two-thirds of the IT staff will be laid off. Would you disclose this information? Schrage answered, "I don't think so."

In another scenario, Schrage asked readers if they would consider deliberately withholding important information from their boss if they knew that its disclosure would provoke his or her immediate counterproductive intervention in an important project. Schrage said he would withhold it. Business involves competing values, he argued, and trade-offs must be made to keep business operations from becoming paralyzed.

Schrage was hit with a barrage of responses accusing him of being dishonorable, shortsighted, and lazy. Other feedback provided new perspectives on his scenarios that Schrage had not considered in his article. For example, an IT manager at Boise State University argued that doing the right thing is good for business. Not disclosing layoffs, she argued, is a trick that only works once. Remaining employees will no longer trust the company and may pursue jobs where they can feel more secure. New job applicants will think twice before joining a company with a reputation for exploiting employees. Other readers responded to that scenario by suggesting that the company could try to maintain loyalty by offering incentives for those who stayed or by providing job-placement services to departing employees.

Addressing the second scenario, another reader, Dewey, suggested that not giving the boss important information could backfire on the employee: "What if your boss finds out the truth? What if you were wrong and the boss could have helped? Once your boss knows that you lied once, will he believe you the next time?"

Another reader had actually worked under an unproductive, reactive, meddling boss. Based on his experience, he suggested speaking to the boss about the problem at an appropriate time and place. In addition, the reader explained that as situations arose that required him to convey important information that might elicit interference, he developed action plans and made firm presentations to his boss. The boss, the reader assured Schrage, will adapt.

Some readers argued that CIOs must consider the company's long-term needs rather than just the current needs of a specific project. Others argued that engaging in unethical behavior, even for the best of purposes, crosses a line that eventually leads to more serious transgressions. Some readers suspected that Schrage had published the article to provoke outrage. Another reader agreed with Schrage, arguing that ethics has to "take a back seat to budgets and schedules" in a large organization. This reader explained, "At the end of the day, IT is business."

Discussion Questions

1. Discuss how a CIO might handle Schrage's scenarios using the suggested process for ethical decision making presented in this chapter.

2. Discuss the possible short-term losses and long-term gains in implementing ethical solutions for each of Schrage's scenarios.

3. Must businesses choose between good ethics and financial benefits? Explain your answer using Schrage's scenarios as examples.

End Notes

[1] Joe Kurtzman, "Business Forum: Executive Ethics; Doing Business, Doing Good," *New York Times*, January 3, 1988.

[2] David A. Kaplan, "Scandal at HP: The Boss Who Spied on Her Board," *Newsweek*, September 10, 2006.

[3] Robert Mullins, "HP Hires Ethics and Compliance Officer," *Computerworld*, October 17, 2006.

[4] Associated Press, "Charges Dropped Against Ex–HP Chairwoman," March 14, 2007.

[5] K. C. Jones, "California Attorney General Attempting Deal Between HP, Pretext Victims," *InformationWeek*, December 8, 2006.

[6] Business Software Alliance, "Fifth Annual BSA and IDC Global Software Piracy Study," May 2008, http://global.bsa.org/idcglobalstudy2007/studies/highlights_globalstudy07.pdf (accessed January 22, 2009).

[7] "FDA: Peanut Plant Knowingly Sent Bad Products," MSNBC.com News Services, February 6, 2009.

[8] W. David Gardner, "Comverse Sues Founder in Options Backdating Scandal," *InformationWeek*, January 17, 2008.

[9] Reuters, "Satyam Seeks Emergency Funding," *InformationWeek*, January 21, 2009.

[10] Wu Zhong, "Ouch: China's IT Scandals," July 4, 2006, www.jucee.org/China/Ouch--China-s-IT-Scandals.html.

[11] Kim Hart, "Tech Firms Make Charitable Giving a Priority," *Washington Post*, June 17, 2006, www.charitenet.com/news/1-10.htm (accessed March 17, 2009).

[12] Kim Hart, "Tech Firms Make Charitable Giving a Priority," *Washington Post*, June 17, 2006, www.charitenet.com/news/1-10.htm (accessed March 17, 2009).

[13] Kim Hart, "Tech Firms Make Charitable Giving a Priority," *Washington Post*, June 17, 2006, www.charitenet.com/news/1-10.htm (accessed March 17, 2009).

[14] Kim Hart, "Tech Firms Make Charitable Giving a Priority," *Washington Post*, June 17, 2006, www.charitenet.com/news/1-10.htm (accessed March 17, 2009).

[15] University of Washington, "Microsoft Giving Program," www.cs.washington.edu/building/givingGuidelines.doc.

[16] SAP America, Inc., "2008/2009 Academic Year Scholarship Program," http://sfa.mst.edu/documents/2008-09_SAP_Scholarship_Program_Description.pdf.

[17] Yahoo!, "Yahoo! Marks Inauguration of "Cyber Giving Week," With Overall Online Donations Expected To Reach Record Highs in 2005," http://yahoo.client.shareholder.com/releasedetail.cfm?ReleaseID=182524.

[18] "Consumer Satisfaction with PCs Drops," *ZDNet UK*, August 14, 2007.

[19] Antone Gonsalves, "Dell Committed Consumer Fraud, N.Y. Judge Rules," *InformationWeek*, May 27, 2008.

[20] *United States v. New York Central & Hudson River R. Co.*, 212 U.S. 509 (1909), http://supreme.justia.com/us/212/509/case.html.

[21] Clifford F. Thies, "The Demise of Arthur Andersen," www.gold-eagle.com/editorials_02/thies041502.html (accessed January 25, 2009).

[22] Paula J. Desio, "Ethics and Compliance Programs May Get Their Day in Court," Ethics Resource Center, www.ethics.org/ethics-today/1208/policy-report.asp (accessed January 25, 2009).

[23] Kenneth W. Johnson, "Federal Sentencing Guidelines: Enterprise Risk Management," Ethics Resource Center, www.ethics.org/erc-publications/staff-articles.asp?aid=787.

[24] "2007 Ethics Resource Center's National Business Ethics Survey," www.ethics.com/research/nbes.asp (accessed January 25, 2009).

[25] "Three Main Responsibilities of an Ethics Officer," Corporate-Ethics.US, www.corporate-ethics.us/ethics_officer.htm (accessed January 25, 2009).

[26] Ethics Resource Center, "National Business Ethics Survey: An Inside View of Private Sector Ethics," *Kiplinger Business Resource Center*, February 2008.

[27] "Performance Reviews Often Skip Ethics, HR Professionals Say," Ethics Resource Center, June 12, 2008, www.ethics.org/about-erc/press-releases.asp?aid=1150 (accessed January 27, 2009).

[28] "Computer Associates Guilty Plea," Corp-Ethics.com, April 10, 2004, www.corp-ethics.com/company/computer_associates/computer-associates-guilty-plea.html.

[29] Karen Kaplan, "CSC Suit Alleges Bribery, Extortion in Merger Try," *Los Angeles Times*, February 4, 1998.

[30] Kim Girard, "CSC, CA Call Truce," *ZDNet Asia*, October 8, 1999.

[31] Linda Harrison, "Wang Wrangles Fatter Bonus," *Software*, July 19, 2000.

[32] "Computer Associates Guilty Plea," Corp-Ethics.com, April 10, 2004, www.corp-ethics.com/company/computer_associates/computer-associates-guilty-plea.html.

[33] "Computer Associates Revises Finances," Corp-Ethics.com, May 6, 2004, www.corp-ethics.com/company/computer_associates/computer-associates-revises-finances.html.

[34] Frank Eltman, "Former CA CEO Kumar Indicted," *InformationWeek*, September 22, 2004.

[35] Dan Neel, "U.S. Strengthens Case Against Ex–CA CEO Kumar," *InformationWeek*, July 1, 2005.

[36] Bill Breen, "Living in Dell Time," Fast Company.com, December 19, 2007, www.fastcompany.com/magazine/88/dell.html.

[37] Bill Breen, "Living in Dell Time," Fast Company.com, December 19, 2007, www.fastcompany.com/magazine/88/dell.html.

[38] Bill Breen, "Living in Dell Time," Fast Company.com, December 19, 2007, www.fastcompany.com/magazine/88/dell.html.

[39] Dell Inc., "Supplier Principles: Supply Chain Management System," www.dell.com/content/topics/global.aspx/about_dell/values/supp_citizen/supply?c=us&l=en&s=gen (accessed January 25, 2009).

[40] Bill Breen, "Living in Dell Time," Fast Company.com, December 19, 2007, www.fastcompany.com/magazine/88/dell.html.

[41] Bill Breen, "Living in Dell Time," Fast Company.com, December 19, 2007, www.fastcompany.com/magazine/88/dell.html.

[42] Bill Breen, "Living in Dell Time," Fast Company.com, December 19, 2007, www.fastcompany.com/magazine/88/dell.html.

CHAPTER 2

ETHICS FOR IT WORKERS AND IT USERS

VIGNETTE

IT Technicians Fired After Reporting Child Porn

In 2002, Dorothea Perry was an employee of Collegis, an IT outsourcing firm that implemented and supported software for colleges and universities. She worked on the IT help desk at New York Law School, a private law school in lower Manhattan founded in 1891. When she arrived at her desk on a Sunday afternoon in June 2002, she had a voice mail from Dr. Edward Samuels. Dr. Samuels, a highly respected professor at the school for over a quarter century, asked that she check his computer for a possible virus. Perry went to Dr. Samuels' empty office to check things out but was unable fix the problem. She left a message for her coworker, Robert Gross—also a Collegis employee—to take a look at the computer on Monday. Gross examined the machine, but he, too, was unable to resolve the problem. So Perry and Gross, following their employer's guidelines, gave the professor a "loaner" machine while they worked on his.[1]

Again, following their employer's procedures, they began backing up Samuels' files to the school's network so that they could transfer them to Samuels' loaner machine. In the process, they uncovered more than two dozen photos of nude, young girls—several in "very sexually provocative poses." Perry immediately reported the incident to Collegis's executive director who in turn alerted her liaison at the school, Fred DeJohn. DeJohn met with Dean Richard Matasar three days later to discuss the photos and determine the appropriate next steps. (Possession of child pornography is both a federal and a New York state crime.) The New York Law School sought advice from counsel, alerted authorities, and cooperated fully in the investigation. A subsequent search of Dr. Samuels' apartment by the New York Police Department uncovered more than 100,000 similar photos—many of them depicting violent sex acts involving young girls. On advice of the district attorney, Dean Matasar did not alert Dr. Samuels of the findings of legal authorities.[2]

Following the investigation, Dr. Samuels was arrested in August 2002. Dean Matasar sent an e-mail to New York Law School faculty, staff, and students, saying, "I'm saddened to report to you that I learned this afternoon that our colleague, Professor Edward Samuels, was arrested on charges relating to possession of child pornographic images. The Law School has placed Professor Samuels on paid administrative leave so that he may attend to his defense … Our hearts go out to Ed and his family as they face the difficult time ahead."[3]

Dr. Samuels was protected by tenure and remained on paid administrative leave for months. Some of the faculty and senior staff were critical of Dean Matasar for turning the photos over to the authorities without first warning Dr. Samuels. Some claimed that Dr. Samuels had committed a victimless crime and there was no proof that he had done anything other than view the photos. Even Dean Matasar said, "When there's no purchase or sale of these materials, I don't know. As a lawyer, I am ambivalent on these issues."[4]

In April 2003, Samuels pled guilty to 100 counts of possessing child pornography. The sentencing judge received 20 testimonial letters expressing great respect for Dr. Samuels' integrity as a professor and stating that the incident did not affect their respect for his professionalism. In June, Samuels was sentenced to six months in jail and 10 years' probation.[5] Samuels eventually served four months.

Perry and Gross were placed on probation for job-performance issues shortly after they reported finding the photos. In October 2002, four months after the discovery, Gross and Perry were fired. Gross had worked at the school just over a year, and his evaluation three months prior to the incident rated him as "fully competent plus." Perry had worked at the school for 12 years, and her previous performance appraisal rated her work as "excellent."[6]

Dean Matasar stated that he had asked Collegis to improve the help desk and that the firing of Gross and Perry was likely a result of this request. Matasar claims he received countless complaints from students and faculty at the school about poor service. The terminations came just prior to the renewal of Collegis's multimillion-dollar contract with the school.[7]

Tom Huber, CEO of Collegis, defended his firm's actions, saying, "Employment of the technicians ended due to issues completely unrelated to this isolated incident. It is the policy of Collegis to treat all of its employees in a fair and impartial manner, and it would never dismiss an employee for doing the right thing. Our employees are instructed and expected to uphold the law and to enforce the policies of the client institutions they serve."[8]

Perry and Gross claimed they had done nothing wrong. They filed a $15 million lawsuit that charged New York Law School and their employer, Collegis, with retaliation for their reporting of the incident. In their suit, the two contend that the child pornography was a form of sexual harassment that violated New York City's "human rights" law. Their attorney stated that Perry and Gross "had a right to a workplace free from degrading and offensive pornography."[9] In October 2004, a judge dismissed some of the claims and ruled that the action did not constitute sexual harassment. However, the judge stated: "That they were terminated shortly after they reported finding child pornography, and despite unblemished employment records, raises a substantial question as to whether the defendants were fired for reporting the professor's alleged criminal activities."[10]

Gross and Perry, after several months, found new jobs in IT support. Gross settled the suit. Perry continued to fight for years, and in October 2008, six years after she was fired, the Appellate Division of the Supreme Court of New York ruled that Perry's employment with Collegis was terminable at will and that Perry had no tenable claim that New York Law School acted for the sole purpose of harming her.

Questions to Consider

1. What message is sent to IT workers by the actions of New York Law School and Collegis—even if unrelated job-performance issues justified their actions in firing Gross and Perry?

2. Since this incident, a number of states have enacted laws that require workers to report immediately any child pornography found while servicing equipment. Most of the laws state that a worker who reports such a discovery is immune from any criminal, civil, or administrative liability. Failure to report the discovery can result in a fine, imprisonment, or both. Do you think such laws will encourage reporting? Why or why not?

LEARNING OBJECTIVES

As you read this chapter, consider the following questions:

1. What key characteristics distinguish a professional from other kinds of workers, and is an IT worker considered a professional?

2. What factors are transforming the professional services industry?

3. What relationships must an IT worker manage, and what key ethical issues can arise in each?

4. How do codes of ethics, professional organizations, certification, and licensing affect the ethical behavior of IT professionals?

5. What are the key tenets of five different codes of ethics that provide guidance for IT professionals?

IT PROFESSIONALS

A **profession** is a calling that requires specialized knowledge and often long and intensive academic preparation. Over the years, the United States government adopted labor laws and regulations that required a more precise definition of what is meant by a *professional* employee. The U.S. Code of Federal Regulations defines a person "employed in a professional capacity" as one who meets these four criteria:

1. One's primary duties consist of the performance of work requiring knowledge of an advanced type in a field of science or learning customarily acquired by a prolonged course of specialized intellectual instruction and study or work.
2. One's instruction, study, or work is original and creative in character in a recognized field of artistic endeavor, the result of which depends primarily on the invention, imagination, or talent of the employee.
3. One's work requires the consistent exercise of discretion and judgment in its performance.
4. One's work is predominantly intellectual and varied in character, and the output or result cannot be standardized in relation to a given period of time.[11]

In other words, professionals such as doctors, lawyers, and accountants require advanced training and experience; they must exercise discretion and judgment in the course of their work; and their work cannot be standardized. Many people would also expect professionals to contribute to society, to participate in a lifelong training program (both formal and informal), to keep abreast of developments in their field, and to assist other professionals in their development. In addition, many professional roles carry special rights and responsibilities. Doctors, for example, prescribe drugs, perform surgery, and request confidential patient information while maintaining doctor–patient confidentiality.

Are IT Workers Professionals?

Many business workers have duties, backgrounds, and training that qualify them to be classified as professionals, including marketing analysts, financial consultants, and IT specialists. A partial list of IT specialists includes programmers, systems analysts, software engineers, database administrators, local area network (LAN) administrators, and chief information officers (CIOs). One could argue, however, that not every IT role requires "knowledge of an advanced type in a field of science or learning customarily acquired by a prolonged course of specialized intellectual instruction and study," to quote again from the U.S. Code of Federal Regulations. From a *legal* perspective, IT workers are not recognized as professionals because they are not licensed by the state or federal government. This distinction is important, for example, in malpractice lawsuits, as many courts have ruled that IT workers are not liable for malpractice because they do not meet the legal definition of a professional.

The Changing Professional Services Industry

Although not legally classified as professionals, IT workers are considered part of the professional services industry, which is experiencing immense changes that impact how members of this industry must think and behave to be successful. Ross Dawson, author and CEO of the consulting firm Advanced Human Technology, identifies seven forces that are changing the nature of professional services: client sophistication, governance, connectivity, transparency, modularization, globalization, and commoditization.[12]

Client Sophistication

Clients are more aware of what they need from service providers, more willing to look outside their own organization to get the best possible services, and better able to drive a hard bargain to get the best possible services at the lowest possible cost.

Governance

Major scandals and tougher laws enacted to avoid future scandals (e.g., Sarbanes-Oxley) have created an environment in which there is less trust and more oversight in client–service provider relationships.

Connectivity

Clients and service providers have built their working relationships on the expectation that they can communicate easily and instantly around the globe through electronic teleconferences, audio conferences, e-mail, and wireless devices.

Transparency

Clients expect to be able to see work-in-progress in real time, and they expect to be able to influence that work. No longer are clients willing to wait until the end product is complete before they weigh in with comments and feedback.

Modularization

Clients are able to break down their business processes into the fundamental steps and decide which they will perform themselves and which they will outsource to service providers.

Globalization

Clients are able to evaluate and choose among service providers around the globe, making the service provider industry extremely competitive.

Commoditization

Clients look at the delivery of low-end services (e.g., staff augmentation to complete a project) as a commodity service for which price is the primary criterion for choosing a service provider. For the delivery of high-end services (e.g., development of an IT strategic plan), clients seek to form a partnership with their service providers.

Professional Relationships That Must Be Managed

IT workers typically become involved in many different relationships, including those with employers, clients, suppliers, other professionals, IT users, and the society at large. In each relationship, an ethical IT worker acts honestly and appropriately. These various relationships are discussed in the following sections.

Relationships Between IT Workers and Employers

IT workers and employers have a critical, multifaceted relationship that requires ongoing effort by both parties to keep it strong. An IT worker and an employer typically agree on fundamental aspects of this relationship before the worker accepts an employment offer. These issues can include job title, general performance expectations, specific work responsibilities, drug-testing requirements, dress code, location of employment, salary, work hours, and company benefits. Many other issues are addressed in the company's policy and procedures manual or in the company's code of conduct, if one exists. These issues include protection of company secrets; vacation policy; time off for a funeral or an illness in the family; tuition reimbursement; and use of company resources, including computers and networks.

Other aspects of the relationship develop over time as the need arises (for example, whether the employee can leave early one day if the time is made up another day). Some aspects are addressed by law—for example, an employee cannot be required to do anything illegal, such as falsify the results of a quality assurance test. Some aspects are specific to the role of the IT worker and are established based on the nature of the work or project—for example, the programming language to be used, the type and amount of documentation to be produced, and the extent of testing to be conducted.

As the stewards of an organization's IT resources, IT workers must set an example and enforce policies regarding the ethical use of IT. IT workers have the skills and knowledge to abuse systems and data or to allow others to do so. Software piracy is an area in which IT workers can be tempted to violate laws and policies.

Although end users often get the blame when it comes to using illegal copies of commercial software, software piracy in a corporate setting is sometimes directly traceable to IT staff members—either they allow it to happen or they actively engage in it, often to reduce IT-related spending. As mentioned in Chapter 1, 38 percent of the world's software was illegally copied in 2007, representing lost revenue of $48 billion. These costs impact the software manufacturers, software distributors, and service providers. In the United States, state and local governments lose sales tax revenue.

The **Business Software Alliance (BSA)** is a trade group that represents the world's largest software and hardware manufacturers. Its mission is to stop the unauthorized copying of software produced by its members (see Table 2-1). More than 100 BSA lawyers and investigators prosecute thousands of cases of software piracy each year. BSA investigations are usually triggered by calls to the BSA hotline (1-888-NO-PIRACY), reports sent to the BSA Web site, and referrals from member companies. Many of these cases are reported by disgruntled employees. In addition, under BSA's "Know It, Report It, Reward It" program, individuals who report software piracy are eligible to receive up to $1 million in cash rewards. Although BSA paid a total of $136,100 in rewards in 2008, most informants opt out of the reward program, citing their motivation to simply "do the right thing."[13] Each year, BSA receives over 2,500 piracy reports.[14] When BSA finds cases of software piracy, it assesses heavy monetary penalties. BSA is funded both through dues based on member companies' software revenues and through settlements from companies that commit piracy.

In 2009, XMCO, a Michigan-based subsidiary of Koniag Development Corporation that writes and produces technical manuals for military equipment and vehicles, paid BSA $70,000 to settle a claim that it had installed unlicensed copies of Adobe, Corel, and

TABLE 2-1 Members of Business Software Alliance (as of January 2009)

Adobe	Apple	Autodesk
Bentley Systems	Borland	CA
Cadence	Cisco Systems	CNC Software-Mastercam
Corel	CyberLink	Dassault Systemes SolidWorks Corporation
Dell	EMC	HP (Hewlett-Packard)
IBM	Intel	Intuit
McAfee	Microsoft	Mindjet
Minitab	Monotype Imaging	Quark
Quest	Rosetta Stone	SAP
Siemens	Sybase	Symantec

Microsoft software on its computers. XMCO agreed to delete all unlicensed copies of software, buy the necessary number of licenses to become compliant, and commit itself to implementing improved practices for managing software licenses.[15]

In December 2008, a 24-year-old college student was sentenced to three years in jail; fined $10,000; and had his $40,000 Porsche, HDTV, and computer equipment seized as part of his sentence for engaging in software piracy. He and his friends operated for-profit Web sites that allowed unsuspecting customers to download pirated copies of Adobe and Macromedia software.[16]

Failure to cooperate with BSA can be extremely expensive. The cost of criminal or civil penalties to a corporation and the people involved can easily be many times more expensive than the cost of "getting legal" by acquiring the correct number of software licenses. Software manufacturers can file a civil suit against software pirates with penalties of up to $100,000 per copyrighted work. Furthermore, the government can criminally prosecute violators and fine them up to $250,000, incarcerate them for up to five years, or both.

Trade secrecy is another area that can present challenges for IT workers and their employers. A **trade secret** is information, generally unknown to the public, that a company has taken strong measures to keep confidential. It represents something of economic value that has required effort or cost to develop and that has some degree of uniqueness or novelty. Trade secrets can include the design of new software code, hardware designs, business plans, the design of a user interface to a computer program, and manufacturing processes. Examples include the Colonel's secret recipe of 11 herbs and spices, the formula for Coke, and Intel's manufacturing process for the i7 quad core processing chip. Employers fear that employees may reveal these secrets to competitors, especially when they leave the company. As a result, they often require employees to sign confidentiality agreements and promise not to reveal the company's trade secrets. However, the IT industry is known for high employee turnover, and things can get complicated when an employee moves on to a competitor.

In September 2006, a vice president of HP's printer division was fired from his new job after just four months. Shortly after his employment began with HP, he e-mailed confidential information from his former employer, IBM, to two senior vice presidents at HP. The information was marked confidential on each page and included product costs and material data that could help the HP sales team understand the goals of IBM. HP quickly investigated, fired the employee, and reported him to IBM and law enforcement authorities. He faces a possible sentence of 10 years' imprisonment and a $250,000 fine.[17]

Another issue that can create friction between employers and IT workers is whistle-blowing. **Whistle-blowing** is an effort by an employee to attract attention to a negligent, illegal, unethical, abusive, or dangerous act by a company that threatens the public interest. Whistle-blowers often have special information based on their expertise or position within the offending organization. For example, an employee of a chip manufacturing company may know that the chemical process used to make the chips is dangerous to employees and the general public. A conscientious employee would call the problem to management's attention and try to correct it by working with appropriate resources within the company. But what if the employee's attempt to correct the problem through internal channels was thwarted or ignored? The employee could then consider becoming a whistle-blower and reporting the problem to people outside the company, including state or federal agencies that have jurisdiction. Obviously, such actions could have negative consequences on the employee's job, perhaps resulting in retaliation and firing.

In May 2005, Oracle Corporation paid $8 million to settle charges that it fraudulently collected fees before providing training for clients and failed to comply with federal travel regulations in billing for travel and expenses. The charges arose from a whistle-blower lawsuit brought by a former Oracle vice president. As a result of the settlement, the whistle-blower received $1.58 million of the $8 million total settlement.

Relationships Between IT Workers and Clients

An IT worker often provides services to clients who either work outside the worker's own organization or are "internal." In relationships between IT workers and clients, each party agrees to provide something of value to the other. Generally speaking, the IT worker provides hardware, software, or services at a certain cost and within a given time frame. For example, an IT worker might agree to implement a new accounts payable software package that meets the client's requirements. The client provides compensation, access to key contacts, and perhaps a work space. This relationship is usually documented in contractual terms—who does what, when the work begins, how long it will take, how much the client pays, and so on. Although there is often a vast disparity in technical expertise between IT workers and their clients, the two parties must work together to be successful.

Typically, the client makes decisions about a project on the basis of information, alternatives, and recommendations provided by the IT worker. The client trusts the IT worker to use his or her expertise and to act in the client's best interests. The IT worker must trust that the client will provide relevant information, listen to and understand what the IT worker says, ask questions to understand the impact of key decisions, and use the information to make wise choices among various alternatives. Thus, the responsibility for decision making is shared between client and IT worker.

One potential ethical problem that can interfere with the relationship between IT workers and their clients involves IT consultants or auditors who recommend their own products and services or those of an affiliated vendor to remedy a problem they have detected. For example, an IT consulting firm might be hired to assess a firm's IT strategic plan. After a few weeks of analysis, the consulting firm might provide a poor rating for the existing strategy and insist that its proprietary products and services are required to develop a new strategic plan. Such findings raise questions about the vendor's objectivity and whether its recommendations can be trusted.

Problems can also arise during a project if IT workers find themselves unable to provide full and accurate reporting of the project's status due to a lack of information, tools, or experience needed to perform an accurate assessment. The project manager may want to keep resources flowing into the project and hope that problems can be corrected before anyone notices. The project manager may also be reluctant to share status information because of contractual penalties for failure to meet the schedule or to develop certain system functions. In this situation, the client may not be informed about the problem until it has become a crisis. After the truth comes out, finger-pointing and heated discussions about cost overruns, missed schedules, and technical incompetence can lead to charges of fraud, misrepresentation, and breach of contract.

Fraud is the crime of obtaining goods, services, or property through deception or trickery. Fraudulent misrepresentation occurs when a person consciously decides to induce another person to rely and act on the misrepresentation. To prove fraud in a court of law, prosecutors must demonstrate the following elements:

- The wrongdoer made a false representation of material fact.
- The wrongdoer intended to deceive the innocent party.
- The innocent party justifiably relied on the misrepresentation.
- The innocent party was injured.

Stein Bagger was a Danish-Norwegian businessman indicted in December 2008 by Danish authorities for fraud based on an advanced leasing scam. Bagger's company, IT Factory, provided hosted applications using IBM and HP hardware and software. Bagger allegedly created fake contracts with companies that did not exist, and then sold the contracts to banks and other investors. Major U.S. IT firms were caught in the fraud, with IBM losing $25 million and HP losing $5 million.[18]

Misrepresentation is the misstatement or incomplete statement of a material fact. If the misrepresentation causes the other party to enter into a contract, that party may have the legal right to cancel the contract or seek reimbursement for damages.

As an example of misrepresentation, an Oracle marketing campaign claimed that retail companies using its software are "49.7 percent more profitable and 61.5 percent more capital efficient than peers." The statement was based on analysis performed by Oracle on data provided by Stratascope, a market research firm. The Stratascope CEO demanded that Oracle issue a correction, claiming, "We have no knowledge of the validity of the criteria and the methodology [Oracle] used, particularly because several of its claims are based on a set of data we do not possess."[19]

Breach of contract occurs when one party fails to meet the terms of a contract. Further, a **material breach of contract** occurs when a party fails to perform certain express or implied obligations, which impairs or destroys the essence of the contract. Because there is no clear line between a minor breach and a material breach, determination is made on a case-by-case basis. "When there has been a material breach of contract, the non-breaching party can either: (1) rescind the contract, seek restitution of any compensation paid under the contract to the breaching party, and be discharged from any further performance under the contract; or (2) treat the contract as being in effect and sue the breaching party to recover damages."[20]

When IT projects go wrong because of cost overruns, schedule slippage, lack of system functionality, and so on, aggrieved parties might charge fraud, fraudulent misrepresentation, or breach of contract. Trials can take years to settle, generate substantial legal fees, and create bad publicity for both parties. As a result, the vast majority of such disputes are settled out of court, and the proceedings and outcomes are concealed from the public. In addition, IT vendors have become more careful about protecting themselves from major legal losses by requiring that contracts place a limit on potential damages.

Most IT projects are joint efforts in which vendors and customers work together to develop a system. Assigning fault when such projects go wrong can be difficult; one side might be partially at fault while the other side is mostly at fault. Consider the following frequent causes of problems in IT projects:

- The customer changes the scope of the project or the system requirements.
- Poor communication between customer and vendor leads to performance that does not meet expectations.
- The vendor delivers a system that meets customer requirements, but a competitor comes out with a system that offers more advanced and useful features.
- The customer fails to reveal information about legacy systems or databases that make the new system extremely difficult to implement.

Who is to blame in such circumstances? For example, the Texas Department of Information Resources and IBM signed a seven-year, $863 million contract in November 2006 for IBM to update the state government's IT infrastructure. The effort involved the consolidation of servers and mainframes of the data centers of 26 state agencies in 1,300 locations to reduce costs and to improve information security and disaster recovery capabilities. In November 2008, the director of the Texas Department of Information Resources, Brian Rawson, said that IBM had "breached its contractual duties and obligations to the state of Texas" by failing to perform the crucial backup of data for more than 20 state agencies. IBM was given a month to fix the backup problems or the state would terminate the contract. The IBM vice president for the Texas project disputed the idea that IBM was at fault for the data backup failures but agreed that it would work to develop a plan to fix the problems.[21]

Relationships Between IT Workers and Suppliers

IT workers deal with many different hardware, software, and service providers. Most IT workers understand that building a good working relationship with suppliers encourages the flow of useful communication as well as the sharing of ideas. Such information can lead to innovative and cost-effective ways of using the supplier's products and services that the IT worker may never have considered.

IT workers should develop good relationships with suppliers by dealing fairly with them and not making unreasonable demands. Threatening to replace a supplier who can't deliver needed equipment tomorrow, when the normal industry lead time is one week, is aggressive behavior that does not help a working relationship.

Suppliers strive to maintain positive relationships with their customers in order to make and increase sales. To achieve this goal, they may sometimes engage in unethical actions—for example, offering an IT worker a gift that is actually intended as a bribe. Clearly, IT workers should not accept a bribe from a vendor, and they must be careful in considering what constitutes a bribe. For example, accepting invitations to expensive dinners or payment of entry fees for a golf tournament may seem innocent to the recipient, but it may be perceived as bribery by an auditor.

Bribery involves providing money, property, or favors to someone in business or government to obtain a business advantage. An obvious example is a software supplier sales representative who offers money to another company's employee to get its business. This type of bribe is often referred to as a kickback or a payoff. The person who offers a bribe commits a crime when the offer is made, and the recipient is guilty of bribery upon accepting the offer. Various states have enacted bribery laws, which have sometimes been used to invalidate contracts involving bribes but have seldom been used to make criminal convictions.

The **U.S. Foreign Corrupt Practices Act (FCPA)** makes it a crime to bribe a foreign official, a foreign political party official, or a candidate for foreign political office. The act applies to any U.S. citizen or company, or to any company with shares listed on any U.S. stock exchange. However, a bribe is not a crime if the payment was lawful under the laws of the foreign country in which it was paid. Penalties for violating the FCPA are severe—corporations face a fine of up to $2 million per violation, and individual violators may be fined up to $100,000 and imprisoned for up to five years.

The FCPA also requires corporations to meet its accounting standards by having an adequate system of internal controls, including maintaining books and records that accurately and fairly reflect their transactions. The goal of these standards is to prevent companies from using slush funds or other means to disguise payments to foreign officials. A firm's business practices and its accounting information systems are frequently audited by both internal and outside auditors to ensure that they meet these standards.

The FCPA permits facilitating payments that are made for "routine government actions," such as obtaining permits or licenses; processing visas; providing police protection; providing phone services, power, or water supplies; or facilitating actions of a similar nature. Thus, it is permissible under the FCPA to pay an official to perform some official function faster (for example, to speed customs clearance) but not to make a different substantive decision (for example, to award business to one's firm).

KBR, a major global engineering, construction, and services company (formerly a part of Halliburton), confessed that it paid $180 million in consulting fees to two agents to bribe Nigerian officials to gain a construction contract. KBR pled guilty to violating the FCPA and paid a $402 million fine. The company stated in a 10-K filing for the SEC that "limitations on our use of agents to comply with applicable laws, including the FCPA, could put us at a competitive disadvantage in pursuing large-scale international projects."[22] KBR may have a point in that 44 percent of U.S. respondents to a survey said they had lost a contract because they did not pay a bribe.[23]

In some countries, gifts are an essential part of doing business. In fact, in some countries, it would be considered rude not to bring a present to an initial business meeting. In the United States, a gift might take the form of free tickets to a sporting event from a personnel agency that wants to get on your company's list of preferred suppliers. At what point does a gift become a bribe, and who decides?

The key distinguishing factor is that no gift should be hidden. A gift may be considered a bribe if it is not declared. As a result, most companies require that all gifts be declared and that everything but token gifts be declined. Some companies have a policy of pooling the gifts received by their employees, auctioning them off, and giving the proceeds to charity.

When it comes to distinguishing between bribes and gifts, the perceptions of the donor and the recipient can differ. The recipient may believe he received a gift that in no way obligates him to the donor, particularly if the gift was not cash. The donor's intentions, however, might be very different. Table 2-2 shows some distinctions between bribes and gifts.

TABLE 2-2 Distinguishing between bribes and gifts

Bribes	Gifts
Are made in secret, as they are neither legally nor morally acceptable	Are made openly and publicly, as a gesture of friendship or goodwill
Are often made indirectly through a third party	Are made directly from donor to recipient
Encourage an obligation for the recipient to act favorably toward the donor	Come with no expectation of a future favor for the donor

Relationships Between IT Workers and Other Professionals

Professionals feel a degree of loyalty to the other members of their profession. As a result, they are quick to help each other obtain new positions but slow to criticize each other in public. Professionals also have an interest in their profession as a whole, because how it is perceived affects how individual members are seen and treated. (For example, politicians are not generally thought to be very trustworthy, but teachers are.) Hence, professionals owe each other an adherence to the profession's code of conduct. Experienced professionals can also serve as mentors and help develop new members of the profession.

A number of ethical problems can arise among members of the IT profession. One of the most common is **résumé inflation**, which involves lying on a résumé and claiming competence in an IT skill that is in high demand. Even though an IT worker might benefit in the short term from exaggerating his or her qualifications, such an action can hurt the profession and the individual in the long run. Customers—and society in general—might become much more skeptical of IT workers as a result. Some studies have shown that around 30 percent of all job applicants exaggerate their accomplishments, while roughly 10 percent "seriously misrepresent" their backgrounds.[24]

Another ethical issue is the inappropriate sharing of corporate information. Because of their roles, IT workers have access to corporate databases of private and confidential information about employees, customers, suppliers, new product plans, promotions, budgets, and so on. As discussed in Chapter 1, this information is sometimes shared inappropriately.

It might be sold to other organizations or shared informally during work conversations with others who have no need to know.

Relationships Between IT Workers and IT Users

The term **IT user** distinguishes the person who uses a hardware or software product from the IT workers who develop, install, service, and support the product. IT users need the product to deliver organizational benefits or to increase their productivity.

IT workers have a duty to understand a user's needs and capabilities and to deliver products and services that best meet those needs—subject, of course, to budget and time constraints. IT workers also have a key responsibility to establish an environment that supports ethical behavior by users. Such an environment discourages software piracy, minimizes the inappropriate use of corporate computing resources, and avoids the inappropriate sharing of information.

Relationships Between IT Workers and Society

Regulatory laws establish safety standards for products and services to protect the public. However, these laws are less than perfect, and they fail to safeguard against all negative side effects of a product or process. Often, professionals can clearly see what effect their work will have and can take action to eliminate potential public risks. Thus, society expects members of a profession to provide significant benefits and to not cause harm through their actions. One approach to meeting this expectation is to establish and maintain professional standards that protect the public.

Clearly, the actions of an IT worker can affect society. For example, a systems analyst may design a computer-based control system to monitor a chemical manufacturing process. A failure or an error in the system may put workers or residents near the plant at risk. As a result, IT workers have a relationship with society members who may be affected by their actions. However, there is currently no single, formal organization of IT workers that takes responsibility for establishing and maintaining standards that protect the public. However, as discussed in the following sections, there are a number of professional organizations that provide useful professional codes of ethics to guide actions that support the ethical behavior of IT workers.

Professional Codes of Ethics

A **professional code of ethics** states the principles and core values that are essential to the work of a particular occupational group. Practitioners in many professions subscribe to a code of ethics that governs their behavior. For example, doctors adhere to varying versions of the 2000-year-old Hippocratic oath, which medical schools offer as an affirmation to their graduating classes. Most codes of ethics created by professional organizations have two main parts: the first outlines what the organization aspires to become, and the second typically lists rules and principles by which members of the organization are expected to abide. Many codes also include a commitment to continuing education for those who practice the profession. (For examples of professional codes of ethics, see Appendices B through F.)

Laws do not provide a complete guide to ethical behavior. Just because an activity is not defined as illegal does not mean it is ethical. You also cannot expect a professional code of ethics to provide an answer to every ethical dilemma—no code can be the definitive collection of behavioral standards. However, following a professional code of ethics can produce many benefits for the individual, the profession, and society as a whole:

- *Ethical decision making*—Adherence to a professional code of ethics means that practitioners use a common set of core values and beliefs as a guideline for ethical decision making.
- *High standards of practice and ethical behavior*—Adherence to a code of ethics reminds professionals of the responsibilities and duties that they may be tempted to compromise to meet the pressures of day-to-day business. The code also defines behaviors that are acceptable and unacceptable to guide professionals in their interactions with others. Strong codes of ethics have procedures for censuring professionals for serious violations, with penalties that can include the loss of the right to practice. Such codes are the exception, however, and few exist in the IT arena.
- *Trust and respect from the general public*—Public trust is built on the expectation that a professional will behave ethically. People must often depend on the integrity and good judgment of a professional to tell the truth, abstain from giving self-serving advice, and offer warnings about the potential negative side effects of their actions. Thus, adherence to a code of ethics enhances trust and respect for professionals and their profession.
- *Evaluation benchmark*—A code of ethics provides an evaluation benchmark that a professional can use as a means of self-assessment. Peers of the professional can also use the code for recognition or censure.

Professional Organizations

No IT professional organization has emerged as preeminent, so there is no universal code of ethics for IT workers. However, the existence of such organizations is useful in a field that is rapidly growing and changing. In order to stay on top of the many new developments in their field, IT workers need to network with others, seek out new ideas, and continually build on their personal skills and expertise. Whether you are a freelance programmer or the CIO of a Fortune 500 company, membership in an organization of IT workers enables you to associate with others of similar work experience, develop working relationships, and exchange ideas. These organizations disseminate information through e-mail, periodicals, Web sites, meetings, and conferences. Furthermore, in recognition of the need for professional standards of competency and conduct, many of these organizations have developed a code of ethics. Five of the most prominent IT-related professional organizations are highlighted in the following sections.

Association for Computing Machinery (ACM)

The Association for Computing Machinery (ACM) is a computing society founded in 1947 with 24,000 student members and 68,000 professional members in more than 100 countries. It offers many publications and electronic forums for technology workers, including *Tech News*, a comprehensive news-gathering service; *Career News*, a career news digest; *RISKS Forum*, a moderated dialogue on risks to the public from computers and related systems; *Queuecasts*, a set of podcasts with IT experts; *eLearn*, an online magazine about online education and training; and *Ubiquit*, a forum and opinion magazine. The organization also offers a substantial digital library of bibliographic information, citations, articles, and journals. The ACM sponsors 34 special-interest groups (SIGs), representing major areas of computing. Each group provides publications, workshops, and conferences for information exchange.

The ACM has a code of ethics and professional conduct with supplemental explanations and guidelines. The ACM code consists of eight general moral imperatives, eight specific professional responsibilities, six organizational leadership imperatives, and two elements of compliance. The complete text of this code is provided in Appendix B.

Association of Information Technology Professionals (AITP)

The Association of Information Technology Professionals (AITP) started in Chicago in 1951, when a group of machine accountants got together and decided that the future was bright for the IBM punched-card tabulating machines they were operating—a precursor of the modern electronic computer. They were members of a local group called the Machine Accountants Association (MAA), which first evolved into the Data Processing Management Association in 1962 and finally the AITP in 1996.

The AITP provides IT-related seminars and conferences, information on IT issues, and forums for networking with other IT workers for about 6,000 members. It has been a leader in the development of model curricula for four-year institutions, with the most current one issued in December 2002. Its mission is to provide superior leadership and education in information technology, and one of its goals is to help members make themselves more marketable within their industry. The AITP also has a code of ethics and standards of conduct, which are presented in Appendix C. The standards of conduct are considered to be rules that no true IT professional should violate.

Institute of Electrical and Electronics Engineers Computer Society (IEEE-CS)

The Institute of Electrical and Electronics Engineers (IEEE) covers the broad fields of electrical, electronic, and information technologies and sciences. The IEEE-CS is one of the oldest and largest IT professional associations, with about 85,000 members. Founded in 1946, the IEEE-CS is one of the largest of the three dozen societies of the IEEE. The IEEE-CS helps meet the information and career development needs of computing researchers and practitioners with technical journals, magazines, conferences, books, conference publications, and online courses. It also offers a Certified Software Development Professional (CSDP) program for experienced professionals and a Certified Software Development Associate (CSDA) for recent college graduates.[25] The society sponsors many conferences, applications-related and research-oriented journals, local and student chapters, technical committees, and standards working groups.

In 1993, the IEEE-CS and the ACM formed a Joint Steering Committee for the Establishment of Software Engineering as a Profession. The initial recommendations of the committee were to define ethical standards, to define the required body of knowledge and recommended practices in software engineering, and to define appropriate curricula to acquire knowledge. The Software Engineering Code of Ethics and Professional Practice (see Appendix D) documents the ethical and professional responsibilities and obligations of software engineers.

Project Management Institute (PMI)

The Project Management Institute (PMI) was established in 1969 and has more than 420,000 members and people who have passed the PMI certification process in more than

170 countries. Its members include project managers from such diverse fields as construction, sales, finance, and production, as well as information systems. The PMI Member Code of Ethics and Professional Conduct is presented in Appendix E.

SysAdmin, Audit, Network, Security (SANS) Institute

The SysAdmin, Audit, Network, Security (SANS) Institute provides information security training and certification for a wide range of individuals, such as auditors, network administrators, and security managers. Each year, its programs train some 12,000 people, and a total of more than 165,000 security professionals around the world have taken one or more of its courses. SANS publishes a weekly news digest (NewsBites), a weekly security vulnerability digest (@Risk), and flash security alerts. The SANS IT Code of Ethics is provided in Appendix F.

At no cost, SANS makes available a collection of some 1,200 research documents about various topics of information security. SANS also operates Internet Storm Center—a program that monitors malicious Internet activity and provides a free early warning service to Internet users—and works with Internet service providers to thwart the malicious attackers.

Certification

Certification indicates that a professional possesses a particular set of skills, knowledge, or abilities, in the opinion of the certifying organization. Unlike licensing, which applies only to people and is required by law, certification can also apply to products (e.g., the Wi-Fi CERTIFIED logo assures that the product has met rigorous interoperability testing to ensure that it will work with other Wi-Fi-certified products) and is generally voluntary. IT-related certifications may or may not include a requirement to adhere to a code of ethics, whereas such a requirement is standard with licensing.

Numerous companies and professional organizations offer certifications, and opinions are divided on their value. Many employers view them as a benchmark that indicates mastery of a defined set of basic knowledge. On the other hand, because certification is no substitute for experience and doesn't guarantee that a person will perform well on the job, some hiring managers are rather cynical about the value of certifications. Most IT employees are motivated to learn new skills, and certification provides a structured way of doing so. For such people, completing a certification provides clear recognition and correlates with a plan to help them continue to grow and advance in their careers. Others view certification as just another means for product vendors to generate additional revenue with little merit attached.

Vendor Certifications

Many IT vendors—such as Cisco, IBM, Microsoft, Sun, SAP, and Oracle—offer certification programs for their products. Workers who successfully complete a program can represent themselves as certified users of a manufacturer's product. Depending on the job market and the demand for skilled workers, some certifications might substantially improve an IT worker's salary and career prospects. Certifications that are tied to a vendor's product are relevant for job roles with very specific requirements or certain aspects of broader roles. Sometimes, however, vendor certifications are too focused on technical details of the vendor's technology and do not address more general concepts.

To become certified, one must pass a written exam. Because of legal concerns about whether other types of exams can be graded objectively, most exams are presented in a multiple-choice format. A few certifications, such as the Cisco Certified Internetwork Expert (CCIE) certification, also require a hands-on lab exam that demonstrates skills and knowledge. It can take years to obtain the necessary experience required for some certifications. Courses and training material are available to help speed up the preparation process, but some training costs can be expensive. Depending on the certification, study materials can cost $1,000 or more, and in-class formal training courses often cost more than $10,000.

Right now, probably no certification carries as much value as a certification in SAP—in terms of employment opportunities and higher salaries. During a six-month period in 2008, some 18,000 professionals earned new SAP-related certifications. In spite of this surge in trained resources, SAP executives fear that the firm may lose potential new customers to its competitors because finding SAP help will be too difficult or costly. The current shortfall is estimated to be 30,000 to 40,000 SAP experts globally. To address that need, SAP has built alliances with some 900 universities around the globe to deliver training and certify additional resources. SAP has so many certifications that it has posted a tool on its Web site to help people identify the appropriate certification for their role. The tool requires you to enter the solution (SAP NetWeaver, SAP ERP, SAP Customer Relationship Management), focus area (applications, development, or technology), and role (architect, application consultant, industry consultant) in which you are interested. The tool then suggests a number of appropriate SAP training opportunities from which to choose.[26]

Industry Association Certifications

There are many industry certifications in a variety of IT-related subject areas (see Table 2-3). Their value varies greatly depending on where people are in their career path, what other certifications they possess, and the nature of the IT job market. For example, according to a 2007 study by Foote Partners LLC, formally certified security professionals are generally paid at least 10 percent more than noncertified individuals.[27]

The requirements for certification generally require that the individual has the prerequisite education and experience, sits for and passes an exam, and commits to and abides by a code of ethics established by the organization providing the certification. In order to remain certified, the individual must typically pay an annual certification fee, earn continuing education credits, and—in some cases—pass a periodic renewal test.

Certifications from industry associations generally require a higher level of experience and a broader perspective than vendor certifications; however, industry associations often lag in developing tests that cover new technologies. The trend in IT certification is to move from purely technical content to a broader mix of technical, business, and behavioral competencies, which are required in today's demanding IT roles. This trend is evident in industry association certifications that address broader roles, such as project management and network security.

TABLE 2-3 IT subject-area certifications (vendor neutral)

Subject area	Organization providing certification	Primary certification(s)
Auditing	Information Systems Audit and Control Association (ISACA)	Certified Information Systems Auditor (CISA)
General	Institute for Certification of Computing Professionals (ICCP)	Certified Computing Professional (CCP)
Security	International Information Systems Security Certification Consortium, Inc. (ISC)^2	Certified Information Systems Security Professional (CISSP)
	SysAdmin, Audit, Network, Security (SANS) Institute	Global Information Assurance Certification (GIAC)
Computer service technician	Computing Technology Industry Association (CompTIA)	CompTIA

Due to the ongoing need for strong project managers, some of the most widely recognized and most sought-after certifications come from the Project Management Institute, which offers certification at several different levels, as summarized in Table 2-4. There are over 300,000 people worldwide who have earned some form of PMI certification.[28]

TABLE 2-4 Available PMI certifications

Certificate	Intended for	Primary eligibility requirement
Project Management Professional (PMP)	Individuals who lead and direct projects	3–5 years of project management experience, depending on level of college education
Certified Associate of Project Management (CAPM)	Individuals who contribute to project teams	1,500 hours of experience or 23 hours of project management education
Program Management Professional (PgMP)	Individuals who achieve organizational objectives through defining and overseeing projects and resources	4–7 years of both project management and program management experience, depending on level of college education
PMI Scheduling Professional (PMI-SP)	Individuals who develop and maintain project schedules	3,500–5,000 hours of project scheduling experience and 30–40 hours of project scheduling education, depending on level of college education
PMI Risk Management Professional (PMI-RMP)	Individuals who assess and identify risks, mitigate threats, and identify and take advantage of unique project opportunities	3,000–4,500 hours of project risk management experience and 30–40 hours of project risk management education, depending on level of college education

PMI certification requires that a person meet specific education and experience requirements, agree to follow the PMI Code of Ethics, and pass the necessary PMI exam for the desired certification. College business majors are well advised to obtain PMI's Certified Associate of Project Management prior to their graduation as a means of opening up future employment opportunities.

Government Licensing

In the United States, a **government license** is government-issued permission to engage in an activity or to operate a business. It is generally administered at the state level and often requires that the recipient pass a test of some kind. Some professionals must be licensed, including certified public accountants (CPAs), lawyers, doctors, various types of medical- and day-care providers, and some engineers.

Various states have enacted legislation to establish licensing requirements and protect public safety in a variety of fields. For example, Texas passed the Engineering Registration Act after a tragic school explosion at New London, Texas, in 1937. Under the act and subsequent revisions, only duly licensed people may legally perform engineering services for the public, and public works must be designed and constructed under the direct supervision of a licensed professional engineer. People cannot call themselves engineers or professional engineers unless they are licensed, and violators are subject to legal penalties. Most states have similar laws.

The Case for Licensing IT Workers

The days of simple, stand-alone information systems are over. Modern systems are highly complex, interconnected, and critically dependent on one another. Highly integrated enterprise resource planning (ERP) systems help multibillion-dollar companies control all of their business functions, including forecasting, production planning, purchasing, inventory control, manufacturing, and distribution. Complex computers and information systems manage and control the nuclear reactors of power plants that generate electricity. Medical information systems monitor the vital statistics of hospital patients on critical life support. Every year, local, state, and federal government information systems are entrusted with generating and distributing millions of checks worth billions of dollars to the public.

As a result of the increasing importance of IT in our everyday lives, the development of reliable, effective information systems has become an area of mounting public concern. This concern has led to a debate about whether the licensing of IT workers would improve information systems. Proponents argue that licensing would strongly encourage IT workers to follow the highest standards of the profession and practice a code of ethics. Licensing would also allow for violators to be punished. Without licensing, there are no requirements for heightened care and no concept of professional malpractice.

Issues Associated with Government Licensing of IT Workers

Australia, Great Britain, and the Canadian provinces of Ontario and British Columbia have adopted licensing for software engineers. The National Council of Examiners for Engineering and Surveying (NCEES) has developed a professional exam for electrical engineers and computer engineers. However, there are many reasons why there are few international or national licensing programs for IT professionals:

- *There is no universally accepted core body of knowledge.* The core **body of knowledge** for any profession outlines agreed-upon sets of skills and abilities that all licensed professionals must possess. At present, however, there are no universally accepted standards for licensing programmers, software engineers, and other IT workers. Instead, various professional societies, state agencies, and federal governments have developed their own standards.

- *It is unclear who should manage the content and administration of licensing exams.* How would licensing exams be constructed, and who would be responsible for designing and administering them? Would someone who passes a license exam in one state or country be accepted in another state or country? In a field as rapidly changing as IT, workers must clearly commit to ongoing, continuous education. If an IT worker's license were to expire every few years (like a driver's license), how often would practitioners be required to prove competence in new practices in order to renew their license? Such questions would normally be answered by the state agency that licenses other professionals.

- *There is no administrative body to accredit professional education programs.* Unlike the American Medical Association for medical schools or the American Bar Association for law schools, no single body accredits professional education programs for IT. Furthermore, there is no well-defined, step-by-step process to train IT workers, even for specific jobs such as programming. There is not even broad agreement on what skills a good programmer must possess; it is highly situational, depending on the computing environment.

- *There is no administrative body to assess and ensure competence of individual workers.* Lawyers, doctors, and other licensed professionals are held accountable to high ethical standards and can lose their license for failing to meet those standards or for demonstrating incompetence. The AITP standards of conduct state that professionals should "take appropriate action in regard to any illegal or unethical practices that come to [their] attention. However, [they should] bring charges against any person only when [they] have reasonable basis for believing in the truth of the allegations and without any regard to personal interest." The AITP code addresses the censure issue much more forcefully than other IT codes of ethics, although it has seldom, if ever, been used to censure practicing IT workers.

IT Professional Malpractice

Negligence has been defined as not doing something that a reasonable person would do, or doing something that a reasonable person would not do. **Duty of care** refers to the obligation to protect people against any unreasonable harm or risk. For example, people have a duty to keep their pets from attacking others and to operate their cars safely. Similarly, businesses must keep dangerous pollutants out of the air and water, make safe products, and maintain safe operating conditions for employees.

The courts decide whether parties owe a duty of care by applying a **reasonable person standard** to evaluate how an objective, careful, and conscientious person would have acted in the same circumstances. Likewise, defendants who have particular expertise or competence are measured against a **reasonable professional standard**. For example, in a medical malpractice suit based on improper treatment of a broken bone, the standard of measure would be higher if the plaintiff were an orthopedic surgeon rather than a general practitioner. In the IT arena, consider a hypothetical negligence case in which an employee inadvertently destroyed millions of customer records in an Oracle database. The standard of measure would be higher if the plaintiff were a licensed, Oracle-certified database administrator (DBA) with 10 years of experience rather than an unlicensed systems analyst with no DBA experience or specific knowledge of the Oracle software.

If a court finds that a defendant actually owed a duty of care, it must determine whether the duty was breached. A **breach of the duty of care** is the failure to act as a reasonable person would act. A breach of duty may consist of an action, such as throwing a lit cigarette into a fireworks factory and causing an explosion, or a failure to act when there is a duty to do so—for example, a police officer not protecting a citizen from an attacker.

Professionals who breach the duty of care are liable for injuries that their negligence causes. This liability is commonly referred to as **professional malpractice**. For example, a CPA who fails to use reasonable care, knowledge, skill, and judgment when auditing a client's books is liable for accounting malpractice. Professionals who breach this duty are liable to their patients or clients, and possibly to some third parties.

Courts have consistently rejected attempts to sue individual parties for computer-related malpractice. Professional negligence can only occur when people fail to perform within the standards of their profession, and software engineering is not a uniformly licensed profession in the United States. Because there are no uniform standards against which to compare a software engineer's professional behavior, he or she cannot be subject to malpractice lawsuits.

IT USERS

Chapter 1 outlined the general topic of how corporations are addressing the increasing risks of unethical behavior. This section focuses on improving employees' ethical use of IT, which is an area of growing concern as more companies provide employees with PCs, access to corporate information systems and data, and the Internet.

Common Ethical Issues for IT Users

This section discusses a few common ethical issues for IT users. Additional ethical issues will be discussed in future chapters.

Software Piracy

As mentioned earlier in this chapter, software piracy in a corporate setting can sometimes be directly traceable to IT professionals—they might allow it to happen, or they might actively engage in it. Corporate IT usage policies and management should encourage users to report instances of piracy and to challenge its practice.

Sometimes IT users are the ones who commit software piracy. A common violation occurs when employees copy software from their work computers for use at home. When confronted, the IT user's argument might be: "I bought a home computer partly so I could take work home and be more productive; therefore, I need the same software on my home computer as I have at work." However, if no one has paid for an additional license to use the software on the home computer, this is still piracy.

Inappropriate Use of Computing Resources

Some employees use their computers to surf popular Web sites that have nothing to do with their jobs, participate in chat rooms, view pornographic sites, and play computer games. These activities eat away at worker productivity and waste time. Furthermore, activities such as viewing sexually explicit material, sharing lewd jokes, and sending hate e-mail could lead to lawsuits and allegations that a company allowed a work environment conducive to racial or sexual harassment. According to a survey by Harris Interactive, 16 percent of men and 8 percent of women with Internet access at work acknowledged that they had seen pornography in the workplace.[29] Companies often fire frequent pornography offenders and take disciplinary action against less egregious offenders. After a monthlong investigation of computer usage habits of Washington, D.C., municipal workers, nine employees were fired and an unspecified number of employees were sanctioned for visiting pornographic Web sites while at work.[30]

Inappropriate Sharing of Information

Every organization stores vast amounts of information that can be classified as either private or confidential. Private data describes individual employees—for example, their salary information, attendance data, health records, and performance ratings. Private data also includes information about customers—credit card information, telephone number, home address, and so on. Confidential information describes a company and its operations, including sales and promotion plans, staffing projections, manufacturing processes, product formulas, tactical and strategic plans, and research and development. An IT user who shares this information with an unauthorized party, even inadvertently, has violated someone's privacy or created the potential that company information could fall into the hands of competitors. For example, if an IT employee saw a coworker's payroll records and then discussed them with a friend, it would be a clear violation of the coworker's privacy.

Supporting the Ethical Practices of IT Users

The growing use of IT has increased the potential for new ethical issues and problems; thus, many organizations have recognized the need to develop policies that protect against abuses. Although no policy can stop wrongdoers, it can set forth the general rights and responsibilities of all IT users, establish boundaries of acceptable and unacceptable behavior, and enable management to punish violators. Adherence to a policy can improve services to users, increase productivity, and reduce costs. Companies can take several of the following actions when creating an IT usage policy.

Establishing Guidelines for Use of Company Software

Company IT managers must provide clear rules that govern the use of home computers and associated software. Some companies negotiate contracts with software manufacturers and provide PCs and software so that IT users can work at home. Other companies help employees buy hardware and software at corporate discount rates. The goal should be to ensure that employees have legal copies of all the software they need to be effective, regardless of whether they work in an office, on the road, or at home.

Defining and Limiting the Appropriate Use of IT Resources

Companies must develop, communicate, and enforce written guidelines that encourage employees to respect corporate IT resources and use them to enhance their job performance. Effective guidelines allow some level of personal use while prohibiting employees from visiting objectionable Internet sites or using company e-mail to send offensive or harassing messages.

Structuring Information Systems to Protect Data and Information

Organizations must implement systems and procedures that limit data access to just those employees who need it. For example, sales managers may have total access to sales and promotion databases through a company network, but their access should be limited to products for which they are responsible. Furthermore, they should be prohibited from accessing data about research and development results, product formulas, and staffing projections if they don't need it to do their jobs.

Installing and Maintaining a Corporate Firewall

A **firewall** is a hardware or software device that serves as a barrier between an organization's network and the Internet; a firewall also limits access to the company's network based on the organization's Internet usage policy. The firewall can be configured to serve as an effective deterrent to unauthorized Web surfing by blocking access to specific objectionable Web sites. Unfortunately, the number of such sites is continually growing, so it is difficult to block them all. The firewall can also serve as an effective barrier to incoming e-mail from certain Web sites, companies, or users. It can even be programmed to block e-mail with certain kinds of attachments (for example, Microsoft Word documents), which reduces the risk of harmful computer viruses.

Table 2-5 presents a manager's checklist that summarizes items to consider when establishing an IT usage policy. The preferred answer in each case is *yes*.

TABLE 2-5   Manager's checklist of items to consider when establishing an IT usage policy

Question	Yes	No
Is there a statement that explains the need for an IT usage policy?		
Does the policy provide a clear set of guiding principles for ethical decision making?		
Is it clear how the policy applies to the following types of workers?		
• Employees		
• Part-time workers		
• Temps		
• Contractors		

TABLE 2-5 Manager's checklist of items to consider when establishing an IT usage policy (continued)

Question	Yes	No
Does the policy address the following issues?		
• Protection of the data privacy rights of employees, customers, suppliers, and others		
• Limits and control of access to proprietary company data and information		
• Use of unauthorized or pirated software		
• Employee monitoring, including e-mail, wiretapping and eavesdropping on phone conversations, computer monitoring, and surveillance by video		
• Respect of the intellectual rights of others, including trade secrets, copyrights, patents, and trademarks		
• Inappropriate use of IT resources, such as Web surfing, personal e-mailing, and other use of computers for purposes other than business		
• The need to protect the security of IT resources through adherence to good security practices, such as not sharing user IDs and passwords, use of "hard-to-guess" passwords, and frequent changing of passwords		
• The use of the computer to intimidate, harass, or insult others through abusive language in e-mails and by other means		
Are disciplinary actions defined for IT-related abuses?		
Is there a process for communicating the policy to employees?		
Is there a plan to provide effective, ongoing training relative to the policy?		
Has a corporate firewall been implemented?		
Is the corporate firewall maintained?		

Summary

- Key characteristics that distinguish professionals from other kinds of workers: (1) they require advanced training and experience, (2) they must exercise discretion and judgment in the course of their work, and (3) their work cannot be standardized.

- A professional is expected to contribute to society, to participate in a lifelong training program, to keep abreast of developments in the field, and to help develop other professionals.

- From a legal standpoint, a professional has passed the state licensing requirements (if they exist) and earned the right to practice there.

- From a legal perspective, IT workers are not recognized as professionals because they are not licensed by the state or federal government. As a result, IT workers are not liable for malpractice.

- As a member of the professional services industry, IT workers must be cognizant of seven major factors that are transforming the professional services industry: (1) increased client sophistication, (2) greater governance requirements, (3) increased connectivity, (4) more transparency, (5) increased need for modularization, (6) growing globalization, and (7) greater commoditization.

- IT professionals typically become involved in many different relationships, each with its own set of ethical issues and potential problems.

- In relationships between IT professionals and employers, important issues include setting and enforcing policies regarding the ethical use of IT, the potential for whistle-blowing, and the safeguarding of trade secrets.

- In relationships between IT professionals and clients, key issues revolve around defining, sharing, and fulfilling each party's responsibilities for successfully completing an IT project.

- A major goal for IT professionals and suppliers is to develop good working relationships in which no action can be perceived as unethical.

- In relationships between IT workers, the priority is to improve the profession through activities such as mentoring inexperienced colleagues and demonstrating professional loyalty.

- Résumé inflation and the inappropriate sharing of corporate information are relevant problems.

- In relationships between IT professionals and IT users, important issues include software piracy, inappropriate use of IT resources, and inappropriate sharing of information.

- When it comes to the relationship between IT professionals and society at large, the main challenge is to practice the profession in ways that cause no harm to society and provide significant benefits.

- A professional code of ethics states the principles and core values that are essential to the work of an occupational group.

- A code serves as a guideline for ethical decision making, promotes high standards of practice and ethical behavior, enhances trust and respect from the general public, and provides an evaluation benchmark.

- Many people believe that the licensing and certification of IT workers would increase the reliability and effectiveness of information systems.

- Licensing and certification raise many issues, including the following: (1) there is no universally accepted core body of knowledge on which to test people; (2) it is unclear who should manage the content and administration of licensing exams; (3) there is no administrative body to accredit professional education programs; and (4) there is no administrative body to assess and ensure competence of individual professionals.

- Several IT-related professional organizations have developed a code of ethics, including ACM, AITP, IEEE-CS, PMI, and SANS.

- These codes have two main parts—the first outlines what the organization aspires to become, and the second typically lists rules and principles that members are expected to live by. They also include a commitment to continuing education for those who practice the profession.

Self-Assessment Questions

The answers to the Self-Assessment Questions can be found in Appendix G.

1. A professional is someone who:
   a. requires advanced training and experience
   b. must exercise discretion and judgment in the course of his or her work
   c. does work that cannot be standardized
   d. all of the above

2. Although end users often get the blame when it comes to using illegal copies of commercial software, software piracy in a corporate setting is sometimes directly traceable to _____.

3. The mission of the Business Software Alliance is to _____.

4. Reporting a trade secret is an effort by an employee to attract attention to a negligent, illegal, unethical, abusive, or dangerous act by a company that threatens the public interest. True or False?

5. _____ is the crime of obtaining goods, services, or property through deception or trickery.

6. Résumé inflation is a usual and customary practice tolerated by employers. True or False?

7. Society expects professionals to act in a way that:
   a. causes no harm to society
   b. provides significant benefits
   c. establishes and maintains professional standards that protect the public
   d. all of the above

8. _____ involves providing money, property, or favors to someone in business or government to obtain a business advantage.

9. _____ is a process that one undertakes voluntarily to prove competency in a set of skills.

   a. Licensing

   b. Certification

   c. Registering

   d. all of the above

10. There are many industry associations and vendor certificate programs for IT workers. True or False?

11. _____ has been defined as not doing something that a reasonable person would do, or doing something that a reasonable person would not do.

12. A _____ states the principles and core values that are essential to the work of a particular occupational group.

Discussion Questions

1. How do you distinguish between a gift and a bribe? Provide an example of a "gift" that falls in the gray area between a gift and a bribe.

2. Discuss the following topic: Laws do not provide a complete guide to ethical behavior. An activity can be legal but not ethical.

3. What is professional malpractice? Can an IT worker ever be sued for professional malpractice? Why or why not?

4. Review the PMI member Code of Ethics in Appendix E. For each point in the code, provide an example of a project manager action that would break the code.

5. What must IT professionals do to ensure that the projects they lead meet the client's expectations and do not lead to charges of fraud, fraudulent misrepresentation, or breach of contract?

6. Should all IT professionals be either licensed or certified? Why or why not?

7. What commonalities do you find among the IT professional codes of ethics discussed in this chapter? What differences are there? Do you think there are any important issues not addressed by these codes of ethics?

What Would You Do?

1. Your old roommate from college was recently let go from his firm during a wave of employee terminations to reduce costs. You two have kept in touch over the six years since school, and he has asked you to help him get a position in the IT organization where you work. You offered to review his résumé, make sure that it gets to the "right person," and even put in a good word for him. However, as you read the résumé, it is obvious that your friend has

greatly exaggerated his accomplishments at his former place of work and even added some IT-related certifications that you are sure he never earned. What would you do?

2. You are a U.S. citizen currently on a three-year assignment as the IT manager for your company's Armenian manufacturing plant. The company is a U.S.-based Fortune 1000 company, and the Armenian plant employs 1,500 workers. The plant budgeted for 250 new computers and associated software to replace all computers over six years old. The cost of the software licenses is $125,000 (USD). Unfortunately, the plant has been hit by the worldwide financial crises, and you have been told that your department must reduce its budget by $100,000 (USD). You have been focusing on two options. The first is to not purchase all the software licenses needed and instead make illegal copies of existing software. (The software piracy rate in Armenia is 93%.) The second is to terminate the employment of two IT employees. You have identified who would be terminated based on their recent job performance and their relatively low level of anticipated future work activity. What would you do?

3. You are the human resources contact for your firm's IT organization. The daughter of the firm's CIO is scheduled to participate in a job interview for an entry-level position in the IT organization next week. It will be your responsibility to meet with the three people who will interview her to form an assessment and make a group decision about whether or not she will be offered the position and, if so, at what salary. How do you handle this situation?

4. You are in charge of awarding all PC service contracts for your employer. In recent e-mails with the company's PC service contractor, you casually exchanged ideas about home landscaping, your favorite pastime. You also mentioned that you would like to have a few Bradford pear trees in your yard. Upon returning from a vacation, you discover three mature trees in your yard, along with a thank-you note in your mailbox from the PC service contractor. You really want the trees, but you didn't mean for the contractor to buy them for you. You suspect that the contractor interpreted your e-mail comment as a hint that you wanted him to buy the trees. You also worry that the contractor still has the e-mail. If the contractor sent your boss a copy, it might look as if you were trying to solicit a bribe. Can the trees be considered a bribe? What would you do?

5. In Italy, *raccomandazione* is the custom of seeking and receiving special treatment from people in power or people close to power. The ability to solicit favors from someone in a higher place, be it through the chief of police or the chief's chauffeur, has been part of the Italian art of getting things done for more than 2000 years. In April 2001, Italy's highest court of appeal ruled that influence peddling is not a crime. The judges did rule, however, that it is a crime to overstate one's power to exert influence.

    Your firm is opening a new sales office in Rome and will be using a local employment agency to identify and screen candidates, who will then undergo employment testing and interviews by members of your organization. What guidelines would you provide to the agency regarding the practice of *raccomandazione* to ensure that the agency operates ethically and effectively?

6. You are an experienced, mid-level manager in your firm's IT organization. One of your responsibilities is to screen résumés for job openings in the organization. You are in the process of reviewing more than 100 résumés you received for a position as an Oracle database administrator. Your goal is to trim the group down to the top 10 candidates to invite to an in-house interview. About half the résumés are from IT workers with less than three years of experience who claim to have one or more Oracle certifications. There are also a few

candidates with over five years of impressive experience but no Oracle certifications listed on their résumés. You were instructed to only include candidates with an Oracle certification in the list of finalists. However, you are concerned about possible résumé inflation and the heavy emphasis on certification versus experience. What would you do?

Cases

1. Google Named in $1 Billion Trade Secret Lawsuit

In February 2007, Google released a suite of software applications called Google Apps to compete against Microsoft's Office (Word, Excel, Outlook). Google Apps are Web-based applications that reside on Google's servers so that users do not need to load large programs onto their computers. At the initial launch of Google Apps, however, Google was missing a key application—there was no solution for Microsoft Outlook users who wanted to move their e-mail, calendar, and contact information to Google Apps.

LimitNone, a Chicago-based software development firm of just four employees, saw the opportunity and developed a product to perform this function that it called MY GRATE. After a successful demo to Google executives in March 2007, Google invited LimitNone to become a part of the Google Enterprise Professional Program to further develop the tool. Under this program, independent, third-party software developers partner with Google to develop solutions together. Such programs are quite common for large software manufacturers, such as Microsoft, SAP, Oracle, IBM, and Apple.

LimitNone claims that Google was clear in saying that it had no interest in developing a competing product. Google asked LimitNone to lower the retail price from $29 to $19 for Google customers and to rename the product gMove; LimitNone did so.

Throughout the rest of 2007, Google promoted gMove on its Web site and got the product in front of its largest customers. According to LimitNone, Google executives continued to reassure the firm that they would not develop a competing product.

In December 2007, Google reversed its position and told LimitNone that it would introduce a competing product (Google Email Uploader), which it would provide at no charge to its Premier customers. Then in May 2008, Google modified its user interface so that gMove no longer worked, and LimitNone was forced to refund its customers.[31]

In June 2008, Kelley Drye & Warren LLP filed a lawsuit on behalf of its client LimitNone, alleging that Google misappropriated trade secrets from LimitNone and committed fraud by enticing LimitNone to share confidential information integral to building Google's competing product. "LimitNone alleges that the Google software is essentially identical to LimitNone's product and that Google could not have developed its competing product without using the information it learned from having access to and studying LimitNone's confidential and proprietary program."[32] The lawsuit also stated: "With gMove priced at $19 per copy and Google's prediction that there were potentially 50 million users, Google deprived LimitNone of a $950 million opportunity....Without Google's knowledge and use of the gMove trade secrets and confidential information, Google would not have been able to solve its long standing Microsoft Outlook to Gmail conversion problem." According to the complaint, a senior Google executive told LimitNone that Google reversed its position because the revenue potential of nearly $1 billion was "just too big to come from someone else."[33]

Kelley Drye & Warren LLP has experience in going up against Google. The firm battled Google in a trademark case against AdWords, Google's online advertising system. After four years of legal haggling, the plaintiffs dropped the AdWords suit before ever going to trial, citing mounting legal costs. Google was able to claim a victory.

Discussion Questions

1. This incident illustrates some of the potential problems for small software developers working with giant software manufacturers to extend or enhance their products. Provide two good reasons why small developers should still consider working closely with large software firms.

2. What measures could LimitNone have taken to better protect itself from Google's alleged actions? What measures could Google have taken to protect itself from this lawsuit?

3. Do research on the Web to find out how this case is proceeding in the courts. Write a short summary of your findings.

2. Waste Management Sues SAP for Alleged Fraud and Breach of Contract

Waste Management, Inc. (WMI) is a provider of integrated waste and environmental services in North America. Through its subsidiaries, it provides collection, transfer, recycling, disposal, and waste-to-energy services. WMI is headquartered in Houston, Texas, and employs over 47,000 people, with a recent annual income of $1 billion based on revenue of $14 billion. The firm's 20 million customers include commercial, industrial, municipal, and residential customers; other waste management companies; electric utilities; and governmental entities.

In December 2005, WMI signed a fixed-fee contract with SAP to transition to a mySAP Business Suite software environment for its order-to-cash process, which includes customer billing, collections, pricing, and customer setup. SAP promised to deliver this system by December 31, 2007. The project goal was to streamline WMI's customer care and billing functions to raise revenue, increase customer satisfaction, and become more competitive.

On March 20, 2008, WMI filed suit in U.S. District Court in Texas. The suit alleged that SAP committed fraudulent inducement, fraud, negligent misrepresentation, and breach of contract by getting WMI to purchase untested software not able to handle the complexities of its business.[34] WMI seeks compensatory and punitive damages.

WMI alleges that SAP assured WMI that its software was mature and represented the industry standard software for the waste industry. WMI alleges that SAP assured the company that mySAP would operate in "the considerably more complex competitive environment" of open pricing for U.S. waste hauling—much different from the government-controlled pricing system in Europe, where SAP software was implemented at dozens of waste and recycling customers—and would not require special customization to meet WMI's needs.[35]

WMI had grown largely through the acquisition of smaller waste-hauling companies—each with its own information systems and business processes. As a result, the firm had a lot of legacy systems that were not well integrated. The promise that SAP could tie together all these disparate systems and processes, with no customization to the standard SAP software, to create an efficient order-to-cash process was enticing. WMI also alleges that it discovered that presale demonstrations of the SAP software were of "fake, mock-up simulations" of software with "false functionality."[36] When WMI conducted an initial pilot of the actual SAP software at a WMI site in New Mexico (unfortunately for WMI, this was after the contract was signed), WMI was dismayed to find that the software "was unable to run [our] most basic revenue management operation."[37] WMI further alleges that SAP wrote tens of thousands of lines of code in a fruitless attempt to address the issues. The pilot was originally scheduled to be completed in December 2006 but was still not complete as of March 2008.

SAP recommended a new development project to provide the functionality needed by WMI. SAP felt that it could finish this new project in 2010 but "without any assurance of success."[38] WMI was not warm to this idea, as it would require the company to "once again act as SAP's guinea pig by agreeing to convert what was supposed to be an 18-month-out-of-the-box implementation into an even more expensive, longer, and highly risky software development project."[39]

SAP attorneys filed a counterclaim to the WMI lawsuit asserting the vendor's innocence. SAP alleged that WMI violated the deal's contract by "failing to timely and accurately define its business requirements; [not providing] sufficient, knowledgeable, decision-empowered users and managers" to complete necessary work on the project; and not successfully migrating data from legacy systems.[40] SAP is seeking millions in maintenance and service fees for the software as well as unspecified compensatory damages.

Meanwhile, the attorneys for both parties are earning their pay. SAP has requested the court delay the proceedings until early 2010 due to the complexity of the case. The firm has also subpoenaed Deloitte Consulting, which served as the system integrator for the project. Waste Management claims that "SAP has sought to delay the case at every turn," and that the trial should begin in April 2009.[41] Both sides are complaining that the other side is providing volumes of meaningless paperwork in the discovery process—WMI has provided nearly 1 million pages of documents, including customer invoices, office building sign-in sheets, and customer addresses, which do not relate specifically to the matter at hand. Meanwhile, SAP has provided over 300,000 pages of documents.[42]

According to Ray Wang at Forrester Research, "large scale implementations are complicated affairs that require alignment among the system integrator, vendor, and client for success. It's a three-legged stool and lawsuits are typically the last resort when any one party faces irreconcilable differences."[43]

Discussion Questions

1. What actions should WMI have taken to lessen the risk of this project and avoid these problems?

2. What sort of losses has WMI incurred from the delay of this project? How has the lack of success on this project affected SAP?

3. Do research on the Web to find out the current status of the lawsuits between WMI and SAP. Write a brief report summarizing your findings.

3. When Certification Is Justified

On June 13, 2005, Don Tennant, editor-in-chief of *Computerworld*, published an editorial in favor of IT certification and was promptly hit with a barrage of angry responses from IT workers.[44] They argued that testable IT knowledge does not necessarily translate into quality IT work. A worker needs good communications and problem-solving skills as well as perseverance to get the job done well. Respondents explained that hardworking IT workers focus on skills and knowledge that are related to their current projects and don't have time for certifications that will quickly become obsolete. They suspected vendors of offering certification as a marketing ploy and a source of revenue. They accused managers without technical backgrounds of using certification as "a crutch, a poor but politically defensible substitute for knowing what and how well one's subordinates are doing."[45]

Any manager would certainly do well to review these insightful points, yet they beg the question: What useful purposes *can* certification serve within an organization?

Some CIOs and vice presidents of technology assert that many employers use certification as a means of training employees and increasing skill levels within the company. Some companies are even using certification as a perk to attract and keep good employees. Such companies may also enhance their employee training programs by offering a job-rotation program through which workers can acquire certification and experience as well.

Employers are also making good use of certification as a hiring gate both for entry-level positions and for jobs that require specific core knowledge. For example, a company with a Windows Server network might run an ad for a systems integration engineer and require a Microsoft Certified Systems Engineer (MCSE) certification. A company that uses Siebel customer relationship management software might require a new hire to have a certification in the latest version of Siebel.

In addition, specific IT fields such as project management and security have a greater need for certification. As the speed and complexity of production increase within the global marketplace, workers in a variety of industries are showing an increasing interest in project management certification. With mottos like "Do It, Do It Right, Do It Right Now," the Project Management Institute has already certified more than 300,000 people. IT industry employers are beginning to encourage and sometimes require project management certification.

Calls for training in the field of security management go beyond certification. The demand for security workers is expected to continue to grow rapidly in the next few years in the face of growing threats. Spam, computer viruses, spyware, botnets, and identity theft have businesses and government organizations worried. They want to make sure that their security managers can protect their data, systems, and resources.

The best recognized security certification is the CISSP, awarded by the International Information Systems Security Certification Consortium (ISC)[2]. Yet the CISSP examination, like so many other IT certification examinations, is multiple choice. Employers and IT workers alike have begun to recognize the limitations of these types of examinations. They want to ensure that examinees not only have core knowledge but also know how to use that knowledge—and a multiple-choice exam, even a six-hour, 250-question exam like the CISSP, can't provide this assurance.

Other organizations are catching on. Sun Microsystems requires the completion of programming or design assignments for some of its certifications. So, while there is no universal call for certification or a uniform examination procedure that answers all needs within the IT profession, certifying bodies are beginning to adapt their programs to better fulfill the evolving needs for certification in IT.

Discussion Questions

1. How can organizations and vendors change their certification programs to test for skills as well as core knowledge? What issues might this introduce?

2. What are the primary arguments against certification, and how can certifying bodies change their programs to overcome these shortcomings?

3. What are the benefits of certification? How might certification programs need to change in the future to better serve the needs of the IT community?

End Notes

[1] Elisabeth Franck, "The Professor and the Porn," *New York*, June 16, 2003.

[2] Elisabeth Franck, "The Professor and the Porn," *New York*, June 16, 2003.

[3] Elisabeth Franck, "The Professor and the Porn," *New York*, June 16, 2003.

[4] Elisabeth Franck, "The Professor and the Porn," *New York*, June 16, 2003.

[5] John Foley, "Law School Professor Sentenced in Child-Porn Case," *InformationWeek*, June 23, 2003.

[6] John Foley, "Troubling Discovery," *InformationWeek*, May 12, 2003.

[7] Elisabeth Franck, "The Professor and the Porn," *New York*, June 16, 2003.

[8] Tom Huber, "Response from Collegis CEO Tom Huber," *InformationWeek*, May 8, 2003.

[9] John Foley, "Troubling Discovery," *InformationWeek*, May 12, 2003.

[10] John Foley, "Work/Life: When Things Go Wrong," *InformationWeek*, August 16, 2004.

[11] U.S. code, Title 5, Part III, Subpart F, Chapter 71, Subchapter 1, Section 7103, www.law. cornell.edu/uscode/5/7103.shtml.

[12] Ross Dawson, "The Seven MegaTrends of Professional Services," 2005, www.rossdawsonblog. com/SevenMegaTrendsofPS.pdf.

[13] Business Software Alliance, "BSA Distributed over $136,000 in Rewards for Software Piracy Tips in 2008," January 15, 2008, www.bsa.org/country/News%20and%20Events/News% 20Archives/en/2009/en-01152009-rewards.aspx.

[14] Business Software Alliance, "Michigan-Based Company Settles with Software Group and Agrees to Pay $70k," January 14, 2009, www.bsa.org/country/News%20and%20Events/News %20Archives/en/2009/en-01142009-xmco.aspx.

[15] Business Software Alliance, "Michigan-Based Company Settles with Software Group and Agrees to Pay $70k," January 14, 2009, www.bsa.org/country/News%20and%20Events/News %20Archives/en/2009/en-01142009-xmco.aspx.

[16] Business Software Alliance, "College Athlete Turned Software Pirate Sentenced to Three Years Behind Bars," December 23, 2008, www.bsa.org/country/News%20and%20Events/ News%20Archives/en/2008/en-12232008-rushing.aspx.

[17] David Kravets, "Former VP for HP Guilty of IBM Trade Secret Theft," *Wired*, July 11, 2008.

[18] "Stein Bagger Accused of $170 ml Fraud," Kaz 911 Rescue Service, December 4, 2008, http://kaz911.blogspot.com/2008/12/stein-bagger-accused-of-170-ml-fraud.html.

[19] Tom Sanders, "Oracle Slapped over Advertising Claims," January 24, 2006, www.vnunet.com/ vnunet/news/2149045/oracle-slapped-false.

[20] Henry R. Cheeseman, *Contemporary Business Law*, 3rd ed. (Upper Saddle River, NJ: Prentice-Hall, 2000), 292.

[21] Emily Ramshaw and Robert T. Garrett, "Texas Gives IBM 30 Days to Fix Computer Glitches," *Dallas Morning News*, November 6, 2008.

[22] Jason Leopold, "KBR Says Abiding by U.S. Laws Puts It at a 'Competitive Disadvantage,'" *Public Record*, March 20, 2009.

[23] Hannah Clark, "Are Anti-Bribe Laws Working?," *Forbes*, October 13, 2006.

24 Ropella, "Hiring Smart; How to Avoid the Top Ten Mistakes," www.ropella.com/index.php/knowledge/articles/hiring_smart.

25 "About the Computer Society," IEEE Computer Society, www.computer.org/portal/site/ieeecs/menuitem.c5efb9b8ade9096b8a9ca0108bcd45f3/index.jsp?&pName=ieeecs_level1&path=ieeecs/about/csinfo&file=index.xml&xsl=generic.xsl&? (accessed February 18, 2009).

26 Marianne Kolbasuk McGee, "What's Hot? SAP Skills and Pay," *InformationWeek*, September 22, 2008.

27 Jaikumar Vijayan, "Salary Premiums for Security Certifications Increasing, Study Shows," *Computerworld*, July 9, 2007.

28 "PMI Credential Overview," Project Management Institute, www.pmi.org/CareerDevelopment/Pages/PMICredentialOverview.aspx (accessed February 8, 2009).

29 Stephanie Armour, "Technology Makes Porn Easier to Access at Work," *USA Today*, October 18, 2007.

30 Jaikumar Vijayan, "D.C. Employees Fired for Viewing Porn Sites at Work," *Computerworld*, January 24, 2008.

31 Stephen Shankland, "Google Faces US $1B Damages in Trade Secret Lawsuit," *CNET News.com*, June 26, 2008.

32 Kelley Drye, "Kelley Drye Represents LimitNone in Lawsuit Against Google," www.kelleydrye.com/news/transaction_matters/0230 (accessed February 6, 2009).

33 "Lawsuit Filed by Kelley Drye & Warren LLP Alleges Google Misappropriated Trade Secrets to Grow Google Apps," *Business Wire*, June 24, 2008.

34 Mary Hayes Weier, "SAP Software a 'Complete Failure,' Lawsuit Claims," *Intelligent Enterprise*, March 2008.

35 SAP America, "Waste Management Selects SAP Safe Passage Program," December 15, 2005.

36 Mary Hayes Weier, "SAP Software a 'Complete Failure,' Lawsuit Claims," *Intelligent Enterprise*, March 2008.

37 Mary Hayes Weier, "SAP Software a 'Complete Failure,' Lawsuit Claims," *Intelligent Enterprise*, March 2008.

38 Chris Kanaracus, "SAP Fires Back at Waste Management," *Network World*, August 14, 2008.

39 Mary Hayes Weier, "SAP Software a 'Complete Failure,' Lawsuit Claims," *Intelligent Enterprise*, March 2008.

40 Chris Kanaracus, "SAP Fires Back at Waste Management," *Network World*, August 14, 2008.

41 Chris Kanaracus, "Accusations Flying in SAP-Waste Management Suit," *InfoWorld*, October 24, 2008.

42 Chris Kanaracus, "Accusations Flying in SAP-Waste Management Suit," *InfoWorld*, October 24, 2008.

43 Chris Kanaracus, "SAP Fires Back at Waste Management," *Network World*, August 14, 2008.

44 Don Tennant, "Certifiably Concerned," *Computerworld*, June 13, 2005.

45 Don Tennant, "Certifiably Mad?" *Computerworld*, June 20, 2005.

CHAPTER 3

COMPUTER AND INTERNET CRIME

VIGNETTE

Trading Scandal at Société Générale

In January 2008, Société Générale (SocGen), France's second largest banking establishment, was a victim of internal fraud carried out by an employee, Jérôme Kerviel. SocGen bank lost €4.9 billion (euros) as an immediate result of the fraud. (At the time of the incident, the euro was worth approximately $1.45.)

In 2007, SocGen was rated the best equity derivatives operation in the world by *Risk* magazine. Its internal control system of checks and balances was world renowned. For example, its trading room had five levels of hierarchy, each of which had a clear set of trading limits and controls, checked daily by a small army of compliance officers.[2] In addition, "the bank also [had] a shock team of internal auditors who descend on a corner of the bank without warning and pull apart its operations to ensure they conform to bank rules."[3]

During the summer of 2000, Kerviel began his employment at the bank—ironically, in its Compliance Department. Five years later, he was promoted to a junior trader in the arbitrage desk, which deals in program trading, exchange-traded funds, swaps, stock index futures trading, and quantitative trading. Kerviel was responsible for generating profits for the bank and its customers by betting on the market's future performance. His first major win came when he shorted stock of German insurer Allianz and earned the bank €55,000.

Thanks to his years of experience in the Compliance Department, Kerviel was an expert in the proprietary information system SocGen used to book trades. He knew that while the Risk-Control Department monitored the bank's overall positions very closely, it did not verify the data that individual traders entered into the system. Kerviel also knew the timing of the nightly reconciliation of the day's trades, so he was able to delete and then reenter unauthorized transactions without getting caught.

On November 7, 2007, SocGen received an e-mail alert from a surveillance officer at Eurex (one of Europe's largest exchanges). The message stated that Kerviel had engaged in several transactions that had set off alarms at the exchange over the past seven months. A SocGen risk-control expert responded two weeks later indicating that there was nothing irregular about the transactions. A week later, Eurex sent a second e-mail alert, stating that they were not satisfied with SocGen's explanation and demanding more details. Following another two-week delay, SocGen provided further details, and both Eurex and SocGen let the matter drop. The SocGen risk-control expert used information provided by Kerviel and his supervisor as well as a compliance officer at a SocGen subsidiary as the basis of both of his replies to Eurex. Kerviel's supervisor stated that there was no anomaly whatsoever.

Following the Eurex warnings, Kerviel took additional steps to cover his tracks by manipulating portions of the internal risk-control system with which he was unfamiliar. This ultimately led to the discovery of his alleged fraud.[4] On January 18, 2008, Kerviel executed a set of trades that set off another alarm.

This time, upon a more thorough investigation, a major problem became apparent. As SocGen risk-control experts carefully reviewed Kerviel's latest transactions, they were shocked to discover that the trades had resulted in a market exposure for the firm of €50 billion (obviously far beyond Kerviel's trading limit), which, when finally cleared, resulted in a loss of more than €4.9 billion.

As of this writing, Kerviel is still under investigation and involved in litigation charging him with using his insider knowledge to falsify records and commit computer fraud. Prosecutors suspect his motivation was to boost his income by making successful trades far beyond his trading limits, thus earning large bonuses (his total salary and bonus for 2007 was a relatively modest €94,000). Kerviel spent five weeks in jail but is currently free on bond. He was hired in February 2008 as a computer consultant by the French firm Lemaire Consultants & Associates; however, he is said to be "traumatized" by his newfound infamy.

Kerviel admits he took trading positions beyond his authorized limit to make transactions involving European index futures. Kerviel told prosecutors, "the techniques I used aren't at all sophisticated and any control that's properly carried out should have caught it."[5] He insists that he did no wrong and that the bank was fully aware of his transactions. Kerviel says that he refuses to be made a scapegoat for the bank's lapses in oversight. He argues that his superiors tacitly approved his activities—as long as they were generating a profit. Kerviel had earned a profit for the bank of nearly €1.5 billion in 2007 by exceeding his trade limit and executing similar, but successful, trades. Meanwhile, the bank says that the fraud was based on simple transactions but was concealed by "sophisticated and varied techniques."[6] If convicted, Kerviel faces up to five years in jail and fines totaling as much as €300,000.[7]

The sterling reputation of SocGen was badly tarnished, and the market value of the firm dropped 50 percent over the course of just a few months. The bank's highly respected CEO and chairman of the board, Daniel Bouton, was put under enormous pressure to step down; this included requests for his resignation from French president Nicholas Sarkozy. Bouton eventually resigned as CEO in May 2008, but he remains chairman of the board.[8] In December 2008, European hedge fund GLG Partners entered into an agreement to acquire the bank in the second half of 2009.[9]

Several internal and external investigations of the bank's operating procedures and internal controls have been completed. The French banking regulator stated that there were "grave deficiencies" in the bank's internal controls and fined it €4 million. The Banking Commission said that SocGen did not focus sufficiently on fraud weaknesses and that there were "significant weaknesses" in the bank's IT security systems. Another report pointed out that Kerviel's direct supervisor was inexperienced and received insufficient support to do his job properly. It also stated that Kerviel's fraudulent transactions were entered by an unnamed assistant trader, thus raising the issue of collusion and indicating even more widespread weaknesses in internal controls.

Pascal Decque, a financial analyst who covers SocGen for Natixis (a leading player in corporate and investment banking), commented, "SocGen was brilliant in [its] achievement, . . . the world leader in derivatives. Maybe when you are that good, you think you will never fail."[10]

Questions to Consider

1. Peter Gumble, European editor for *Fortune* magazine, comments, "Kerviel is a stunning example of a trader breaking the rules, but he's by no means alone. One of the dirty little secrets of trading floors around the world is that every so often, somebody is caught concealing a position and is quickly—and quietly—dismissed…. [This] might be shocking for people unfamiliar with the macho, high-risk, high-reward culture of most trading floors, but consider this: the only way banks can tell who will turn into a good trader and who won't is by giving every youngster it hires a chance to show his mettle. That means allowing even the most junior traders to take aggressive positions. This leeway is supposed to be matched by careful controls, but clearly they aren't foolproof."[11] What is your reaction to this statement by Mr. Gumble?

2. What explanation can there be for the failure of SocGen's internal control system to detect Kerviel's transactions while Eurex detected many suspicious transactions?

LEARNING OBJECTIVES

As you read this chapter, consider the following questions:

1. What key trade-offs and ethical issues are associated with the safeguarding of data and information systems?
2. Why has there been a dramatic increase in the number of computer-related security incidents in recent years?
3. What are the most common types of computer security attacks?
4. Who are the primary perpetrators of computer crime, and what are their objectives?
5. What are the key elements of a multilayer process for managing security vulnerabilities based on the concept of reasonable assurance?
6. What actions must be taken in response to a security incident?

IT SECURITY INCIDENTS: A MAJOR CONCERN

The security of information technology used in business is of utmost importance. Confidential business data and private customer and employee information must be safeguarded, and systems must be protected against malicious acts of theft or disruption. Although the necessity of security is obvious, it must often be balanced against other business needs and issues. Business managers, IT professionals, and IT users all face a number of ethical decisions regarding IT security:

- If their firm is a victim of a computer crime, should they pursue prosecution of the criminals at all costs, maintain a low profile to avoid the negative publicity, inform their affected customers, or take some other action?
- How much effort and money should be spent to safeguard against computer crime? (In other words, how safe is safe enough?)
- If their firm produces software with defects that allow hackers to attack customer data and computers, what actions should they take?
- What should be done if recommended computer security safeguards make life more difficult for customers and employees, resulting in lost sales and increased costs?

Unfortunately, the number of IT-related security incidents is increasing—not only in the United States but around the world. Table 3-1 lists the most common computer security incidents according to the "2008 CSI Computer Crime and Security Survey." The figures shown in the table represent the percentage of organizations responding to the survey that experienced such an incident during the specified year. According to the survey, 53 percent of the responding organizations spend 5 percent or less of their overall IT budget on information security.[12]

TABLE 3-1   Most common security incidents

Type of security incident	2004	2005	2006	2007	2008
Virus	78%	74%	65%	52%	50%
Insider abuse	59%	48%	42%	59%	44%
Laptop theft	49%	48%	47%	50%	42%
Unauthorized access	37%	32%	32%	25%	29%
Denial of service	39%	32%	25%	25%	21%
Instant messaging abuse				25%	21%
Bots				21%	20%

Source: "2008 CSI Computer Crime and Security Survey."

Why Computer Incidents Are So Prevalent

Around the world, organizations are reporting security incidents such as those listed in Table 3-1. In today's computing environment of increasing complexity, higher user expectations, expanding and changing systems, and increased reliance on software with known vulnerabilities, it is no wonder that the number, variety, and impact of security incidents are increasing dramatically.

Increasing Complexity Increases Vulnerability

The computing environment has become enormously complex. Networks, computers, operating systems, applications, Web sites, switches, routers, and gateways are interconnected and driven by hundreds of millions of lines of code. This environment continues to increase in complexity every day. The number of possible entry points to a network expands continually as more devices are added, increasing the possibility of security breaches.

Higher Computer User Expectations

Today, time means money, and the faster computer users can solve a problem, the sooner they can be productive. As a result, computer help desks are under intense pressure to respond very quickly to users' questions. Under duress, help desk personnel sometimes forget to verify users' identities or to check whether they are authorized to perform a requested action. In addition, even though they have been warned against doing so, some computer users share their login ID and password with other coworkers who have forgotten their own passwords. This can enable workers to gain access to information systems and data for which they are not authorized.

Expanding and Changing Systems Introduce New Risks

Business has moved from an era of stand-alone computers, in which critical data was stored on an isolated mainframe computer in a locked room, to an era in which personal computers connect to networks with millions of other computers, all capable of sharing information. Businesses have moved quickly into e-commerce, mobile computing, collaborative work groups, global business, and interorganizational information systems. Information technology has become ubiquitous and is a necessary tool for organizations to achieve their goals. However, it is increasingly difficult to keep up with the pace of technological change, successfully perform an ongoing assessment of new security risks, and implement approaches for dealing with them.

Increased Reliance on Commercial Software with Known Vulnerabilities

In computing, an **exploit** is an attack on an information system that takes advantage of a particular system vulnerability. Often this attack is due to poor system design or implementation. Once the vulnerability is discovered, software developers quickly create and issue a "fix," or patch, to eliminate the problem. Users of the system or application are responsible for obtaining and installing the patch, which they can usually download from the Web. (These fixes are in addition to other maintenance and project work that software developers perform.) Any delay in installing a patch exposes the user to a security breach.

Estimates of the rate at which software vulnerabilities are discovered by organizations around the world vary widely; the daily rate has been estimated to be as low as seven vulnerabilities[13] and as high as 382.[14] All these bugs and potential vulnerabilities create a serious work overload for developers, who are responsible for security fixes. Clearly, it can be difficult to keep up with all the required patches. A **zero-day attack** takes place before the security community or software developer knows about the vulnerability or has been able to repair it. Although the potential for damage from zero-day exploits is great, few such attacks have been documented as of this writing.

U.S. companies increasingly rely on commercial software with known vulnerabilities. Even when vulnerabilities are exposed, many corporate IT organizations prefer to use already installed software "as is" rather than implement security fixes that will either make the software harder to use or eliminate "nice-to-have" features suggested by current users or potential customers that will help sell the software.

Types of Exploits

There are numerous types of computer attacks, with new varieties being invented all the time. This section will discuss some of the more common attacks, including the virus, worm, Trojan horse, botnet, distributed denial-of-service, rootkit, spam, and phishing.

While we usually think of such exploits being aimed at computers, smartphones such as Apple's iPod and Research In Motion's BlackBerry are becoming increasingly computer capable. They can store personal identity information, including credit card numbers and bank account numbers. They can be used to surf the Web and transact business electronically. The more people use their smartphones for these purposes, the more attractive these devices become as a target for cyberthieves. Some IT security experts warn that it will not be long before we see exploits directed at these devices to steal users' data or turn them into remote-controlled bots.[15]

Viruses

Computer virus has become an umbrella term for many types of malicious code. Technically, a **virus** is a piece of programming code, usually disguised as something else, that causes a computer to behave in an unexpected and usually undesirable manner. Often a virus is attached to a file, so that when the infected file is opened, the virus executes. Other viruses sit in a computer's memory and infect files as the computer opens, modifies, or creates them. Most viruses deliver a "payload," or malicious software that causes the computer to perform in an unexpected way. For example, the virus may be programmed to display a certain message on the computer's display screen, delete or modify a certain document, or reformat the hard drive.

A true virus does not spread itself from computer to computer. A virus is spread to other machines when a computer user opens an infected e-mail attachment, downloads an infected program, or visits infected Web sites. In other words, it takes action by the "infected" computer user to spread a virus.

Macro viruses have become a common and easily created form of virus. Attackers use an application macro language (such as Visual Basic or VBScript) to create programs that infect documents and templates. After an infected document is opened, the virus is executed and infects the user's application templates. Macros can insert unwanted words, numbers, or phrases into documents or alter command functions. After a macro virus infects a user's application, it can embed itself in all future documents created with the application.

Worms

Unlike a computer virus, which requires users to spread infected files to other users, a **worm** is a harmful program that resides in the active memory of the computer and duplicates itself. Worms differ from viruses in that they can propagate without human intervention, sending copies of themselves to other computers by e-mail or Internet Relay Chat (IRC).

The negative impact of a worm attack on an organization's computers can be considerable—lost data and programs, lost productivity due to workers being unable to use their computers, additional lost productivity as workers attempt to recover data and programs, and lots of effort for IT workers to clean up the mess and restore everything to as close to normal as possible. The cost to repair the damage done by each of the Code Red, SirCam, and Melissa worms was estimated to exceed $1 billion, with that of the Storm and ILOVEYOU worms totaling well above that, as shown in Table 3-2.

TABLE 3-2 Cost impact of worms

Name	Year released	Worldwide economic impact
Storm	2007	> $10 billion (est.)
ILOVEYOU	2000	$8.75 billion
Code Red	2001	$2.62 billion
SirCam	2001	$1.15 billion
Melissa	1999	$1.10 billion

Source: Pelin Aksoy and Laura Denardis, *Information Technology in Theory* (Boston: Cengage Learning, 2007), 299–301.

Trojan Horses

A **Trojan horse** is a program in which malicious code is hidden inside a seemingly harmless program. The program's harmful payload can enable the hacker to destroy hard drives, corrupt files, control the computer remotely, launch attacks against other computers, steal passwords or Social Security numbers, and spy on users by recording keystrokes and transmitting them to a server operated by a third party.

A Trojan horse can be delivered as an e-mail attachment, downloaded from a Web site, or contracted via a removable media device such as a CD/DVD or USB memory stick. Once an unsuspecting user executes the program that hosts the Trojan horse, the malicious payload is automatically launched as well—with no telltale signs. Common host programs include screen savers, greeting card systems, and games.

iWork is the suite of applications created by Apple that includes word processing, desktop publishing, presentation preparation software, and a spreadsheet application. Some pirated copies of this software contain a Trojan horse, iServices.a, which launches when the user begins installation of the pirated software. When installed, the Trojan horse "phones home" to the hacker's server to confirm the Mac is infected and awaits further instructions.[16]

Users are often tricked into installing Trojan horses. For example, the Opanki Trojan horse disguised itself as a file coming from Apple's popular online iTunes music service. It was distributed via an instant message that read "This picture never gets old." An unsuspecting user who clicked a link in the message would install the Trojan horse. Another type of Trojan horse is a **logic bomb**, which executes when it is triggered by a specific event. For example, logic bombs can be triggered by a change in a particular file, by typing a specific series of keystrokes, or by a specific time or date.

Botnets

A **botnet** is a large group of computers controlled from one or more remote locations by hackers, without the knowledge or consent of their owners. Botnets are frequently used to distribute spam and malicious code. The collective processing capacity of some botnets exceeds that of the world's most powerful supercomputers. Cutwail, a large botnet, controlled approximately one million active bots at one time.[17] It is estimated that about one in four personal computers in the United States is part of a botnet.[18] In 2008, about 90 percent of spam was distributed by botnets, including the notorious Storm, Srizbi, and Cutwail botnets.[19] Dealing with "bot" computers within an organization's network can be quite expensive. Survey respondents who had been attacked by bots pegged the average cost to repair the damage at $350,000.[20]

Distributed Denial-of-Service (DDoS) Attacks

A **distributed denial-of-service attack (DDoS)** is one in which a malicious hacker takes over computers on the Internet and causes them to flood a target site with demands for data and other small tasks. A distributed denial-of-service attack does not involve infiltration of the targeted system. Instead, it keeps the target so busy responding to a stream of automated requests that legitimate users cannot get in—the Internet equivalent of dialing a telephone number repeatedly so that all other callers hear a busy signal. The targeted machine "holds the line open" while waiting for a reply that never comes, and eventually the requests exhaust all resources of the target.

The software to initiate a denial-of-service attack is simple to use and readily available at hacker sites. A tiny program is downloaded surreptitiously from the attacker's computer to dozens, hundreds, or even thousands of computers all over the world. Based on a command by the attacker or at a preset time, these computers (called zombies) go into action, each sending a simple request for access to the target site again and again—dozens of times per second.

The zombies involved in a denial-of-service attack are often seriously compromised and are left with more enduring problems than their target. As a result, zombie machines need to be inspected to ensure that the attacker software is completely removed from the system. In addition, system software must often be reinstalled from a reliable backup to reestablish the system's integrity, and an upgrade or patch must be implemented to eliminate the vulnerability that allowed the attacker to enter the system.

The Republic of Estonia is a small country (population 1.4 million) in the Baltic region of northern Europe. It was occupied by the Soviet Union following World War II and gained its independence in 1991. During April and May of 2007, a global botnet of compromised home computers was used to launch hundreds of coordinated distributed denial-of-service attacks, which disrupted the Web sites of numerous Estonian government agencies, financial institutions, and media outlets. It appears that pro-Russian activists led the attacks in retaliation for the Estonian government's decision to move a Soviet World War II memorial.[21]

Rootkits

A **rootkit** is a set of programs that enables its user to gain administrator level access to a computer without the end user's consent or knowledge. Once installed, the attacker can gain full control of the system and even obscure the presence of the rootkit from legitimate system administrators. Attackers can use the rootkit to execute files, access logs, monitor user activity, and change the computer's configuration. Rootkits are one part of a blended threat, consisting of the dropper, loader, and rootkit. The dropper code gets the rootkit installation started and can be activated by clicking on a link to a malicious Web site in an e-mail or opening an infected .pdf file. The dropper launches the loader program and then deletes itself. The loader loads the rootkit into memory; at that point the computer has been compromised. Rootkits are designed so cleverly that it is difficult to even discover if they are installed on a computer. The fundamental problem with trying to detect a rootkit is that the operating system currently running cannot be trusted to provide valid test results. Here are some symptoms of rootkit infections:

- The computer locks up or fails to respond to input from the keyboard or mouse.
- The screen saver changes without any action on the part of the user.
- The taskbar disappears.
- Network activities function extremely slowly.

When it is determined that a computer has been infected with a rootkit, there is little to do but reformat the disk; reinstall the operating system and all applications; and reconfigure the user's settings, such as mapped drives. This can take hours, and the user may be left with a basic working machine, but all locally held data and settings may be lost.

Spam

E-mail spam is the abuse of e-mail systems to send unsolicited e-mail to large numbers of people. Most spam is a form of low-cost commercial advertising, sometimes for questionable products such as pornography, phony get-rich-quick schemes, and worthless stock. Spam is also an extremely inexpensive method of marketing used by many legitimate organizations. For example, a company might send e-mail to a broad cross section of potential customers to announce the release of a new product in an attempt to increase initial sales. Spam may also be used to deliver harmful worms or other malware.

The cost of creating an e-mail campaign for a product or service is several hundred to a few thousand dollars, compared to tens of thousands of dollars for direct-mail campaigns. In addition, e-mail campaigns take only a couple of weeks to develop, compared with three months or more for direct-mail campaigns, and the turnaround time for feedback averages 48 hours for e-mail as opposed to weeks for direct mail. However, the benefits of spam to companies can be largely offset by the public's generally negative reaction to receiving unsolicited ads.

Spam forces unwanted and often objectionable material into e-mail boxes, detracts from the ability of recipients to communicate effectively due to full mailboxes and relevant e-mails being hidden among many unsolicited messages, and costs Internet users and service providers millions of dollars annually. It takes users time to scan and delete spam e-mail, a cost that can add up if they pay for Internet connection charges on an hourly basis. It also costs money for ISPs and online services to transmit spam, which is reflected in the rates charged to all subscribers.

The Controlling the Assault of Non-Solicited Pornography and Marketing (CAN-SPAM) Act went into effect in January 2004. The act says that it is legal to spam, provided the messages meet a few basic requirements—spammers cannot disguise their identity by using false return addresses, there must be a label in the message specifying that it is an ad or a solicitation, and such e-mails must include a way for recipients to indicate that they do not want future mass mailings. From 2005 to 2008, the average global proportion of spam in e-mail traffic shrunk from 92 percent to 81 percent.[22]

Several companies (Microsoft, Yahoo!) offer free e-mail services. Spammers often seek to use e-mail accounts from such major, free, and reputable Web-based e-mail service providers, as their spam can be sent at no charge and is less likely to be blocked. At the start of 2008, about 6.5 percent of spam was sent from such accounts. By the end of the year, the percentage had increased to 12 percent. Spammers can defeat the registration process of the free e-mail services by launching a coordinated bot attack that can sign up for thousands of e-mail accounts. These accounts are then used by the spammers to send thousands of untraceable e-mail messages for free. A third of the world's spam is generated from compromised computers in North America (21%), Russia (8%), and China (4%).[23]

A partial solution to this problem is the use of CAPTCHA to ensure that only humans obtain free accounts. **Completely Automated Public Turing Test to Tell Computers and Humans Apart (CAPTCHA)** software generates and grades tests that humans can pass but all but the most sophisticated computer programs cannot. For example, humans can read the distorted text in Figure 3-1, but simple computer programs cannot.

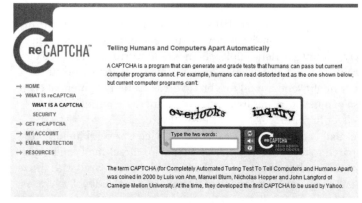

Source: Permission needed from http://recaptcha.net/captcha.html

FIGURE 3-1   Example of CAPTCHA

For nearly five years, Edward Davidson ran a spamming operation out of his home, where he managed a large network of computers that sent hundreds of thousands of spam e-mails. The spam operation promoted the sale of watches, perfume, and other items for nearly two dozen companies. Davidson and his subcontractors sent e-mail messages with header information that concealed the actual sender from the recipient of the e-mail—a violation of the federal CAN-SPAM Act. In April 2008, Edward Davidson was sentenced to serve 21 months in federal prison for violation of the CAN-SPAM Act. He was also ordered to pay $714,139 in restitution to the IRS for taxes on income from the operation that he failed to report.[24]

Phishing

Phishing is the act of using e-mail fraudulently to try to get the recipient to reveal personal data. In a phishing scam, con artists send legitimate looking e-mails urging the recipient to take action to avoid a negative consequence or to receive a reward. The requested action may involve clicking on a link to a Web site or opening an e-mail attachment. These e-mails, such as the one shown in Figure 3-2, lead consumers to counterfeit Web sites designed to trick them into divulging personal data. Savvy users often become suspicious and refuse to enter data into the fake Web sites; however, sometimes just accessing the Web site can trigger an automatic and unnoticeable download of malicious software to a computer. eBay, PayPal, and Citibank are among the Web sites that phishers spoof most frequently.

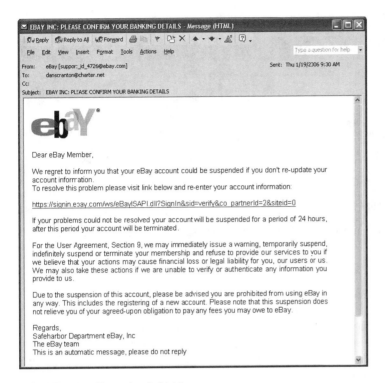

FIGURE 3-2 Example of phishing

Spear-phishing is a variation of phishing in which the phisher sends fraudulent e-mails to a certain organization's employees. The phony e-mails are designed to look like they came from high-level executives within the organization. Employees are again directed to a fake Web site and then asked to enter personal information, such as name, Social Security number, and network passwords.

Phishers have grown quite sophisticated from the days when the subject lines of their e-mail messages read, "Photo of boss at party!" During the worldwide financial crisis that started during late 2008, ruthless phishers took advantage of a panic-stricken public by using subject lines that mentioned "acquisition," "merger," or the solvency of major organizations in an attempt to get people to read and respond to their messages. In 2008, phishing activity peaked in February at a level of 1 percent of all e-mails.[25] Botnets have become the primary means for distributing spam, malware, and phishing scams.

Types of Perpetrators

The people who launch these kinds of computer attacks include thrill seekers wanting a challenge, common criminals looking for financial gain, industrial spies trying to gain a competitive advantage, and terrorists seeking to cause destruction to further their cause. Each type of perpetrator has different objectives and access to varying resources, and each is willing to accept different levels of risk to accomplish his or her objective. Knowing the profile of each set of likely attackers, as shown in Table 3-3, is the first step toward establishing effective countermeasures.

TABLE 3-3 Classifying perpetrators of computer crime

Type of perpetrator	Typical motives
Hacker	Test limits of system and/or gain publicity
Cracker	Cause problems, steal data, and corrupt systems
Malicious insider	Gain financially and/or disrupt company's information systems and business operations
Industrial spy	Capture trade secrets and gain competitive advantage
Cybercriminal	Gain financially
Hacktivist	Promote political ideology
Cyberterrorist	Destroy infrastructure components of financial institutions, utilities, and emergency response units

Hackers and Crackers

Hackers test the limitations of information systems out of intellectual curiosity—to see whether they can gain access and how far they can go. They have at least a basic understanding of information systems and security features, and much of their motivation comes from a desire to learn even more. The term *hacker* has evolved over the years, leading to its negative connotation today rather than the positive one it used to have. While there is a vocal minority who believe that hackers perform a service by identifying security weaknesses, most people now believe that a hacker no longer has the right to explore public or private networks.

Some hackers are smart and talented, but many are technically inept and are referred to as **lamers** or **script kiddies** by more skilled hackers. Surprisingly, hackers have a wealth of available resources to hone their skills—online chat groups, Web sites, downloadable hacker tools, and even hacker conventions (such as DEFCON, an annual gathering in Las Vegas).

The microblogging Web site Twitter has been hacked numerous times. One hacker took advantage of a vulnerability to force victims to join his Twitter follow list automatically. Other hackers created a Twitter account under the name of Vint Cerf (the person most often called the father of the Internet) and used it for spamming. Hackers gained access to several high-profile accounts (Barack Obama, Britney Spears, and CNN's Rick Sanchez) and sent out fake updates in their names.[26]

In a more serious example of hacking that borders on cyberterrorism, Chinese hackers have repeatedly hacked into systems to intercept e-mails between U.S. and UK government officials. Fortunately, the compromised computer network carried only unclassified communications. A separate, more secure network used to carry classified communications has not yet been compromised. Foreign-government-sponsored hackers are a growing concern because they have access to millions of dollars, the most knowledgeable people, and the best equipment to attempt to hack into U.S.-based Web sites.[27]

Cracking is a form of hacking that is clearly criminal activity. **Crackers** break into other people's networks and systems to cause harm—defacing Web pages, crashing

computers, spreading harmful programs or hateful messages, and writing scripts and automated programs that let other people do the same things. For example, crackers defaced a
CERN (the European Organization for Nuclear Research) Web page, disparaging CERN's IT
security staff as a "bunch of school kids" and saying they had no plan to disrupt CERN's
operations but simply wanted to highlight the lab's security problems. The crackers came
very close to gaining access to a computer that controlled one of the 12,500 magnets that
control the Large Hadron Collider built to perform particle physics experiments.[28]

Malicious Insiders

A major security concern for companies is the malicious insider—an ever present and
extremely dangerous adversary, as shown in the opening vignette. Companies are exposed
to a wide range of fraud risks, including diversion of company funds, theft of assets, fraud
connected with bidding processes, invoice and payment fraud, computer fraud, and credit
card fraud. Not surprisingly, fraud that occurs within an organization is usually due to weaknesses in its internal control procedures. As a result, many frauds are discovered by chance
and by outsiders—via tips, through resolving payment issues with contractors or suppliers,
or during a change of management—rather than through control procedures. Often, frauds
involve some form of **collusion**, or cooperation, between an employee and an outsider.
For example, an employee in Accounts Payable may engage in collusion with a company
supplier. Each time the supplier submits an invoice, the Accounts Payable employee adds
$1,000 to the amount approved for payment. The inflated payment is received by the
supplier, and the two split the extra money.

Insiders are not necessarily employees; they can also be consultants and contractors.
However, "the typical employee who commits fraud has many years with the company, is
an authorized user, is in a nontechnical position, has no record of being a problem employee,
uses legitimate computer commands to commit the fraud, and does so mostly during business
hours."[29] The risk tolerance of these employees depends on whether they are motivated by
financial gain, revenge on their employers, or publicity.

Malicious insiders are extremely difficult to detect or stop because they are often authorized to access the very systems they abuse. Although insiders are less likely to attack systems than outside hackers or crackers are, the company's systems are far more vulnerable
to them. Most computer security measures are designed to stop external attackers but are
nearly powerless against insiders. Insiders have knowledge of individual systems, which
often includes the procedures to gain access to login IDs and passwords. Insiders know how
the systems work and where the weak points are. Their knowledge of organizational structure and security procedures helps them avoid investigation of their actions.

There are several steps organizations can take to reduce the potential for attacks from
insiders, including the following:

- Perform a thorough background check as well as psychological and drug testing
  of candidates for sensitive positions.
- Establish an expectation of regular and ongoing psychological and drug testing
  as a normal routine for people in sensitive positions.
- Carefully limit the number of people who can perform sensitive operations, and
  grant only the minimum rights and privileges necessary to perform essential
  duties.

- Define job roles and procedures so that it is not possible for the same person to both initiate and approve an action.
- Periodically rotate employees in sensitive positions so that any unusual procedures can be detected by the replacement.
- Immediately revoke all rights and privileges required to perform old job responsibilities when someone in a sensitive position moves to a new position.
- Implement an ongoing audit process to review key actions and procedures.

Industrial Spies

Industrial spies use illegal means to obtain trade secrets from competitors of their sponsor. Trade secrets are protected by the Economic Espionage Act of 1996, which makes it a federal crime to use a trade secret for one's own benefit or another's benefit. Trade secrets are most often stolen by insiders, such as disgruntled employees and ex-employees.

Competitive intelligence uses legal techniques to gather information that is available to the public. Participants gather and analyze information from financial reports, trade journals, public filings, and printed interviews with company officials. Industrial espionage involves using illegal means to obtain information that is not available to the public. Participants might place a wiretap on the phones of key company officials, bug a conference room, or break into a research and development facility to steal confidential test results. An unethical firm may spend a few thousand dollars to hire an industrial spy to steal trade secrets that can be worth a thousand times that amount. The industrial spy avoids taking risks that would expose his employer, as the employer's reputation (an intangible but valuable item) would be considerably damaged if the espionage were discovered.

Industrial espionage can involve the theft of new product designs, production data, marketing information, or new software source code. For example, Shekhar Verma was employed by Geometric Software Solutions Ltd. (GSSL), an Indian company that provides outsourcing services, including software development. GSSL was awarded a contract to debug the source code of SolidWorks 2001 Plus, a popular computer-aided design software package. Verma was eventually fired from GSSL; he allegedly stole the source code and offered it to several of SolidWorks' U.S. competitors for $200,000. (The value of the source code has been estimated to exceed $50 million.) A competitor contacted the FBI, a sting was set up, and Verma was arrested. However, Indian law at the time did not recognize misappropriation of trade secrets, so technically Verma did not steal from his employer, as the source code belonged to SolidWorks. Prosecutors were forced to charge Verma with simple theft; four years after those charges, he is still free and making a living as a programmer in India.[30]

Cybercriminals

Information technology provides a new and highly profitable venue for cybercriminals, who are attracted to the use of information technology for its ease in reaching millions of potential victims. **Cybercriminals** are motivated by the potential for monetary gain and hack into corporate computers to steal, often by transferring money from one account to another to another—leaving a hopelessly complicated trail for law enforcement officers to follow. Cybercriminals also engage in all forms of computer fraud—stealing and reselling credit card numbers, personal identities, and cell phone IDs. Because the potential for monetary gain is high, they can afford to spend large sums of money to buy the technical expertise and access they need from unethical insiders.

The use of stolen credit card information is a favorite ploy of computer criminals. Fraud rates are highest for merchants who sell downloadable software or expensive items such as electronics and jewelry (because of their high resale value). Credit card companies are so concerned about making consumers feel safe while shopping online that many are marketing new and exclusive zero-liability programs, although the Fair Credit Billing Act limits consumer liability to only $50 of unauthorized charges. When a charge is made fraudulently in a retail store, the bank that issued the credit card must pay the fraudulent charges. For fraudulent credit card transactions over the Internet, the Web merchant absorbs the cost.

A high rate of disputed transactions, known as chargebacks, can greatly reduce a Web merchant's profit margin. However, the permanent loss of revenue caused by lost customer trust has far more impact than the costs of fraudulent purchases and bolstering security. Most companies are afraid to admit publicly that they have been hit by online fraud or hackers because they don't want to hurt their reputations.

To reduce the potential for online credit card fraud, most e-commerce Web sites use some form of encryption technology to protect information as it comes in from the consumer. Some also verify the address submitted online against the one the issuing bank has on file, although the merchant may inadvertently throw out legitimate orders as a result—for example, a consumer might place a legitimate order but request shipment to a different address because it is a gift. Another security technique is to ask for a card verification value (CVV), the three-digit number above the signature panel on the back of a credit card. This technique makes it impossible to make purchases with a credit card number stolen online. An additional security option is transaction-risk scoring software, which keeps track of a customer's historical shopping patterns and notes deviations from the norm. For example, say that you have never been to a casino and your credit card turns up at Caesar's Palace at 2 a.m. Your transaction-risk score would go up dramatically, so much so that the transaction might be declined.

Some card issuers are issuing debit and credit cards in the form of **smart cards**, which contain a memory chip that is updated with encrypted data every time the card is used. This encrypted data might include the user's account identification and the amount of credit remaining. To use a smart card for online transactions, consumers must purchase a card reader that attaches to their personal computers and enter a personal identification number to gain access to the account. Although smart cards are used widely in Europe, they are not as popular in the United States because of the changeover costs for merchants.

Hacktivists and Cyberterrorists

Hacktivism, a combination of the words *hacking* and *activism*, is hacking to achieve a political or social goal. A **cyberterrorist** launches computer-based attacks against other computers or networks in an attempt to intimidate or coerce a government in order to advance certain political or social objectives. Cyberterrorists are more extreme in their goals than hacktivists although there is no clear demarcation line. Because of the Internet, cyberattacks can easily originate from foreign countries, making detection and retaliation much more difficult.

Three years before the September 11, 2001, terrorist attacks, the U.S. government considered the threat of cyberterrorism serious enough that it established the National

Infrastructure Protection Center, which served as the focal point for threat assessment, warning, investigation, and response for threats or attacks against American infrastructures. (This function was later transferred to the Homeland Security Department's Information Analysis and Infrastructure Protectorate.) These infrastructures include telecommunications, energy, banking and finance, water, government operations, and emergency services. Specific targets might include telephone-switching systems, an electric power grid that serves major portions of a geographic region, or an air traffic control center that ensures airplanes can take off or land safely. Successful cyberattacks on such targets could cause widespread and massive disruptions to society. Some computer security experts believe that cyberterrorism attacks could be used to further complicate matters following a major act of terrorism by reducing the ability of fire and emergency teams to respond.

Cyberterrorists seek to cause harm rather than gather information, and they use techniques that destroy or disrupt services. They are extremely dangerous, consider themselves to be at war, have a very high acceptance of risk, and seek maximum impact. In early 2009, Israeli hacktivists made available malware dubbed Patriot. When downloaded to computers of Israeli sympathizers, this malware converts those computers into zombies, which launch a distributed denial-of-service attack intended to silence Hamas Web sites. Meanwhile, anti-Israeli hacktivists were also on the offensive. Bruce Jenkins, a consultant from the application security firm Fortify Security, states, "Our observations suggest that a large number of Web sites have been defaced by a variety of hacker groups from Iran, Lebanon, Morocco and Turkey and the trend is accelerating."[31]

Federal Laws for Prosecuting Computer Attacks

Computers came into use in the 1950s. Initially, there were no laws that pertained strictly to computer-related crimes. For example, if a group of criminals entered a bank and stole money at gunpoint, they could be captured and charged with robbery—the crime of seizing property through violence or intimidation. However, by the mid-1970s, it was possible to access a bank's computer remotely using a terminal (a keyboard and monitor), modem, and telephone line. A knowledgeable person could then transfer money (in the form of computer bits) from accounts in that bank to an account in another bank. This act did not fit the definition of robbery, and the traditional laws were no longer adequate to punish criminals who used computer modems.

Over the years, several laws have been enacted to help prosecute those responsible for computer-related crime; these are summarized in Table 3-4. For example, the USA Patriot Act defines cyberterrorism as hacking attempts that cause $5,000 in aggregate damage in one year, damage to medical equipment, or injury to any person. Those convicted of cyberterrorism are subject to a prison term of five to 20 years. (The $5,000 threshold is quite easy to exceed, and as a result, many young people who have been involved in what they consider to be minor computer pranks have found themselves meeting the criteria to be tried as cyberterrorists.)

TABLE 3-4 Federal laws that apply to computer attacks

Federal law	Subject area
USA Patriot Act	Defines cyberterrorism and penalties
Computer Fraud and Abuse Act	Malicious code and unauthorized access to computers
Identity Theft and Assumption Deterrence Act	Identity theft
U.S. Code Title 18, Part I, Chapter 47, Section 1030	Fraud and related activities in association with computers: • Accessing a computer without authorization or exceeding authorized access • Transmission of a program, code, or command that causes harm to a computer • Trafficking of computer passwords • Threats to cause damage to a protected computer
U.S. Code Title 18, Part I, Chapter 121, Section 2701	Unlawful access to stored communications to obtain, alter, or prevent authorized access to a wire or electronic communication while it is in electronic storage

Now that we have discussed various types of computer exploits, the people who perpetrate these exploits, and the laws under which they can be prosecuted, we will discuss how organizations can take steps to implement a trustworthy computing environment to defend against such attacks.

IMPLEMENTING TRUSTWORTHY COMPUTING

Trustworthy computing is a method of computing that delivers secure, private, and reliable computing experiences based on sound business practices; this is what organizations worldwide are demanding today. Everyone who provides computing services (software and hardware manufacturers, consultants, programmers) knows that this is a priority for their customers. For example, Microsoft has pledged to deliver on a trustworthy computing initiative designed to improve trust in its software products, as summarized in Figure 3-3 and Table 3-5.[32]

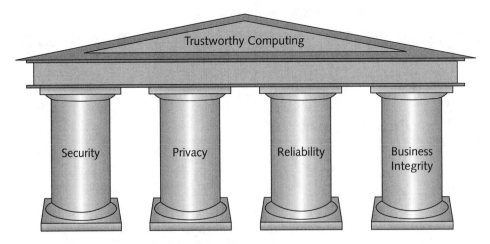

FIGURE 3-3 Microsoft's Four Pillars of Trustworthy Computing

TABLE 3-5 Actions taken by Microsoft to support trustworthy computing

Pillar	Actions taken by Microsoft to support trustworthy computing
Security	Invest in the expertise and technology required to create a trustworthy environment.
	Work with law enforcement agencies, industry experts, academia, and private sectors to create and enforce secure computing.
	Develop trust by educating consumers on secure computing.
Privacy	Make privacy a priority in the design, development, and testing of products.
	Contribute to standards and policies created by industry organizations and government.
	Provide users with a sense of control over their personal information.
Reliability	Build systems so that (1) they continue to provide service in the face of internal or external disruptions; (2) in the event of a disruption, they can be easily restored to a previously known state with no data loss; (3) they provide accurate and timely service whenever needed; (4) required changes and upgrades do not disrupt them; (5) on release, they contain minimal software bugs; and (6) they work as expected or promised.
Business integrity	Be responsive—take responsibility for problems and take action to correct them.
	Be transparent—be open in dealings with customers, keep motives clear, keep promises, and make sure customers know where they stand in dealing with the company.

The security of any system or network is a combination of technology, policy, and people and requires a wide range of activities to be effective. "Society ultimately expects computer systems to be trustworthy—that is, that they do what is required and expected of them despite environmental disruption, human user and operator errors, and attacks by hostile parties, and that they not do other things."[33] A strong security program begins by assessing threats to the organization's computers and network, identifying actions that address the most serious vulnerabilities, and educating end users about the risks involved and the actions they must take to prevent a security incident. The IT security group must lead the effort to prevent security breaches by implementing security policies and procedures, as well as effectively employing available hardware and software tools. However, no security system is perfect, so systems and procedures must be monitored to detect a possible intrusion. If an intrusion occurs, there must be a clear reaction plan that addresses notification, evidence protection, activity log maintenance, containment, eradication, and recovery. The following sections will discuss these activities.

Risk Assessment

A **risk assessment** is the process of assessing security-related risks to an organization's computers and networks from both internal and external threats. Such threats can prevent an organization from meeting its key business objectives. The goal of risk assessment is to identify which investments of time and resources will best protect the organization from its most likely and serious threats. In the context of an IT risk assessment, an asset is any hardware, software, information system, network, or database that is used by the organization to achieve its business objectives. A loss event is any occurrence that has a negative impact on an asset, such as a computer contracting a virus or a Web site undergoing a distributed denial-of-service attack.

Figure 3-4 illustrates a general security risk assessment process. The steps shown in this figure are discussed in the following section.

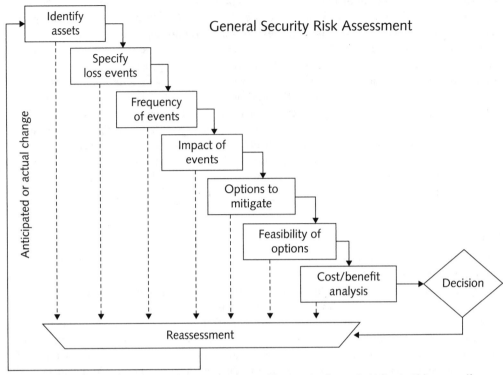

Source: General Security Risk Assessment Guideline, ASIS International, www.asisonline.org/guidelines/guidelinesgsra.pdf.

FIGURE 3-4   General Security Risk Assessment

Step 1. Identify the set of IT assets about which the organization is most concerned. Priority is typically given to those assets that support the organization's mission and the meeting of its primary business goals.

Step 2. Identify the loss events or the risks or threats that could occur, such as a distributed denial-of-service attack or insider fraud.

Step 3. Assess the frequency of events or the likelihood of each potential threat; some threats, such as insider fraud, are more likely to occur than others.

Step 4. Determine the impact of each threat occurring. Would the threat have a minor impact on the organization, or could it keep the organization from carrying out its mission for a lengthy period of time?

Step 5. Determine how each threat can be mitigated so that it becomes much less likely to occur or, if it does occur, has less of an impact on the organization. For example, installing virus protection on all computers makes it much less likely for a computer to contract a virus. Due to time and resource limitations, most organizations choose to focus on those threats that have a high (relative to all other threats) frequency and a high (relative to all other threats) impact. In other words, first address those threats that are likely to occur and that would have a high negative impact on the organization.

Step 6. Assess the feasibility of implementing the mitigation options.

Step 7. Perform a cost-benefit analysis to ensure that your efforts will be cost effective. No amount of resources can guarantee a perfect security system, so organizations must balance the risk of a security breach with the cost of preventing one. The concept of **reasonable assurance** recognizes that managers must use their judgment to ensure that the cost of control does not exceed the system's benefits or the risks involved.

Step 8. Make the decision on whether or not to implement a particular counter-measure. If you decide against implementing a particular countermeasure, you need to reassess if the threat is truly serious and, if so, identify a less costly countermeasure.

Table 3-6 illustrates a risk assessment for a hypothetical organization.

TABLE 3-6   Risk assessment for hypothetical company

Risk	Business objective threatened	Estimated probability of such an event occurring	Estimated cost of a successful attack	Probability × cost = expected cost	Assessment of current level of protection	Relative priority to be fixed
Distributed denial-of-service attack	24/7 operation of B2C Web site	40%	$500,000	$200,000	Poor	1
E-mail attachment with harmful worm	Rapid and reliable communications among employees and suppliers	70%	$200,000	$140,000	Poor	2
Harmful virus	Employees' use of personal productivity software	90%	$50,000	$45,000	Good	3
Invoice and payment fraud	Reliable cash flow	10%	$200,000	$20,000	Excellent	4

A completed risk assessment identifies the most dangerous threats to a company and helps focus security efforts on the areas of highest payoff.

Establishing a Security Policy

A **security policy** defines an organization's security requirements, as well as the controls and sanctions needed to meet those requirements. A good security policy delineates responsibilities and the behavior expected of members of the organization. A security policy outlines *what* needs to be done but not *how* to do it. The details of how to accomplish the goals of the policy are provided in separate documents and procedure guidelines.[34] In a recent survey of over 500 security professionals, 68 percent of respondents said that their organizations had a formal information security policy while 18 percent said that they were currently developing such a policy.[35]

The National Institute of Standards and Technology (NIST) is a nonregulatory federal agency within the U.S. Department of Commerce. Its Computer Security Division develops security standards and technology against threats to the confidentiality, integrity, and availability of information and services.[36] The Computer Security Division has published the NIST SP 800 series of documents, which provides useful definitions, policies, standards, and guidelines related to computer security. These may be found at the Computer Security Division's Computer Security Resource Center Web site at *http://csrc.nist.gov*.

Whenever possible, automated system rules should mirror an organization's written policies. Automated system policies can often be put into practice using the configuration options in a software program. For example, if a written policy states that passwords must be changed every 30 days, then all systems should be configured to enforce this policy automatically. When applying system security restrictions, there are some trade-offs between ease of use and increased security; however, when a decision is made to favor ease of use, security incidents sometimes increase. As security techniques continue to advance in sophistication, they become more transparent to end users.

The use of e-mail attachments is a critical security issue that should be addressed in every organization's security policy. Sophisticated attackers can try to penetrate a network via e-mail attachments, regardless of the existence of a firewall and other security measures. As a result, some companies have chosen to block any incoming mail that has a file attachment, which greatly reduces their vulnerability. Some companies allow employees to receive and open e-mail with attachments, but only if the e-mail is expected and from someone known by the recipient. Such a policy can be risky, however, because worms often use the address book of their victims to generate e-mails to a target audience.

Another growing area of concern is the use of wireless devices to access corporate e-mail, store confidential data, and run critical applications, such as inventory management and sales force automation. Mobile devices such as smartphones can be susceptible to viruses and worms. However, the primary security threat for mobile devices continues to be loss or theft of the device. Wary companies have begun to include special security requirements for mobile devices as part of their security policies. In some cases, users of laptops and mobile devices must use a virtual private network to gain access to their corporate network. A **virtual private network (VPN)** works by using the Internet to relay communications; it maintains privacy through security procedures and tunneling protocols, which encrypt data at the sending end and decrypt it at the receiving end. An additional level of security involves encrypting the originating and receiving network addresses. Because of the ease of loss or theft, it is also vital to encrypt all sensitive corporate data stored on handhelds and laptops. Unfortunately, it is hard to apply a single, simple approach to securing all handheld devices because so many manufacturers and models exist.[37]

Educating Employees, Contractors, and Part-Time Workers

According to a recent survey, one of the major security problems for U.S. companies in 2007 was creating and enhancing user awareness of security policies.[38] Employees, contractors, and part-time workers must be educated about the importance of security so that they will be motivated to understand and follow the security policies. This can often be accomplished by discussing recent security incidents that affected the organization. Users must understand that they are a key part of the security system and that they have certain responsibilities. For example, users must help protect an organization's information systems and data by doing the following:

- Guarding their passwords to protect against unauthorized access to their accounts
- Prohibiting others from using their passwords
- Applying strict access controls (file and directory permissions) to protect data from disclosure or destruction
- Reporting all unusual activity to the organization's IT security group

Prevention

No organization can ever be completely secure from attack. The key is to implement a layered security solution to make computer break-ins so difficult that an attacker eventually gives up. In a layered solution, if an attacker breaks through one layer of security, there is another layer to overcome. These layers of protective measures are explained in more detail in the following sections.

Installing a Corporate Firewall

Installation of a corporate firewall is the most common security precaution taken by businesses. A **firewall** stands guard between an organization's internal network and the Internet, and it limits network access based on the organization's access policy (see Figure 3-5). Firewalls can be established through the use of software, hardware, or a combination of both. Any Internet traffic that is not explicitly permitted into the internal network is denied entry. Similarly, most firewalls can be configured so that internal network users can be blocked from gaining access to certain Web sites based on such content as sex and violence. Most firewalls can also be configured to block instant messaging, access to newsgroups, and other Internet activities.

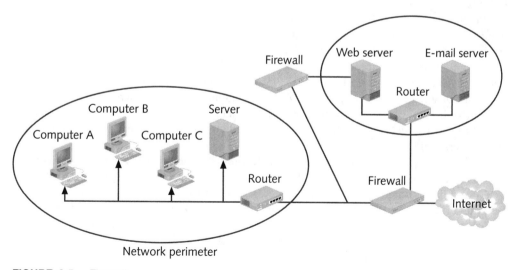

FIGURE 3-5   Firewall

Installing a firewall can lead to another serious security issue—complacency. For example, a firewall cannot prevent a worm from entering the network as an e-mail attachment. Most firewalls are configured to allow e-mail and benign-looking attachments to reach their intended recipient.

Table 3-7 lists some of the top-rated firewall software used to protect personal computers. Typically, firewall software sells for $30 to $60 for a single user license.

TABLE 3-7   Popular firewall software for personal computers

Software	Vendor
Norton Personal Firewall	Symantec
Comodo	Comodo Security Solutions, Inc.
Online Armor	Tall Emu Pty Ltd
ZoneAlarm Pro	Check Point Software Technologies
Personal Firewall	McAfee

Intrusion Prevention Systems

Intrusion prevention systems (IPSs) work to prevent an attack by blocking viruses, malformed packets, and other threats from getting into the protected network. The IPS sits directly behind the firewall and examines all the traffic passing through it. A firewall and a network IPS are complementary. Most firewalls can be configured to block everything except what you explicitly allow through; most IPSs can be configured to let through everything except what you explicitly specify should be blocked.

Installing Antivirus Software on Personal Computers

Antivirus software should be installed on each user's personal computer to scan a computer's memory and disk drives regularly for viruses. Antivirus software scans for a specific sequence of bytes, known as a **virus signature**, that indicates the presence of a specific virus. If it finds a virus, the antivirus software informs the user, and it may clean, delete, or quarantine any files, directories, or disks affected by the malicious code. Good antivirus software checks vital system files when the system is booted up, monitors the system continuously for virus-like activity, scans disks, scans memory when a program is run, checks programs when they are downloaded, and scans e-mail attachments before they are opened. Two of the most widely used antivirus software products are Norton AntiVirus from Symantec and Personal Firewall from McAfee.

The United States Computer Emergency Readiness Team (US-CERT) is a partnership between the Department of Homeland Security and the public and private sectors, established in 2003 to protect the nation's Internet infrastructure against cyberattacks. US-CERT has long served as a clearinghouse for information on new viruses, worms, and other computer security topics (over 500 new viruses and worms are developed each month). According to US-CERT, most of the virus and worm attacks that the team analyzes use already known programs. Thus, it is crucial that antivirus software be continually updated with the latest virus signatures. In most corporations, the network administrator is responsible for monitoring network security Web sites frequently and downloading updated antivirus software as needed. Many antivirus vendors recommend—and provide for—automatic, frequent updates.

Implementing Safeguards Against Attacks by Malicious Insiders

User accounts that remain active after employees leave a company are potential security risks. To reduce the threat of attack by malicious insiders, IT staff must promptly delete the computer accounts, login IDs, and passwords of departing employees and contractors.

Organizations also need to define employee roles carefully and separate key responsibilities properly, so that a single person is not responsible for accomplishing a task that has high security implications. For example, it would not make sense to allow an employee to initiate as well as approve purchase orders. That would allow an employee to input large invoices on behalf of a "friendly vendor," approve the invoices for payment, and then disappear from the company to split the money with the vendor. In addition to separating duties, many organizations frequently rotate people in sensitive positions to prevent potential insider crimes.

Another important safeguard is to create roles and user accounts so that users have the authority to perform their responsibilities and nothing more. For example, members of the Finance Department should have different authorizations from members of Human Resources. An accountant should not be able to review the pay and attendance records of an employee, and a member of Human Resources should not know how much was spent to modernize a piece of equipment. Even within one department, not all members should be given the same capabilities. Within the Finance Department, for example, some users may be able to approve invoices for payment, but others may only be able to enter them. An effective administrator will identify the similarities among users and create profiles associated with these groups.

Addressing the Most Critical Internet Security Threats

The overwhelming majority of successful computer attacks are made possible by taking advantage of well-known vulnerabilities. Computer attackers know that many organizations are slow to fix problems, which makes scanning the Internet for vulnerable systems an effective attack strategy. The rampant and destructive spread of worms, such as Blaster, Slammer, and Code Red, was made possible by the exploitation of known but unpatched vulnerabilities.

Both the SANS (SysAdmin, Audit, Network, Security) Institute and US-CERT regularly update a summary of the most frequent, high-impact vulnerabilities being reported to them. You can read these summaries at *www.sans.org/top20/* and *www.us-cert.gov/current*, respectively. The actions required to address these issues include installing a known patch to the software, and keeping applications and operating systems up to date. Those responsible for computer security must make it a priority to prevent attacks using these vulnerabilities.

Conducting Periodic IT Security Audits

Another important prevention tool is a **security audit** that evaluates whether an organization has a well-considered security policy in place and if it is being followed. For example, if a policy says that all users must change their passwords every 30 days, the audit must check how well the policy is being implemented. The audit should also review who has access to particular systems and data and what level of authority each user has. It is not unusual for an audit to reveal that too many people have access to critical data and that many people have capabilities beyond those needed to perform their jobs. One result of a good audit is a list of items that need to be addressed in order to ensure that the security policy is being met.

A thorough security audit should also test system safeguards to ensure that they are operating as intended. Such tests might include trying the default system passwords that are active when software is first received from the vendor. The goal of such a test is to ensure that all such known passwords have been changed.

Some organizations will also perform a penetration test of their defenses. This entails assigning individuals to try to break through the measures and identify vulnerabilities that still need to be addressed. The individuals used for this test are often contractors rather than employees. The contractors may possess special skills or knowledge and are likely to take unique approaches in testing the security measures.

One organization that employs security audits is the U.S. government. U.S. government agencies must maintain security for their information systems and data to prevent data tampering, disruptions in critical operations, fraud, and the inappropriate disclosure of sensitive information. Title III of the E-Government Act, entitled the Federal Information Security Management Act (FISMA), "requires each federal agency to develop, document, and implement an agency-wide program to provide security for the information and information systems that support the operations and assets of the agency, including those provided or managed by another agency, contractor, or other source."[39]

The annual Federal Computer Security Report Card is based on evaluations defined in FISMA and compiled by the House Government Reform Committee from information provided by each agency's inspector general. Results for selected agencies are shown in Table 3-8.[40,41] The important thing to note is that significant improvements in security can sometimes take years and are not easily achieved. The overall security of federal government computer systems earned only a C average in the 2007 security report card.

TABLE 3-8 Selected federal agencies' computer security report card for 2004 to 2007

Federal agency	2007	2006	2005	2004
Department of Homeland Security	B+	D	F	F
Department of Justice	A+	A–	D	B–
Nuclear Regulatory Commission	F	F	D–	B+
Department of State	C	F	F	D+
Department of the Treasury	F	F	D–	D+
Department of Defense	D–	F	F	D
NASA	C	D–	B–	D–
Department of Energy	B+	C–	F	F
Government-wide grade	C	C–	D+	D+

Detection

Even when preventive measures are implemented, no organization is completely secure from a determined attack. Thus, organizations should implement detection systems to catch intruders in the act. Organizations often employ an intrusion detection system to minimize the impact of intruders.

An **intrusion detection system** is software and/or hardware that monitors system and network resources and activities, and notifies network security personnel when it identifies possible intrusions from outside the organization or misuse from within the organization. Two fundamentally different approaches to intrusion detection are knowledge-based approaches and behavior-based approaches. Knowledge-based intrusion detection systems contain information about specific attacks and system vulnerabilities, and watch for attempts to exploit these vulnerabilities, such as repeated failed login attempts or recurring attempts to download a program to a server. When such an attempt is detected, an alarm is triggered. A behavior-based intrusion detection system models normal behavior of a system and its users from reference information collected by various means. The intrusion detection system compares current activity to this model and generates an alarm if it finds a deviation. Examples include unusual traffic at odd hours or a user in the Human Resources Department who accesses an accounting program that she has never before used.

Response

An organization should be prepared for the worst—a successful attack that defeats all or some of a system's defenses and damages data and information systems. A response plan should be developed well in advance of any incident and be approved by both the organization's legal department and senior management. A well-developed response plan helps keep an incident under technical and emotional control.

In a security incident, the primary goal must be to regain control and limit damage, not to attempt to monitor or catch an intruder. Sometimes system administrators take the discovery of an intruder as a personal challenge and lose valuable time that should be used to restore data and information systems to normal.

Incident Notification

A key element of any response plan is to define who to notify and who *not* to notify. Questions to cover include the following: Within the company, who needs to be notified, and what information does each person need to have? Under what conditions should the company contact major customers and suppliers? How does the company inform them of a disruption in business without unnecessarily alarming them? When should local authorities or the FBI be contacted?

Most security experts recommend against giving out specific information about a compromise in public forums, such as news reports, conferences, professional meetings, and online discussion groups. All parties working on the problem need to be kept informed and up to date without using systems connected to the compromised system. The intruder may be monitoring these systems and e-mail to learn what is known about the security breach.

Protection of Evidence and Activity Logs

An organization should document all details of a security incident as it works to resolve the incident. Documentation captures valuable evidence for a future prosecution and provides data to help during the incident eradication and follow-up phases. It is especially important to capture all system events, the specific actions taken (what, when, and who), and all external conversations (what, when, and who) in a logbook. Because this may become court evidence, an organization should establish a set of document handling procedures using the legal department as a resource.

Incident Containment

Often it is necessary to act quickly to contain an attack and to keep a bad situation from becoming even worse. The response plan should clearly define the process for deciding if an attack is dangerous enough to warrant shutting down or disconnecting critical systems from the network. How such decisions are made, how fast they are made, and who makes them are all elements of an effective response plan.

Eradication

Before the IT security group begins the eradication effort, it must collect and log all possible criminal evidence from the system, and then verify that all necessary backups are current, complete, and free of any virus. Creating a forensic disk image of each compromised system on write-only media both for later study and as evidence can be very useful. After virus eradication, the group must create a new backup. Throughout this process, a log should be kept of all actions taken. This will prove helpful during the follow-up phase and ensure that the problem does not recur. It is imperative to back up critical applications and data regularly. Many organizations, however, have implemented inadequate backup processes and found that they could not fully restore original data after a security incident. All backups should be created with enough frequency to enable a full and quick restoration of data if an attack destroys the original. This process should be tested to confirm that it works.

Incident Follow-Up

Of course, an essential part of follow-up is to determine how the organization's security was compromised so that it does not happen again. Often the fix is as simple as getting a software patch from a product vendor. However, it is important to look deeper than the immediate fix to discover why the incident occurred. If a simple software fix could have prevented the incident, then why wasn't the fix installed *before* the incident occurred?

A review should be conducted after an incident to determine exactly what happened and to evaluate how the organization responded. One approach is to write a formal incident report that includes a detailed chronology of events and the impact of the incident. This report should identify any mistakes so that they are not repeated in the future. The experience from this incident should be used to update and revise the security incident response plan.

Creating a detailed chronology of all events will also document the incident for later prosecution. To this end, it is critical to develop an estimate of the monetary damage. Potential costs include loss of revenue, loss in productivity, and the salaries of people working to address the incident, along with the cost to replace data, software, and hardware.

Another important issue is the amount of effort that should be put into capturing the perpetrator. If a Web site was simply defaced, it is easy to fix or restore the site's HTML (Hypertext Markup Language—the code that describes to your browser how a Web page should look). What if the intruders inflicted more serious damage, however, such as erasing proprietary program source code or the contents of key corporate databases? What if they stole company trade secrets? Expert crackers can conceal their identity, and tracking them down can take a long time as well as a tremendous amount of corporate resources.

The potential for negative publicity must also be considered. Discussing security attacks through public trials and the associated publicity has not only enormous potential costs in public relations but real monetary costs as well. For example, a brokerage firm might lose customers who learn of an attack and think their money or records aren't secure. Even if a company decides that the negative publicity risk is worth it and goes after the perpetrator, documents containing proprietary information that must be provided to the court could cause even greater security threats in the future. On the other hand, an organization must decide if it has an ethical or a legal duty to inform customers or clients of a cyberattack that may have put their personal data or financial resources at risk.

Table 3-9 recommends a set of actions an organization can take to implement a successful IT security initiative. The appropriate answer to each question is *yes*.

TABLE 3-9 Checklist to evaluate an organization's readiness for a security incident

Question	Yes	No
Has a risk assessment been performed to identify investments in time and resources that can protect the organization from its most likely and most serious threats?		
Have senior management and employees involved in implementing security measures been educated about the concept of reasonable assurance?		
Has a security policy been formulated and broadly shared throughout the organization?		
Have automated systems policies been implemented that mirror written policies?		
Does the security policy address: • E-mail with executable file attachments? • Wireless networks and devices? • Use of smartphones deployed as part of corporate rollouts as well as those bought by end users?		
Is there an effective security education program for employees, contractors, and part-time employees?		
Has a layered security solution been implemented to prevent break-ins?		
Has a firewall been installed?		

(continued)

TABLE 3-9 Checklist to evaluate an organization's readiness for a security incident (*continued*)

Question	Yes	No
Is antivirus software installed on all personal computers?		
Is the antivirus software frequently updated?		
Have precautions been taken to limit the impact of malicious insiders?		
Are the accounts, passwords, and login IDs of former employees promptly deleted?		
Is there a well-defined separation of employee responsibilities?		
Are individual roles defined so that users have authority to perform their responsibilities and nothing more?		
Is it a requirement to review at least quarterly the most critical Internet security threats and implement safeguards against them?		
Has it been verified that backup processes for critical software and databases work correctly?		
Has an intrusion detection system been implemented to catch intruders in the act—both in the network and on critical computers on the network?		
Has an intrusion prevention system been implemented to thwart intruders in the act?		
Are periodic IT security audits conducted?		
Has a comprehensive incident response plan been developed?		
Has the security plan been reviewed and approved by legal and senior management?		
Does the plan address all of the following areas: • Incident notification? • Protection of evidence and activity logs? • Incident containment? • Eradication? • Incident follow-up?		

Summary

- The security of information technology used in business is of the utmost importance, but it must be balanced against other business needs and issues.

- Increasing complexity, higher user expectations, expanding and changing systems, and increased reliance on software with known vulnerabilities has caused a dramatic increase in the number, variety, and impact of security incidents.

- Viruses, worms, Trojan horses, botnets, distributed denial-of-service attacks, root-kits, spam, and phishing are among the most common computer exploits.

- A successful computer exploit aimed at several organizations can have a cost impact of more than $1 billion.

- There are many different kinds of people who launch computer attacks, including the hacker, cracker, malicious insider, industrial spy, cybercriminal, hacktivist, and cyberterrorist. Each type has a different motivation.

- Trustworthy computing is a method of computing that delivers secure, private, and reliable computing experiences based on sound business practices.

- The security of any system is a combination of technology, policy, and people, and it requires a wide range of activities to be effective.

- A strong security program begins by assessing threats to the organization's computers and network, identifying actions that address the most serious vulnerabilities, and educating users about the risks involved and the actions they must take to prevent a security incident.

- The IT security group must lead the effort to implement security policies and procedures, along with hardware and software tools to help prevent security breaches.

- The key to prevention of a computer security incident is to implement a layered security solution to make computer break-ins so difficult that an attacker eventually gives up.

- No security system is perfect, so systems and procedures must be monitored to detect a possible intrusion.

- If an intrusion occurs, there must be a clear reaction plan that addresses notification, evidence protection, activity log maintenance, containment, eradication, and recovery.

Self-Assessment Questions

The answers to the Self-Assessment Questions can be found in Appendix G.

1. According to the "2008 CSI Computer Crime and Security Survey," which of the following was the most common security incident?
   a. instant messaging abuse
   b. distributed denial-of-service attacks
   c. laptop theft
   d. virus attack

2. A virus does not spread itself from computer to computer but must be spread through infected e-mail document attachments, infected programs, or infected Web sites. True or False?

3. An attack on an information system that takes advantage of a vulnerability is called a(n) _____.

4. A group of computers controlled centrally from one or more remote locations by hackers without the knowledge of their owners is called a(n) _____.

5. A set of programs that enables a hacker to gain administrative level access to a computer without the end user's consent or knowledge is called a(n):

   a. Trojan horse

   b. logic bomb

   c. rootkit

   d. worm

6. _____ forces unwanted and often objectionable materials into e-mail boxes, detracts from the ability of Internet users to communicate effectively, and costs Internet users and service providers millions of dollars annually.

7. Software that generates and grades tests that humans can pass but that all but the most sophisticated computer programs cannot is called _____.

8. A person who attacks computers and information systems in order to capture trade secrets and gain a competitive advantage is called a cyberterrorist. True or False?

9. To date, there are no documented cases of cyberterrorism. True or False?

10. A type of attacker that is extremely difficult to detect or stop because he or she is often authorized to access the very systems being abused is called a(n) _____.

11. Concern over potential cyberterrorism began well before the attacks of 9/11. True or False?

12. _____ is a method of computing that delivers secure, private, and reliable computing experiences.

13. The process of assessing security-related risks from both internal and external threats to an organization's computers and networks is called a(n) _____.

14. The written statement that defines an organization's security requirements as well as the controls and sanctions used to meet those requirements is known as a:

   a. risk assessment

   b. security policy

   c. firewall

   d. none of the above

15. Implementation of a strong firewall provides adequate security for almost any network. True or False?

16. A device that works to prevent an attack by blocking viruses, malformed packets, and other threats from getting into the company network is called a(n):

   a. firewall

   b. honeypot

c.   intrusion prevention system

d.   intrusion detection system

Discussion Questions

1.   Identify and briefly discuss four reasons why computer incidents have become more prevalent.

2.   A successful distributed denial-of-service attack requires downloading software that turns unprotected computers into zombies under the control of the malicious hacker. Should the owners of the zombie computers be fined as a means of encouraging people to better safeguard their computers? Why or why not?

3.   Do you believe that spam is actually harmful? Why or why not?

4.   How can installation of a firewall give an organization a false sense of security?

5.   Some IT security personnel believe that their organizations should always employ whatever resources are necessary to capture and prosecute computer criminals. Do you agree? Why or why not?

6.   You have been assigned to be a computer security trainer for all of your firm's 2,000 employees, contractors, and part-time workers. What are the key topics you would cover in your training program?

7.   What do you think motivates a hacker to attempt to break into computers to probe their defenses?

8.   Why and how do spammers seek to set up e-mail accounts with major, free, and reputable Web-based e-mail service providers?

9.   Why might a nonprofit charity decide not to involve law enforcement officials in a computer incident that resulted in the loss of personal data about its donors?

What Would You Do?

1.   You are a member of the IT security support group of a large manufacturing company. You have been awakened late at night and informed that someone has defaced your organization's Web site and also attempted to gain access to computer files containing information about a new product currently under development. What are your next steps? How much effort would you spend in tracking down the identity of the hacker?

2.   You are a member of the Human Resources group of a three-year-old software manufacturer that has several products and annual revenue in excess of $500 million. You've just received a request from the manager of software development to hire three notorious crackers to probe your company's software products in an attempt to identify any vulnerabilities. The reasoning is that if anyone can find a vulnerability in your software, they can. This will give your firm a head start on developing patches to fix the problems before anyone can exploit them. You're not sure, and feel uneasy about hiring people with criminal records and connections to unsavory members of the hacker/cracker community. What would you do?

3. You have just been hired as an IT security consultant to "fix the security problem" at Acme United Global Manufacturing. The company has been hacked mercilessly over the last six months, with three of the attacks making headlines for the negative impact they have had on the firm and its customers. You have been given 90 days and a budget of $1 million. Where would you begin, and what steps would you take to fix the problem?

4. You are the CFO of a sporting goods manufacturer and distributor. Your firm has annual sales exceeding $500 million, with roughly 25 percent of your sales coming from online purchases. Today your firm's Web site was not operational for about an hour. The IT group informed you that the site was the target of a distributed denial-of-service attack. You are shocked by an anonymous call later in the day in which a man tells you that your site will be attacked unmercifully unless you pay him $250,000 to stop the attacks. What do you say to the blackmailer?

5. You are the CFO of a midsized manufacturing firm. You have heard nothing but positive comments about the new CIO you hired three months ago. As you watch her outline what needs to be done to improve the firm's computer security, you are impressed with her energy, enthusiasm, and presentation skills. However, your jaw drops when she states that the total cost of the computer security improvements will be $300,000. This seems like a lot of money for security, given that your firm has had no major incident. Several other items in the budget will either have to be dropped or trimmed back to accommodate this project. In addition, the $300,000 is above your spending authorization and will require approval by the CEO. This will force you to defend the expenditure, and you are not sure how to do this. You wonder if this much spending on security is really required. How can you sort out what really needs to be done without appearing to be micromanaging or discouraging the new CIO?

6. It appears that someone is using your firm's corporate directory—which includes job titles and e-mail addresses—to contact senior managers and directors via e-mail. The message requests that they click on a URL that takes them to a Web site that looks as if it were designed by your HR organization. Once at this phony Web site, they are asked to confirm the bank and account number to be used for electronic deposit of their annual bonus check. You are a member of IT security for the firm. What can you do?

7. Your friend just told you that he is developing a worm to attack the administrative systems at your college. The worm is "harmless" and will simply cause a message—"Let's party!"—to be displayed on all workstations on Friday afternoon at 3 p.m. By 4 p.m., the virus will erase itself and destroy all evidence of its presence. What would you say or do?

8. You are a member of the application development organization for a small but rapidly growing software company that produces patient billing applications for doctors' offices. During work on the next release of your firm's one and only software product, you discover a small programming glitch in the current release that could pose a security risk to users. The probability of the problem being discovered is low, but if exposed, the potential impact on your firm's 100 or so customers could be substantial: hackers could access private patient data and change billing records. The problem will be corrected in the next release, scheduled to come out in three months, but you are concerned about what should be done for the users of the current release.

The problem has come at the worst possible time. The firm is seeking approval for a $10 million loan to raise enough cash to continue operations until revenue from the sales of its just-released product offsets expenses. In addition, the effort to develop and distribute the patch, to communicate with users, and to deal with any fallout will place a major drain on your small development staff, delaying the next software release by at least two months. You have your regularly scheduled quarterly meeting with the manager of application development this afternoon; what will you say about this problem?

Cases

1. Wildcats Fight Cyberterrorists

A team of computer scientists from the University of Arizona (UA), whose mascot is the wildcat, is tracking and analyzing the use of the Internet by terrorists to recruit new members, train supporters in terrorist tactics and methods, and spread propaganda. This research project, dubbed Dark Web, is supported by the National Science Foundation and other federal agencies. Its goal is to collect and analyze all terrorist content on the Internet, including e-mail, Internet forums, and terrorist-related Web sites (now estimated to exceed 50,000).[42] The project team is made up of roughly one dozen UA professors and graduate students.

At its initiation, the Dark Web was only able to collect and analyze Internet communications in English, Arabic, and Spanish. German and French were quickly added. Chinese, Farsi (spoken in Iran, Afghanistan, and Tajikistan), Dutch, and other languages are also being added.

UA received a $1.5 million grant to study how terrorists use the Internet to train their followers to build, plant, and explode improvised explosive devices (IEDs). These deadly roadside bombs have killed and maimed thousands of people in Iraq and other countries. "Our young soldiers, before they're deployed, will know the IEDs through the enemies' eyes," said Hsinchun Chen, director of UA's Artificial Lab, home of the Dark Web project.[43] The ability to study this information and to know where it has been downloaded has led to countermeasures that protect soldiers and civilians.

Another research area is the identification of the kinds of individuals who are more susceptible to recruitment by extremist groups, and what messages are more effective in recruiting people. A surprising discovery by the Dark Web project team is that the terrorist Web sites are much richer in multimedia content (audio and video) and much more effective in creating an active community of frequent visitors than are U.S. government Web sites.[44]

The Dark Web project team has also developed a technique called Writeprint, which can be used to determine the author of anonymous postings and e-mail. It does this by matching new anonymous postings to the lingual, structural, and semantic features in hundreds of thousands of online postings by known authors. The accuracy rate is estimated to exceed 95 percent.[45]

Market researchers in many organizations are experimenting with the use of sentiment analysis to determine the attitude of a writer or speaker with respect to a certain subject. It is used in market research to evaluate how people feel about a new product or a just-released movie. One approach to sentiment analysis is to separate objective (fact-based) from subjective (opinion-based) sentences and to separate positive from negative comments. In fighting cyberterrorists, Dark Web researchers apply it to find "emotions like hate and rage in an attempt to tease apart the social activists from the suicide bombers."[46]

Terrorism experts point out that terrorists are trained to maintain a wide separation between the political wing (recruiters, propagandists) and the action wing (bomb makers, murderers). Given this separation, it is likely that the Dark Web is collecting and analyzing the communications of the members of the political wings of terrorist groups. Dark Web is not likely to come across the communications of the action wings, as these communications do not flow through the Internet (unless carefully encrypted). Thus, the Dark Web is not likely to intercept communications about planned terrorist actions.[47]

Discussion Questions

1. There are some who think that the techniques and tools used in the Dark Web project could be used in a way that could negatively affect our way of life. The executive director of the Electronic Privacy Information Center, Marc Rotenberg, fears "the very same tools that can be used to track terrorists can also be used to track political opponents."[48] What are some of the negative ways such tools can be used against U.S. citizens? Do you think that this is a reasonable fear?

2. Identify three positive ways that this technology could be used to improve the performance of organizations or to help us in our daily lives.

3. Do a Web search to find current information about the Dark Web research being carried out at UA, including information about the budget for this effort, the number of researchers currently employed, and any new tools or techniques they've developed. Write a brief summary of your findings.

2. The Storm Worm

The Storm worm is a backdoor Trojan horse that attacks computers running the Windows operating systems. It began its attacks in Europe and the United States in January 2007 using an e-mail with the subject line "230 dead as storm batters Europe," referring to an actual weather disaster that occurred that month. In July 2007, over 42 million Storm-related messages were sent out in a single day.[49] These spam messages, sent to trick people into unknowingly installing the Storm worm, included links to happycards2008.com and newyearcards2008.com. Users who clicked on these links were asked to download files named happy2008.exe or happynewyear.exe. If they did so, their computer became infected with the Storm worm.

Recently, nearly 80 percent of the junk e-mails distributed by the Storm botnet advertise online pharmacy brands. Prescription drugs are sold to people who were not first examined by a doctor. Pharmacological testing has found that only one-third of the shipments included the correct dosage of an active ingredient.

A personal computer infected with the Storm worm will either be used to send out millions of junk e-mails advertising Web links that when clicked attempt to download a copy of the worm, or will serve as a destination for that link—essentially hosting the latest copy of the worm for its download.

The Storm worm is one of the most sophisticated worms of recent years. Each compromised computer is merged into a botnet with no centralized control; instead, it connects to two to three dozen other compromised machines, which act as hosts. No one compromised computer— including those that serve as a host machine—is ever aware of more than a small subset of the computers in the total botnet. This makes it impossible to even know the number of computers in the botnet let alone take down the botnet. In September 2007, the size of the botnet was estimated to be somewhere between one and 10 million computers.[50]

The Storm worm authors are quite devious and know how to evade most traditional antivirus tools in order to reach their targets. One technique they use is to create hundreds of variants of the worm to thwart signature recognition antivirus software from detecting it.[51]

Worm authors are clever about continuously changing their tactics to take advantage of people's interest in natural disasters and celebrities to trick them into installing the worm. On April 1, 2008, a Storm worm was released onto the Internet with subject lines related to April Fool's Day. In June 2008, Storm authors resorted to using such fabricated subject lines as "Eiffel Tower damaged by massive earthquakes," "Donald Trump missing, feared kidnapped," and "F.B.I. vs. Facebook."[52]

According to Joshua Corman of IBM Internet Security Systems, "Storm has been a tremendous financial success because it has created a botnet of compromised machines that could be used to launch profitable spam attacks."[53] Attackers get paid for sending out the spam and also receive a slice of the profits generated by the spam.

The Storm worm even has its own conspiracy theorists who claim that both Russian and U.S. government law enforcement agencies know who the worm's authors are and where they live, but they cannot be arrested due to connections with the Russian domestic security service and high level officials, including Vladimir Putin, the prime minister of Russia. More likely, a multinational group, perhaps based in Russia, is behind the Storm worm.[54]

In late 2008, German researchers developed an idea about how to destroy the Storm worm botnet. Their solution involved sending an update to each Storm node, commanding it to download and execute a file that would break the connection between the computers in the botnet. There were a couple of problems with this idea. First, there was the potential that other damage could be caused to users' computers. Second, all this activity would occur without the users' consent, which would violate the computer abuse laws of many countries. Meanwhile, a new, even more sophisticated version of the Storm worm—Waldac—has been uncovered and is spreading.

Discussion Questions

1. What harm does a worm cause to the public, Internet service providers, and e-mail service providers? Are there other reasons why many people object so strongly to the spread of worms?

2. What would you say or do if you received an e-mail that guaranteed you $20 per month if you agreed to permit your computer to be used to send e-mail during hours you are not using it?

3. Do you feel it is objectionable to let others use your computer to execute files without your prior knowledge or consent? What if it were for a good cause, such as destroying a worldwide botnet, such as that created by the Storm worm, or running an application that analyzes research data related to finding a cure for cancer?

3. Whistle-Blower Divides IT Security Community

As a member of the X-Force, Mike Lynn analyzed online security threats for Internet Security Systems (ISS), a company whose clients include businesses and government agencies across the world. In early 2005, Lynn began investigating a flaw in the Internet operating system (IOS) used by Cisco routers. Through reverse engineering, he discovered that it was possible to create a network worm that could propagate itself as it attacked and took control of routers across the Internet. Lynn's discovery was momentous, and he decided that he had to speak out and let IT security professionals and the public know about the danger. "What politicians are talking about when they talk about the Digital Pearl Harbor is a network worm," Lynn said during a presentation. "That's what we could see in the future, if this isn't fixed."

Lynn had informed ISS and Cisco of his intentions to talk at a Black Hat conference—a popular meeting of computer hackers—and all three parties entered discussions with the conference managers to decide what information Lynn would be allowed to convey. Two days before the presentation, Cisco and ISS pulled the plug. Cisco employees tore out 10 pages from the conference booklet, and ISS asked that Lynn speak on a different topic—Voice over Internet Protocol (VoIP) security.

In a dramatic move, Lynn resigned from ISS on the morning of the conference and decided to give the presentation as originally planned. Within a few hours of his presentation, Cisco had filed suit against Lynn, claiming that he had stolen information and violated Cisco's intellectual property rights. "I feel I had to do what's right for the country and the national infrastructure," Lynn explained.

Lynn's words might have held more credibility had his presentation *not* been titled "The Holy Grail: Cisco IOS Shellcode and Remote Execution" and had Lynn *not* chosen a Black Hat annual conference as the venue for his crucial revelation.

Rather than speak to a gathering of Cisco users, who would have responded to the revelation by installing Cisco's patch and putting pressure on Cisco to find additional solutions, Lynn chose an audience that may well have included hackers who viewed the search for the flaw as a holy crusade. Black hats are crackers who break into systems with malicious intent. By contrast, white hats are hackers who reveal vulnerabilities to protect systems. Black Hat is a company that provides IT security consulting, briefings, and training. The CEO of Black Hat, Jeff Moss, also founded DEFCON, an annual meeting of underground hackers who gather together to drink, socialize, and talk shop. During the DEFCON conference, which followed the Black Hat conference, hackers worked late into the night trying to find the flaw.

"What Lynn ended up doing was describing how to build a missile without giving all the details. He gave enough details so people could understand how a missile could be built, and they could take their research from there," said one DEFCON hacker.

Once well defined, the line between white hat and black hat has become blurry. Security professionals, law enforcement officials, and other white hats have infiltrated the ranks of the black-garbed renegades at DEFCON annual conferences. IT companies hire hackers as IT security experts. Microsoft has declared that it plans to host annual hacker conferences that it will call Blue Hat conferences. Respectable IT giants such as IBM, Microsoft, and Hewlett-Packard have invited the black hats into the industry, and they have accepted the invitation in large part.

Yet Cisco's handling of Lynn had the black hats up in arms. "The whole attempt at security through obscurity is amazing, especially when a big company like Cisco tries to keep a researcher quiet," exclaimed Marc Maiffret, chief hacker for eEye Digital Security. Maiffret felt that Cisco would have to mend some bridges with the IT security community.

White hats, in the meantime, bombarded the IT media with opinion pieces reminding people that a similar Black Hat disclosure about Microsoft precipitated the creation of the Blaster worm, which tore across the Internet and cost billions of dollars in damage.

Discussion Questions

1. Do you think that Mike Lynn acted in a responsible manner? Why or why not?

2. Do you think that Cisco and ISS were right to pull the plug on Lynn's presentation at the Black Hat conference? Why or why not?

3. Outline a more reasonable approach toward communicating the flaw in the Cisco routers that would have led to the problem being promptly addressed without stirring up animosity among the parties involved.

Endnotes

[1] Robert Richardson, "2008 CSI Computer Crime and Security Survey," http://i.zdnet.com/blogs/csisurvey2008.pdf.

[2] Peter Gumbel, "4 Things I Learned at Société Générale," *Fortune*, February 1, 2008.

[3] Peter Gumbel, "4 Things I Learned at Société Générale," *Fortune*, February 1, 2008.

[4] Nelson D. Schwartz and Katrin Bennhold, "Société Générale Scandal: 'A Suspicion That This Was Inevitable,'" *International Herald Tribune*, February 5, 2008.

[5] Peter Gumbel, "4 Things I Learned at Société Générale," *Fortune*, February 1, 2008.

[6] "Rogue Trader to Cost SocGen $7b," *BBS News*, January 24, 2008.

[7] Nicola Clark and James Kanter, "Decision Delayed on Releasing Ex-trader at Center of Société Générale Inquiry," *International Herald Tribune*, March 14, 2008.

[8] "Société Boss Burton to Step Down," *BBC News*, April 17, 2008.

[9] "GLG to Acquire SocGen Long Only Operation," *Hedge Funds Review*, December 24, 2008.

[10] Nelson D. Schwartz and Katrin Bennhold, "Société Générale Scandal: 'A Suspicion That This Was Inevitable,'" *International Herald Tribune*, February 5, 2008.

[11] Peter Gumbel, "4 Things I Learned at Société Générale," *Fortune*, February 1, 2008.

[12] Robert Richardson, "2008 CSI Computer Crime and Security Survey," http://i.zdnet.com/blogs/csisurvey2008.pdf.

[13] Caroline Gabriel, "Symantec Flexes Its Muscles with Zero-Day Products for an Increasingly Hostile Web," *Rethink It*, May 1, 2004.

[14] Sharon Gaudin, "Public in the Dark about 95% of Software Bugs, IBM Says," *InformationWeek*, June 5, 2007.

[15] Jaikumar Vijayan, "Up Next: Cellular Botnets, Cybermilitias," *Computerworld*, October 17, 2008.

[16] Gregg Keizer, "Trojan Hides in Pirated Copies of Apple's iWork '09," *Computerworld*, January 22, 2009.

[17] "Symantec Announces MessageLabs Intelligence 2008 Annual Security Report," *EmailWire.com*, January 22, 2009.

[18] "Computer Crime a Growing Threat, Warns FBI," *New Scientist*, October 16, 2008.

[19] "Symantec Announces MessageLabs Intelligence 2008 Annual Security Report," *EmailWire.com*, January 22, 2009.

[20] Robert Richardson, "2008 CSI Computer Crime and Security Survey," http://i.zdnet.com/blogs/csisurvey2008.pdf.

[21] Carolyn Duffy Marsan, "How Close Is World War 3.0?" *Network World*, August 22, 2007.

[22] "Symantec Announces MessageLabs Intelligence 2008 Annual Security Report," *EmailWire.com*, January 22, 2009.

[23] "Russia Emerges as Spam Superpower," *SC Magazine*, February 2, 2008.

[24] Kieren Nicholoson, "Missing 'Spam King' Kills Family, Self," *Denver Post*, December 2, 2008.

[25] "Symantec Announces MessageLabs Intelligence 2008 Annual Security Report," *EmailWire.com*, January 22, 2009.

[26] Thomas Claburn, "Twitter Has Security Meltdown," *InformationWeek*, January 5, 2009.

[27] George Hulme, "Chinese Hackers Repeatedly Hack White House Network," *InformationWeek*, November 7, 2008.

[28] Thomas Claburn, "Hackers Deface CERN's 'Big Bang' Particle Accelerator Site," *InformationWeek*, September 15, 2008.

[29] Intellinx Ltd., "A Break Through in Insider Threat and Prevention," Intellinx, www.intellinx-sw.com/resources_insider_threat.asp.

[30] Kenneth Wong, "Tech Trends, What Happens in Delhi, Stays in Delhi," *Cadalyst*, September 1, 2006.

[31] John Lyden, "Hacktivist Tool Targets Hamas," *Register*, January 9, 2009.

[32] "Trustworthy Computing," Microsoft, www.microsoft.com/mscorp/twc/default.mspx (accessed March 31, 2009).

[33] Seymour E. Goodman and Herbert S. Lin, eds, "Toward a Safer and More Secure Cyberspace," National Academy of Sciences, www.cyber.st.dhs.gov/docs/Toward_a_Safer_and_More_Secure_Cyberspace-Full_report.pdf (accessed March 9, 2009).

[34] Marc Gartenberg, "How to Develop an Enterprise Security Policy," *Computerworld*, January 13, 2005.

[35] Robert Richardson, "2008 CSI Computer Crime and Security Survey," http://i.zdnet.com/blogs/csisurvey2008.pdf.

[36] "2007 Computer Security Division Report," National Institute of Standards and Technology, http://csrc.nist.gov/publications/nistir/ir7442/NIST-IR-7442_2007CSDAnnualReport.pdf (accessed February 2, 2009).

[37] Jaikumar Vijayan, "Handheld Risks Prompt Push for Usage Policy," *Computerworld*, February 21, 2005.

[38] Larry Greenemeier, "The Threat Within: Employees Pose the Biggest Security Risk," *InformationWeek*, July 16, 2007.

[39] "Background, FISMA Implementation Project," http://csrc.nist.gov/sec-cert/ca-background.html, August 13, 2005.

[40] May 2008 Federal Security Report Card, http://republicans.oversight.house.gov/media/PDFs/Reports/FY2007FISMAReportCard.pdf (accessed December 9, 2008).

[41] FISMA Grades 2005, http://republicans.oversight.house.gov/FISMA (accessed December 9, 2008).

[42] Eric Swedlund, "UA Effort Sifting Web for Terror-Threat Data," *Arizona Daily Star*, September 24, 2007.

[43] Eric Swedlund, "UA Effort Sifting Web for Terror-Threat Data," *Arizona Daily Star*, September 24, 2007.

[44] Eric Swedlund, "UA Effort Sifting Web for Terror-Threat Data," *Arizona Daily Star*, September 24, 2007.

[45] Eric Swedlund, "UA Effort Sifting Web for Terror-Threat Data," *Arizona Daily Star*, September 24, 2007.

[46] Stephen Kotler, "'Dark Web' Project Takes on Cyber-Terrorism," *Fox News.com*, October 12, 2007.

[47] Stephen Kotler, "'Dark Web' Project Takes on Cyber-Terrorism," *Fox News.com*, October 12, 2007.

[48] Stephen Kotler, "'Dark Web' Project Takes on Cyber-Terrorism," *Fox News.com*, October 12, 2007.

[49] Sharon Gaudin, "Storm Worm Erupts into Worst Virus Attack in 2 Years," *InformationWeek*, July 2, 2007.

[50] Clement James, "Storm Malware Still Blowing Strong," *Secure Computing*, January 29, 2008.

[51] Gregg Kizer, "Storm Switches Tactics Third Time, Adds Rootkit," *Computerworld*, December 27, 2007.

[52] Shaun Nichols, "Storm Tries Its Hand at Fiction," *SC Magazine*, June 23, 2008.

[53] George Hulme, "Storm Worm Makers Reaping Millions a Day in Profit," *InformationWeek*, February 10, 2008.

[54] John Leyden, "Storm Worm Turns One," *Security*, January 18, 2008.

PRIVACY

VIGNETTE

Privacy Concerns Abound with New IRS Systems

In the summer of 2007, Josa'lyn Johnson, a clerk with the Philadelphia Service Center of the Internal Revenue Service (IRS), pled guilty to charges that she had provided unauthorized access to certain taxpayers' personal information. Johnson sold the names, addresses, and Social Security numbers of 24 individuals in return for about $1,500. She was sentenced to three years' probation and was required to pay a small fine.[2]

Although the IRS opened over 500 investigations into alleged unauthorized accessing of taxpayers' records by IRS employees in 2007, most of these breaches of privacy resulted from simple curiosity—a clerk snooping into the records of a friend, a neighbor, or an ex-spouse.[3] These incidents are not widespread and involve only a tiny fraction of IRS employees, who are disciplined and, in some cases, prosecuted. Overall, the IRS has handled privacy concerns with great care—that is, until recently, when a new threat to personal tax-related data emerged.

Over 130 million individuals pay federal taxes and file private information with the IRS each year. The data collected by the IRS includes not only names and addresses, but also Social Security numbers and financial information. This data is a tempting target for identity thieves. Moreover, the money the IRS collects funds the federal government. Disrupting the processing of tax returns and tax payments could significantly affect the nation's economy. As a result, the federal government has issued stringent security requirements for storing, accessing, and manipulating this data.

In 1999, the IRS began a $1 billion project to modernize its computer-based master filing system. The new system, the Customer Account Data Engine (CADE), has been implemented gradually, with new functionalities rolling out twice a year; the project is scheduled to be completed in 2012. In October 2007, the IRS began deploying Account Management Services (AMS), a system through which IRS employees can quickly access data to respond to taxpayer inquiries. The IRS has also been developing a Modernized e-File (MeF) system through which taxpayers can file taxes, forms, and schedules over the Internet. The MeF system deployment began in 2004 and is scheduled to be completed in 2020.

In September 2008, the Treasury Inspector General for Tax Administration (TIGTA) issued a report summarizing major security flaws in both the CADE and AMS.[4] The auditors found that those given privileged-user status, including system administrators and others, could gain access to and modify tax-payer data undetected. Moreover, of the hundreds of contractors and employees with privileged-user status, nearly 40 percent had never been properly authorized to be granted such status. In addition, timeout mechanisms did not function properly, so that if an authorized user walked away from a computer, others might gain access—again, undetected.

The auditors also found that the disaster recovery plan was deficient and that the systems were vulnerable in other ways. Backup tapes stored at an off-site location were not being tested regularly to check that data had been retrieved correctly and could be used to restore the system if necessary.

When CADE was deployed, the IRS failed to implement rigorous protections in the system against malicious codes, such as computer viruses. These and other vulnerabilities made the new IRS systems a weaker target for hackers and thieves.

In 2008, CADE—the system so prone to hacking and other failures—handled approximately 20 percent of all taxpayer returns.[5]

On the last business day of 2008, TIGTA issued another report, this one on the MeF system. The auditors found even more problems with unauthorized data and system access in this system. Any IRS employee with Intranet access could log on to the MeF system management console and change security settings. The procedures for creating and checking users' identification did not meet federal security standards. Users who repeatedly entered invalid passwords were not locked out of the system. Several users had established passwords that were not in compliance with regulations, making it easier for hackers to break in. Finally, users had access to more data than they needed; in fact, database administrator privileges were provided to some people who were not in that role and who should not have been granted those privileges.[6]

All these vulnerabilities were known when the system was deployed in 2007. To date, no known major privacy violations have occurred, and the IRS has agreed to the recommendations made by TIGTA to remedy the situation. Yet the deployment of these new information systems greatly increases the risk of privacy breaches by malicious hackers. The new threats dwarf the privacy concerns of the past, which arose primarily from unchecked curiosity.

Questions to Consider

1. What information about you is being held, who is holding it, and what is this information being used for?
2. What measures are being taken to safeguard this information, and what happens if it is inadvertently disclosed or deliberately stolen?

PRIVACY PROTECTION AND THE LAW

The use of information technology in business requires balancing the needs of those who use the information that is collected against the rights and desires of the people whose information is being used.

On the one hand, information about people is gathered, stored, analyzed, and reported because organizations can use it to make better decisions. Some of these decisions, including whether or not to hire a job candidate, approve a loan, or offer a scholarship, can profoundly affect people's lives. In addition, the global marketplace and intensified competition have increased the importance of knowing consumers' purchasing habits and financial condition. Companies use this information to target marketing efforts to consumers who are most likely to buy their products and services. Organizations also need basic information about customers to serve them better. It is hard to imagine an organization having productive relationships with its customers without having data about them. Thus, organizations want systems that collect and store key data from every interaction they have with a customer.

On the other hand, many people object to the data collection policies of government and business on the basis that they strip individuals of the power to control their own personal information. For these people, the existing hodgepodge of privacy laws and practices fails to provide adequate protection; rather, it causes confusion that promotes distrust and skepticism, which are further fueled by the disclosure of threats to privacy, such as those detailed in the opening vignette on IRS systems. Indeed, "one of the key factors affecting the growth of e-commerce is the lack of Internet users' confidence in online information privacy."[7]

A combination of approaches—new laws, technical solutions, and privacy policies—is required to balance the scales. Reasonable limits must be set on government and business access to personal information; new information and communication technologies must be designed to protect rather than diminish privacy; and appropriate corporate policies must be developed to set baseline standards for people's privacy. Education and communication are also essential.

This chapter helps you understand the right to privacy, and presents an overview of developments in information technology that could impact this right. The chapter also addresses a number of ethical issues related to gathering data about people.

First, it is important to gain a historical perspective on the right to privacy. During the debates on the adoption of the United States Constitution, some of the drafters expressed concern that a powerful government would intrude on the privacy of individual citizens. After the Constitution went into effect in 1789, several amendments were proposed that would spell out the rights of individuals. Ten of these proposed amendments were ultimately ratified and became known as the Bill of Rights. So although the Constitution does not contain the word *privacy*, the U.S. Supreme Court has ruled that the concept of privacy is protected by a number of amendments in the Bill of Rights. For example, the Supreme Court has stated that American citizens are protected by the Fourth Amendment when there is a "reasonable expectation of privacy."

The Fourth Amendment is as follows:

> The right of the people to be secure in their persons, houses, papers, and effects, against unreasonable searches and seizures, shall not be violated, and no Warrants shall issue, but upon probable cause, supported by Oath or affirmation, and particularly describing the place to be searched, and the persons or things to be seized.

However, the courts have ruled that *without* a reasonable expectation of privacy, there is no privacy right.

Today, in addition to protection from government intrusion, people need privacy protection from private industry. Few laws actually provide such protection, and most people assume that they have greater privacy rights than the law actually provides. Some people believe that only those with something to hide should be concerned about the loss of privacy; however, everyone should be concerned. As the Privacy Protection Study Commission noted in 1977, when the computer age was in its infancy: "The real danger is the gradual erosion of individual liberties through the automation, integration, and interconnection of many small, separate record-keeping systems, each of which alone may seem innocuous, even benevolent, and wholly justifiable."[8]

Information Privacy

A broad definition of the right of privacy is "the right to be left alone—the most comprehensive of rights, and the right most valued by a free people."[9] Another concept of privacy that is particularly useful in discussing the impact of IT on privacy is the term *information privacy*, first coined by Roger Clarke, director of the Australian Privacy Foundation. **Information privacy** is the combination of communications privacy (the ability to communicate with others without those communications being monitored by other persons or organizations) and data privacy (the ability to limit access to one's personal data by other individuals and organizations in order to exercise a substantial degree of control over that data and its use).[10] The following sections will cover concepts and principles related to information privacy, beginning with a summary of the most significant privacy laws, their applications, and related court rulings.

Privacy Laws, Applications, and Court Rulings

This section outlines a number of legislative acts passed over the past 40 years that affect a person's privacy. Note that most of these actions address invasion of privacy by the government. Legislation that protects people from data privacy abuses by corporations is almost nonexistent.

Although a number of independent laws and acts have been implemented over time, no single, overarching national data privacy policy has been developed in the United States. Nor is there an established advisory agency that recommends acceptable privacy practices to businesses. Instead, there are laws that address potential abuses by the government, with little or no restrictions for private industry. As a result, existing legislation is sometimes inconsistent or even conflicting. You can track the status of privacy legislation in the United States at the Electronic Privacy Information Center's Web site (*www.epic.org*).

The discussion will be broken into the following topics: financial data, health information, children's personal data, electronic surveillance, export of personal data, and access to government records.

Financial Data

Individuals must reveal much of their personal financial data in order to take advantage of the wide range of financial products and services available, including credit cards, checking and savings accounts, loans, payroll direct deposit, and brokerage accounts. To access many of these financial products and services, individuals must use a personal logon name, password, account number, or PIN. The inadvertent loss or disclosure of this personal financial data carries a high risk of loss of privacy and potential financial loss. Individuals should be concerned about how this personal data is protected by businesses and other organizations and whether or not it is shared with other people or companies.

Fair Credit Reporting Act (1970)

The **Fair Credit Reporting Act** of 1970 regulates the operations of credit-reporting bureaus, including how they collect, store, and use credit information. The act, enforced by the U.S. Federal Trade Commission, is designed to ensure the accuracy, fairness, and privacy of information gathered by the credit-reporting companies (such as Experian, Equifax, and TransUnion) and to check those systems that gather and sell information about people. The act outlines who may access your credit information, how you can find out what is in your file, how to dispute inaccurate data, how long data is retained, and so on. The manual procedures as well as the information systems of the credit-reporting bureaus must implement and support all of these regulations.

Gramm-Leach-Bliley Act (1999)

The **Gramm-Leach-Bliley Act (GLBA)**, also known as the Financial Services Modernization Act of 1999, was a bank deregulation law that repealed a Depression-era law known as Glass-Steagall. Glass-Steagall prohibited any one institution from offering investment, commercial banking, and insurance services; individual companies were only allowed to offer one of those types of financial service products. GLBA enabled such entities to merge. The emergence of new corporate conglomerates, such as Bank of America, Citigroup, and JPMorgan Chase, soon followed. These one-stop financial supermarkets owned bank branches, sold insurance, bought and sold stocks and bonds, and engaged in mergers and acquisitions. Some place partial blame for the financial crisis that began in 2008 on the passage of GLBA and the loosening of banking restrictions.[11]

GLBA also included three key rules that affect personal privacy:

- *Financial Privacy Rule*—This rule established mandatory guidelines for the collection and disclosure of personal financial information by financial organizations. Under this provision, financial institutions must provide a privacy notice to each consumer that explains what data about the consumer is gathered, with whom that data is shared, how the data is used, and how the data is protected. The notice must also explain the consumer's right to **opt out**— to refuse to give the institution the right to collect and share personal data with unaffiliated parties. Anytime the privacy policy is changed, the consumer must be contacted again and given the right to opt out. The privacy notice must be provided to the consumer at the time the consumer relationship is formed and once each year thereafter. Customers who take no action automatically **opt in** and give financial institutions the right to share personal data, such as annual earnings, net worth, employers, personal investment information, loan amounts, and Social Security numbers, to other financial institutions.
- *Safeguards Rule*—This rule requires each financial institution to document a data security plan describing its preparation and plans for the ongoing protection of clients' personal data.
- *Pretexting Rule*—This rule addresses attempts by people to access personal information without proper authority by such means as impersonating an account holder or phishing. GLBA encourages financial institutions to implement safeguards against pretexting.

Initially, financial institutions resorted to mass mailings to contact their customers with privacy-disclosure forms. As a result, many people received a dozen or more similar-looking forms—one from each financial institution with which they did business. However, most people did not take the time to read the long forms, which were printed in small type and full of legalese. Rather than making it easy for customers to opt out, the documents required that consumers send one of their own envelopes to a specific address and state in writing that they wanted to opt out—all this rather than sending a simple prepaid postcard that allowed customers to check off their choice. As a result, most customers threw out the forms without grasping their full implications and thus, by default, agreed to opt in to the collection and sharing of their personal data.

Health Information

The use of electronic medical records and the subsequent interlinking and transferring of this electronic information among different organizations has become widespread. Individuals are rightly concerned about the erosion of privacy of data concerning their health. They fear intrusions into their health data by employers, schools, insurance firms, law enforcement agencies, and even marketing firms looking to promote their products and services. The primary law addressing these issues is the Health Insurance Portability and Accountability Act.

Health Insurance Portability and Accountability Act of 1996

The **Health Insurance Portability and Accountability Act of 1996 (HIPAA)** was designed to improve the portability and continuity of health insurance coverage; to reduce fraud, waste, and abuse in health insurance and healthcare delivery; and to simplify the administration of health insurance.

To these ends, HIPAA requires healthcare organizations to employ standardized electronic transactions, codes, and identifiers to enable them to fully digitize medical records, thus making it possible to exchange medical data over the Internet. The Department of Health and Human Services developed over 1,500 pages of specific rules governing exchange of such data. The regulations affect more than 1.5 million healthcare providers, 7,000 hospitals, and 2,000 healthcare plans.[12]

Under the HIPAA provisions, healthcare providers must obtain written consent from patients prior to disclosing any information in their medical records. Thus, patients need to sign a HIPAA disclosure form each time they are treated at a hospital and such a form must be kept on file with their primary care physician. In addition, healthcare providers are required to keep track of everyone who received information from a patient's medical file.

For their part, healthcare companies must appoint a privacy officer to develop privacy policies and procedures as well as train employees on how to handle sensitive patient data. These actions must address the potential for unauthorized access to data by outside hackers as well as the more likely threat of internal misuse of data. For example, the pharmaceutical giant Eli Lilly and Company built a Web site for Prozac, its antidepression drug; Lilly promoted the site as "Your Guide to Evaluating and Recovering from Depression." Consumers who visited the site were able to subscribe to Lilly's Medi-messenger service, which enabled consumers to compose and schedule personal medication-reminder e-mail messages to be sent from Lilly. When Lilly decided to terminate the Medi-messenger service in June 2001, an employee sent an e-mail message, including all of the current participants of the service in the "To" field of the message, thus disclosing the names of all 669 participants. At the time of this incident, HIPAA was not yet fully in effect, so no fines or penalties were levied against Lilly; however, such unintended revelation of personal patient data in relation to electronic medical records is a major concern that HIPAA is designed to address.[13]

HIPAA assigns responsibility to healthcare organizations, as the originators of individual medical data, for certifying that their business partners (billing agents, insurers, debt collectors, research firms, government agencies, and charitable organizations) also comply with HIPAA security and privacy rules. This provision of HIPAA has healthcare executives especially concerned, as they do not have direct control over the systems and procedures that their partners implement. Those who misuse data may be fined $250,000 and serve up to 10 years in prison.

As the full details of HIPAA are becoming clearer, both medical personnel and privacy advocates are becoming concerned. Some fear that between the increasing demands for disclosure of patient information and the inevitable complete digitization of medical records, patient confidentiality will be lost. Many think that HIPAA provisions are too complicated and that rather than achieving the original objective of reducing medical industry costs, HIPAA will instead increase costs and paperwork for doctors without improving medical care. All agree that the medical industry must make a substantial investment to achieve compliance. The second case at the end of this chapter expands on these issues.

Children's Personal Data

Internet use by children continues to climb; a recent report out of the United Kingdom found that teenagers spend an average of 31 hours per week online.[14] As a concerned society, many of us feel that there is a need to protect children from being exposed to inappropriate material and online predators; becoming the target of harassment; divulging personal data; and becoming involved in gambling or other inappropriate behavior. To date, only a few laws have been implemented to protect children online, and most of these have been ruled unconstitutional under the First Amendment and its protection of freedom of expression.

Children's Online Privacy Protection Act (1998)

According to the **Children's Online Privacy Protection Act (COPPA)**, any Web site that caters to children must offer comprehensive privacy policies, notify parents or guardians about its data-collection practices, and receive parental consent before collecting any personal information from children under 13 years of age. COPPA is meant to give parents control over the collection, use, and disclosure of their children's personal information; it does not cover the dissemination of information to children.

The law has had a major impact, requiring many companies to spend hundreds of thousands of dollars to make their sites compliant, while others eliminated preteens as a target audience. A handful were identified to be in violation of COPPA and were fined by the Federal Trade Commission (FTC). For example, in 2006, the FTC charged Bonzi Software, Inc., and UMG Recordings, Inc., with collecting personal information from children online without their parents' consent, and fined them $75,000 and $400,000, respectively.[15] Bonzi Software, which distributes a free software download called BonziBUDDY, was the first company charged for privacy violations over a download. UMG Recordings, which operates music-related Web sites, was charged with collecting birth dates from children through its online registration process. The social networking Web site Xanga.com was also fined $1 million for allowing preteens to sign up for the service without gaining a parent's consent.[16]

While COPPA is a U.S. law, the Federal Trade Commission has stated that the requirements of COPPA will apply to any foreign-operated Web site that is "directed to children in the United States or knowingly collects information from children in the United States."[17] A few other countries, including Canada and Australia, have enacted similar laws to protect preteens.

Electronic Surveillance

This section covers laws that address government surveillance, including various forms of electronic surveillance. New laws have been added and old laws amended in recent years in reaction to worldwide terrorist activities and the development of new communication technologies.

Communications Act of 1934

The **Communications Act of 1934** established the Federal Communications Commission and gave it responsibility for regulating all non-federal-government use of radio and television broadcasting and all interstate telecommunications—including wire, satellite, and cable—as well as all international communications that originate or terminate in the United States. The act also restricted the government's ability to secretly intercept communications.

Title III of the Omnibus Crime Control and Safe Streets Act (1968, amended 1986)

Title III of the Omnibus Crime Control and Safe Streets Act, also known as the Wiretap Act, regulates the interception of wire (telephone) and oral communications. It allows state and federal law enforcement officials to use wiretapping and electronic eavesdropping, but only under strict limitations. Under this act, a warrant must be obtained from a judge to conduct a wiretap. The judge will approve the warrant only if "there is probable cause [to believe] that an individual is committing, has committed, or is about to commit a particular offense ... [and that] normal investigative procedures have been tried and have failed or reasonably appear to be unlikely if tried or to be too dangerous."[18]

One of the driving forces behind the passage of this act was the case of *Katz v. United States*. Katz was convicted of illegal gambling based on recordings by the FBI of Katz's side of various telephone calls made from a public phone booth; the recordings were made using a device attached to the phone booth. Katz challenged the conviction based on a violation of his Fourth Amendment rights. In 1967, the case finally made it to the Supreme Court, which agreed with Katz. The court ruled that "the Government's activities in electronically listening to and recording the petitioner's words violate the privacy upon which he justifiably relied while using the telephone booth and thus constituted a 'search and seizure' within the meaning of the Fourth Amendment."[19] This ruling helped form the basis for the requirement that there be a reasonable expectation of privacy for the Fourth Amendment to apply.

Title III court orders must describe the duration and scope of the surveillance, the conversations that may be captured, and the efforts to be taken to avoid capture of innocent conversations. The number of approved Title III wiretaps increased from 870 in 1991 to 2,208 in 2007. During this 16-year period, less than three dozen wiretap requests were denied.[20]

Since the Wiretap Act was passed, it has been significantly amended by several new laws, including FISA, ECPA, CALEA, and the USA PATRIOT Act. These acts will now be discussed.

Foreign Intelligence Surveillance Act (1978)

The **Foreign Intelligence Surveillance Act (FISA)** of 1978 describes procedures for the electronic surveillance and collection of foreign intelligence information in communications between foreign powers and the agents of foreign powers. **Foreign intelligence** is information relating to the capabilities, intentions, or activities of foreign governments or agents of foreign governments or foreign organizations. The act allows surveillance, without court order, within the United States for up to a year unless the "surveillance will acquire the contents of any communication to which a U.S. person is a party."[21] If a U.S. citizen is involved, judicial authorization is required within 72 hours after surveillance begins.

The act also created the FISA court, which meets in secret to hear applications for orders approving electronic surveillance anywhere within the United States. Each application for a surveillance warrant is made before an individual judge of the court. Such applications are rarely turned down—there were 22,990 applications for warrants made between 1979 and 2006, and only five were denied.

The act also specifies that the U.S. attorney general may request a specific communications common carrier to furnish information, facilities, or technical assistance to accomplish the electronic surveillance.

FISA makes it illegal to intentionally engage in electronic surveillance under appearance of an official act or to disclose or use information obtained by electronic surveillance under appearance of an official act knowing that it was not authorized by statute; this is punishable with a fine of up to $10,000 or up to five years in prison, or both.

FISA was amended by the USA PATRIOT Act in 2001 to include anyone involved in terrorism on behalf of groups not backed by a foreign government in its definition of agents of foreign powers and thus covered by FISA.

Foreign Intelligence Surveillance Amendments Act (2008)

The **Foreign Intelligence Surveillance Amendments Act** contains several amendments that both revised many of the FISA procedures for gathering foreign intelligence and implemented legal protections for electronic communications service providers who previously provided consumer data to the National Security Agency (NSA) and the CIA. (A number of lawsuits have been filed against these companies, alleging that they illegally monitored the phone calls and e-mails of millions of people.[22]) These amendments were highly controversial, and the American Civil Liberties Union (ACLU) filed a lawsuit challenging the constitutionality of the act the same day that it was enacted into law.[23]

Electronic Communications Privacy Act (1986)

The **Electronic Communications Privacy Act of 1986 (ECPA)** deals with three main issues: (1) the protection of communications while in transfer from sender to receiver; (2) the protection of communications held in electronic storage; and (3) the prohibition of devices to record dialing, routing, addressing, and signaling information without a search warrant. ECPA was passed as an amendment to Title III of the Omnibus Crime Control and Safe Streets Act.

Title I of ECPA extends the protections offered under the Wiretap Act to electronic communications, such as e-mail, fax, and messages sent over the Internet. The government is prohibited from intercepting such messages unless it obtains a court order based on probable cause (the same restriction as the Wiretap Act for telephone calls).

Title II of ECPA (also called the Stored Communications Act) prohibits unauthorized access to stored wire and electronic communications, such as the contents of e-mail Inboxes, instant messages, message boards, and social networking sites. However, the law only applies if the stored communications are not readily accessible to the general public. Webmasters who desire protection for their subscribers under this act must take careful measures to limit public access through the use of logon procedures, passwords, and other methods. Under this act, the FBI director or someone acting on his behalf may issue a National Security Letter (NSL) to an Internet service provider to provide various data and records about a service subscriber.

A third part of ECPA establishes a requirement for court-approved law enforcement use of a **pen register**—a device that records electronic impulses to identify the numbers dialed for outgoing calls—or a **trap and trace**—a device that records the originating number of incoming calls for a particular phone number. A recording of every telephone number dialed and the source of every call received can provide an excellent profile of a person's associations, habits, contacts, interests, and activities. A similar type of surveillance has also been applied to e-mail communications to gather e-mail addresses, e-mail header information, and Internet provider addresses. To obtain approval for a pen-register order or a trap-and-trace order, the law enforcement agency only needs to certify that "the information likely to be obtained is relevant to an ongoing criminal investigation." (This requirement is much lower than the probable cause required to obtain a court order to intercept an electronic communication.) A prosecutor does not have to justify the request, and judges are required to approve every request.

Communications Assistance for Law Enforcement Act (1994)

The **Communications Assistance for Law Enforcement Act (CALEA)** was passed by Congress in 1994 and amended both the Wiretap Act and ECPA. CALEA was a hotly debated law because it required the telecommunications industry to build tools into its products that federal investigators could use—after obtaining a court order—to eavesdrop on conversations and intercept electronic communications. The court order can only be obtained if it is shown that a crime is being committed, that communications about the crime will be intercepted, and that the equipment being tapped is being used by the suspect in connection with the crime.

A provision in the act covering radio-based data communication grew from a realization that the Electronic Communications Privacy Act of 1986 failed to cover emerging technologies, such as wireless modems, radio-based electronic mail, and cellular data networks. The ECPA statute outlawed the unauthorized interception of wire-based digital traffic on commercial networks, but the law's drafters did not foresee the growing interest in wireless data networks. Section 203 of CALEA corrected that oversight by effectively covering all publicly available "electronic communication."

With CALEA, the Federal Communications Commission responded to appeals from the Justice Department and other law enforcement officials by requiring providers of Internet phone services and broadband services to ensure that their equipment accommodated the use of law enforcement wiretaps. This equipment includes Voice over Internet Protocol (VoIP) technology, which shifts calls away from the traditional phone network of wires and switches to technology based on converting sounds into data and transmitting them over the Internet. The decision has created a controversy among many who fear that opening VoIP to access by law enforcement agencies will create additional points of attack and security holes that hackers can exploit.

USA PATRIOT Act (2001)

The **USA PATRIOT Act** (Uniting and Strengthening America by Providing Appropriate Tools Required to Intercept and Obstruct Terrorism) of 2001 was passed just after the terrorist attacks of September 11, 2001. It gave sweeping new powers both to domestic law enforcement and international intelligence agencies, including increasing the ability of law enforcement agencies to search telephone, e-mail, medical, financial, and other records. It also eased restrictions on foreign intelligence gathering in the United States.

Although the act was more than 340 pages long and quite complex (it changed more than 15 existing statutes), it was passed into law just five weeks after being introduced. Legislators rushed to get the act approved in the House and Senate, arguing that law enforcement authorities needed more power to help track down terrorists and prevent future attacks. Critics have argued that the law removes many checks and balances that previously gave courts the opportunity to ensure that law enforcement agencies did not abuse their powers. Critics also argue that many of its provisions have nothing to do with fighting terrorism. Table 4-1 summarizes the key provisions of the PATRIOT Act as they affect the privacy of individuals.

TABLE 4-1   Key provisions of the USA PATRIOT Act

Section	Issue addressed	Summary
201	Wiretapping in terrorism cases	Added several crimes for which federal courts may authorize wiretapping of people's communications
202	Wiretapping in computer fraud and felony abuse cases	Added computer fraud and abuse to the list of crimes the FBI may obtain a court order to investigate under Title III of the Wiretap Act
203(b)	Sharing wiretap information	Allows the FBI to disclose evidence obtained under Title III to other federal officials, including "law enforcement, intelligence, protective, immigration, national defense, [and] national security" officials
203(d)	Sharing foreign intelligence information	Provides for disclosure of threat information obtained during criminal investigations to "appropriate" federal, state, local, or foreign government officials for the purpose of responding to the threat
204	FISA pen-register/trap-and-trace exceptions	Exempts foreign intelligence surveillance from statutory prohibitions against the use of pen-register or trap-and-trace devices, which capture "addressing" information about the sender and recipient of a communication; it also exempts the U.S. government from general prohibitions against intercepting electronic communications and allows stored voice-mail communication to be obtained by the government through a search warrant rather than more stringent wiretap orders
206	FISA roving wiretaps	Expands FISA to permit "roving wiretap" authority, which allows the FBI to intercept any communications to or by an intelligence target without specifying the telephone line, computer, or other facility to be monitored
207	Duration of FISA surveillance of non-U.S. agents of a foreign power	Extends the duration of FISA wiretap orders relating to an agent of a foreign power from 90 days to 120 days, and allows an extension in 1-year intervals instead of 90-day increments
209	Seizure of voice-mail messages pursuant to warrants	Enables the government to obtain voice-mail messages under Title III using just a search warrant rather than a wiretap order, which is more difficult to obtain; messages stored on an answering machine tape, however, remain outside the scope of this section
212	Emergency disclosure of electronic surveillance	Permits providers of communication services (such as telephone companies and Internet service providers) to disclose consumer records to the FBI if they believe immediate danger of serious physical injury is involved; communication providers cannot be sued for such disclosure
214	FISA pen-register/trap-and-trace authority	Allows the government to obtain a pen-register/trap-and-trace device "for any investigation to gather foreign intelligence information"; it prohibits the use of FISA pen-register/trap-and-trace surveillance against a U.S. citizen when the investigation is conducted "solely on the basis of activities protected by the First Amendment"

(continued)

Privacy

TABLE 4-1 Key provisions of the USA PATRIOT Act (*continued*)

Section	Issue addressed	Summary
215	FISA access to tangible items	Permits the FBI to compel production of any record or item without showing probable cause; people served with a search warrant issued under FISA rules may not disclose, under penalty of law, the existence of the warrant or the fact that records were provided to the government. It prohibits investigation of a U.S. citizen when it is conducted solely on the basis of activities protected by the First Amendment.
217	Interception of computer-trespasser communications	Creates a new exception to Title III that permits the government to intercept the "communications of a computer trespasser" if the owner or operator of a "protected computer" authorizes it; it defines a protected computer as any computer "used in interstate or foreign commerce or communication" (because of the Internet, this effectively includes almost every computer)
218	Purpose for FISA orders	Expands the application of FISA to situations in which foreign intelligence gathering is merely a significant purpose rather than the sole purpose
220	Nationwide service of search warrants for electronic evidence	Expands the geographic scope in which the FBI can obtain search warrants or court orders for electronic communications and customer records
223	Civil liability and discipline for privacy violations	Provides that people can sue the government for unauthorized disclosure of information obtained through surveillance
225	Provider immunity for FISA wiretap assistance	Provides immunity from lawsuits for people who disclose information to the government pursuant to a FISA wiretap order, a physical search order, or an emergency wiretap or search
505	Authorizes use of National Security Letters (NSLs) to gain access to personal records	Authorizes the attorney general or a delegate to compel holders of your personal records to turn them over to the government simply by writing a National Security Letter, which is not subject to judicial review or oversight; NSLs can be used against anyone, including U.S. citizens, even if they are not suspected of espionage or criminal activity

Source: "Key Provisions of USA Patriot Act," NewsMax.com, December 22, 2005, http://archive.newsmax. com/archives/articles/2005/12/22/113858.shtml.

One of the more contentious aspects of the USA PATRIOT Act has been the guidelines issued for the use of National Security Letters (NSLs). Before the PATRIOT Act was enacted, the FBI could issue an NSL for information about someone only if it had reason to believe the person was a foreign spy. Under the PATRIOT Act, the FBI can issue an NSL to compel banks, Internet service providers, and credit reporting companies to turn over information about their customers without a court order simply on the basis that the information is needed for an ongoing investigation. The FBI issued nearly 200,000 NSLs between 2003 and 2006 under these relaxed requirements.[24] The ACLU has challenged the use of NSLs by the FBI in court several times. These lawsuits are in various stages of hearings and appeals. In one lawsuit, *Doe v. Holder*, the Court of the Southern District of New York and, upon appeal, the Second Circuit Court of Appeals ruled that the NSL gag provision—which prohibits NSL recipients from informing anyone, even the person who is the subject of the NSL request, that the government has secretly requested his or her records—violates the First Amendment.

Recognizing that the PATRIOT Act was both quickly written and very broad in scope, Congress designated several "sunset" provisions that were to terminate on December 31, 2005. A **sunset provision** terminates or repeals a law or portions of it after a specific date unless further legislative action is taken to extend the law. It was March 2006 before Congress decided on the sunset provisions. Only two of the provisions in Table 4-1 were modified:

- Section 215 was extended for four years and altered slightly so that recipients of FISA subpoenas for record searches would have the right to consult with a lawyer to challenge the request in court.
- Section 505 was modified so that the recipient of an NSL could challenge the nondisclosure requirement but no sooner than one year after receiving the National Security Letter. Recipients are not required to disclose the names of any attorneys they consulted, but they are required to tell the FBI if they consulted anyone other than legal counsel.

Export of Personal Data

Various organizations have developed guidelines to ensure that the flow of personal data across national boundaries (transborder data flow) does not result in the unlawful storage of personal data, the storage of inaccurate personal data, or the abuse or unauthorized disclosure of such data. Two sets of these guidelines will now be discussed.

Organisation for Economic Co-operation and Development Fair Information Practices (1980)
Another agency concerned with privacy is the Organisation for Economic Co-operation and Development (OECD), an international organization consisting of 30 member countries, including Australia, Canada, France, Germany, Italy, Japan, Mexico, New Zealand, the United Kingdom, and the United States. Its goal is to set policy and come to agreement on topics for which multilateral consensus is necessary for individual countries to make progress in a global economy. Dialogue, consensus, and peer pressure are essential to make these policies and agreements stick. The 1980 privacy guidelines set by OECD—also known as the **Fair Information Practices**—are often held up as the model of ethical treatment of consumer data for organizations to adopt. These guidelines are composed of the eight principles summarized in Table 4-2.

TABLE 4-2   Summary of the 1980 OECD privacy guidelines

Principle	Guideline
Collection limitation	Limit the collection of personal data; all such data must be obtained lawfully and fairly with the subject's consent and knowledge
Data quality	Personal data should be accurate, complete, current, and relevant to the purpose for which it is used
Purpose specification	The purpose for which personal data is collected should be specified and should not be changed
Use limitation	Personal data should not be used beyond the specified purpose without a person's consent or by authority of law
Security safeguards	Personal data should be protected against unauthorized access, modification, or disclosure
Openness principle	Data policies should exist, and a data controller should be identified
Individual participation	People should have the right to review their data, to challenge its correctness, and to have incorrect data changed
Accountability	A data controller should be responsible for ensuring that the above principles are met

Source: Organisation for Economic Co-operation and Development, "OECD Guidelines on the Protection of Privacy and Transborder Flows of Personal Data," www.oecd.org/document/18/0,3343,en_2649_34255_1815186_1_1_1_1,00.html.

European Union Data Protection Directive (1998)

The **European Union Data Protection Directive** requires any company doing business within the borders of 15 western European nations to implement a set of privacy directives on the fair and appropriate use of information. Basically, this directive requires member countries to ensure that data transferred to non–European Union (EU) countries is protected. It also bars the export of data to countries that do not have data privacy protection standards comparable to the European Union's. The following list summarizes these European data privacy principles:

- *Notice*—Tell all customers what is done with their information.
- *Choice*—Give customers a way to opt out of marketing.
- *Onward transfer*—When data is transferred to suppliers or other business partners, companies must observe the notice and choice principles mentioned above and require all recipients of such data to provide at least the same level of protection for such data.
- *Access*—Give customers access to their information.
- *Security*—Protect customer information from unauthorized access.
- *Data integrity*—Ensure that information is accurate and relevant.
- *Enforcement*—Independently enforce the privacy policy.

Initially, EU countries were concerned that the largely voluntary system of data privacy in the United States did not meet the EU directive's stringent standards. Eventually, the U.S. Department of Commerce worked out an agreement with the European Union; American companies that could certify adherence to certain safe harbor principles were allowed to import and export personal data. An EU organization can ensure that it is sending data to a U.S. organization participating in the safe harbor by visiting *www.export.gov/ safeharbor*, where a list of such organizations is maintained.

BBB Online and TRUSTe

The European philosophy of addressing privacy involves strict government regulation— including enforcement by a set of commissioners—and differs greatly from the U.S. philosophy of having no federal privacy policy. The United States instead relies on self-regulation, which is overseen by the Department of Commerce, the Better Business Bureau Online (BBB*OnLine*), and TRUSTe.

BBB*OnLine* and TRUSTe are independent initiatives that favor an industry-regulated approach to data privacy. They are both concerned that strict government regulation could have a negative impact on the Internet's use and growth, and that such regulation would be costly to implement and difficult to change or repeal. A Web site operator can apply for a BBB*OnLine* reliability seal or a TRUSTe data privacy seal to demonstrate that his or her site adheres to a high level of data privacy. Having one of the seals can increase consumer confidence in the site operator's desire and ability to manage data responsibly. The seals also help users make more informed decisions about whether to release personal information (such as phone numbers, addresses, and credit card numbers) to a Web site.

An organization must join the Better Business Bureau (BBB) and pay an annual fee ranging from $200 to $7,000, depending on annual sales, before applying for the BBB*OnLine* Reliability Program seal (see Figure 4-1). The BBB*OnLine* seal program identifies online businesses that honor their own stated privacy protection policies; therefore, an accredited Web site must also have adopted and posted an online privacy notice. There are currently 51,600 Web sites that are accredited as meeting the BBB*OnLine* standards.[25]

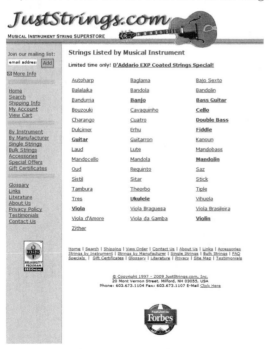

FIGURE 4-1 JustStrings.com displays the BBB*OnLine Reliability Program* seal

For a Web site to receive the TRUSTe seal (see Figure 4-2), its operators must demonstrate that it adheres to established privacy principles. They must also agree to comply with TRUSTe's oversight and consumer resolution process, and pay an annual fee. The privacy principles require the Web site to openly communicate what personal information it gathers, how it will be used, with whom it will be shared, and whether the user has an option to control its dissemination. After operating as a nonprofit organization for 10 years, TRUSTe converted to for-profit status in mid-2008, selling approximately $10 million in newly created stock to the venture capital firm Accel Partners—the same firm that backed eBay and Facebook. Some observers have criticized TRUSTe because it rarely tells users when a Web site has been removed from its program and doesn't require its member Web sites to give users a choice to opt out should the site sell or trade personal data about users to another company.[26] There are currently 1,500 Web sites accredited by TRUSTe.

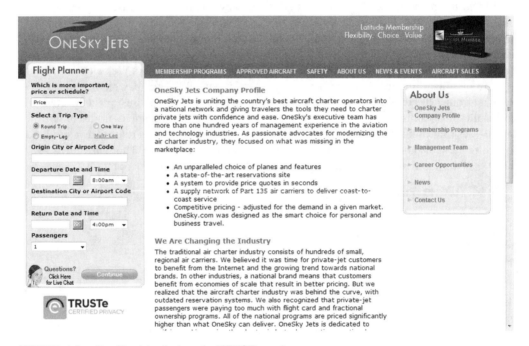

FIGURE 4-2   OneSky Jets displays the TRUSTe seal

Access to Government Records

The government has a great capacity to store data about each and every one of us and about the proceedings of its various organizations. The Freedom of Information Act enables the public to gain access to certain government records, and the Privacy Act prohibits the government from concealing the existence of any personal data record-keeping systems.

Freedom of Information Act (FOIA) (1966, amended 1974)

The **Freedom of Information Act (FOIA)**, passed in 1966 and amended in 1974, grants citizens the right to access certain information and records of the federal government upon request. FOIA is a powerful tool for journalists and the public to acquire information that the government is reluctant to release. The well-defined FOIA procedures have been used to uncover previously unrevealed details about President Kennedy's assassination, determine when and how many times members of Congress or certain lobbyists visited the White House, obtain budget and spending data about a government agency, and even request information on the "UFO incident" at Roswell in 1947. In 2005, U.S. citizens filed 2.7 million requests for government data using FOIA procedures.[27] The act is often used by whistle-blowers to obtain records that they would otherwise be unable to get. Citizens have also used FOIA to find out what information the government has about them.

There are two basic requirements for filing a FOIA request: (1) the request must not require wide-ranging, unreasonable, or burdensome searches for records, and (2) the request must be made according to agency procedural regulations published in the *Federal Register*. A typical FOIA request includes the requester's statement: "pursuant to the Freedom of Information Act, I hereby request"; a reasonably described record; and a statement of willingness to pay for reasonable processing charges. (The fees can be several hundred dollars and include the cost to search for the documents, the cost to review documents to see if they should be disclosed, and the cost of duplication.) FOIA requests are sent to the FOIA officer for the responding agency.[28]

Agencies receiving a request must acknowledge that the request has been received and indicate when the request will be fulfilled. The act requires an initial response within 10 working days. An agency may take an additional 10 days to respond if there are unusual circumstances (e.g., there is a need to consult with another agency). In reality, most requests take much longer. The courts have ruled that this is acceptable as long as the agency treats each request sequentially on a first-come, first-served basis.[29]

If the request is denied, the responding agency must provide the reasons for the denial along with the name and title of each denying officer. The agency must also notify the requester of his or her right to appeal the denial and provide the address to which an appeal should be sent.[30]

Exemptions to the FOIA bar disclosure of information that could compromise national security or interfere with an active law enforcement investigation. Another exemption prevents disclosure of records if it would invade someone's privacy. In this case, a balancing test is applied to evaluate whether the privacy interests at stake are outweighed by competing public interests. Some legislators have called for an additional exemption to protect information shared between the government and private industry regarding attacks against computer and information systems. The sharing of such information, they say, is important for the federal government to protect the nation's critical IT infrastructure from attack; however, private companies are reluctant to share such data. They fear that the FOIA doesn't offer enough assurances that their proprietary data will remain secret.

Privacy Act of 1974

The **Privacy Act of 1974** prohibits U.S. government agencies from concealing the existence of any personal data record-keeping system. Under this law, any agency that maintains such a system must publicly describe both the kinds of information in it and the manner in which the information will be used. The law also outlines 12 requirements that each record-keeping agency must meet, including issues that address openness, individual access, individual participation, collection limitation, use limitation, disclosure limitation, information management, and accountability. The purpose of the act is to provide safeguards for people against invasion of personal privacy by federal agencies. The CIA and law enforcement agencies are excluded from this act; in addition, it does not cover the actions of private industry.

KEY PRIVACY AND ANONYMITY ISSUES

The rest of this chapter discusses a number of current and important privacy issues, including identity theft, consumer profiling, treating consumer data responsibly, workplace monitoring, and advanced surveillance technology.

Identity Theft

Identity theft occurs when someone steals key pieces of personal information to impersonate a person. This information may include such data as name, address, date of birth, Social Security number, passport number, driver's license number, and mother's maiden name. Using this information, an identity thief may apply for new credit or financial accounts, rent an apartment, set up utility or phone service, and register for college courses—all in someone else's name.

Identity theft is recognized as one of the fastest growing forms of fraud in the United States. Recent research indicates that incidents of identity fraud increased from 3.6 percent to 4.3 percent of the overall U.S. adult population from 2007 to 2008, with a resulting increase of $3 billion in losses for a total of $48 billion. The mean cost of identity theft to the consumer was $496 per incident. Interestingly, thieves are increasingly obtaining identity data from stolen personal belongings and via telephone calls rather than online.[31]

Consumers and organizations are becoming more vigilant and proactive in fighting this threat. Many consumers know they can request one free credit report per year from each of the three major credit bureaus, and are also aware that they can put fraud alerts on their credit reports and sign up for credit monitoring services. Many consumers are savvy enough to recognize obvious phishing attempts to capture personal data. Organizations and individuals have put into place improved systems and practices to ward off identity thieves, including spyware and antivirus software. (Table 4-3 offers additional recommendations for safeguarding your identity data.) Still, over 225 million electronic records containing personal data were stolen between January 2005 and May 2008.[32]

TABLE 4-3 Recommendations for safeguarding your identity data

Recommendation	Explanation
Completely and irrevocably destroy digital identity data on used equipment	As it is possible to undelete files and recover data, take necessary actions to ensure that all data is destroyed when you dispose of used computers and data storage devices; consider the use of special software, such as Shred XP
Shred everything	Identity thieves are not above "dumpster diving"—going through your garbage to find financial statements and bills in order to obtain confidential personal information
Require retailers to request a photo ID when accepting your credit card	Writing "Request Photo ID" on the back of your credit cards should prompt retailers to request a photo ID before accepting your card
Beware shoulder surfing	Ensure that nobody can look over your shoulder when you enter or write down personal information—at an ATM, filling out forms in public places, and so on
Minimize personal data shown on checks	Do not include a Social Security number or driver's license number on your checks
Minimize time that mail is in your mailbox	Do not leave paid bills in your mailbox for postal pickup; collect mail from your mailbox as soon as possible after it is delivered
Do not use debit cards to pay for online purchases	Victims of credit card fraud are liable for no more than $50 in losses; debit card users can have their entire checking account wiped out
Treat your credit card receipts safely	Always take your credit card receipts from the retailer and keep them for reconciliation purposes; dispose of them by shredding them
Use hard-to-guess passwords and PINs	Do not use names or words in passwords; include a mix of capital and small letters with at least one special character ($, #, *)

Taking on another identity can be easy and extremely lucrative, enabling the thief to use a victim's credit cards, siphon off money from bank accounts, and even obtain Social Security benefits. The victim is left with a credit history in shambles.

Four approaches are frequently used by identity thieves to capture the personal data of their victims: (1) create a data breach to steal hundreds, thousands, or even millions of personal records; (2) purchase personal data from criminals; (3) use phishing to entice users to willingly give up personal data; and (4) install spyware capable of capturing the keystrokes of victims.

Data Breaches

An alarming number of identity theft incidents involve breaches of large databases to gain personal identity information. The breach may be caused by hackers breaking into the database or, more often than one would suspect, by carelessness or failure to follow proper security procedures. For example, a laptop computer containing the unencrypted Social

Security numbers of 26.5 million U.S. veterans was stolen from the home of a Veterans Affairs (VA) analyst. The analyst violated existing VA policy by removing the data from his workplace. Table 4-4 identifies the largest U.S. data breaches since 2005.

TABLE 4-4   Largest reported data breaches since 2005

Organization	Date	Number of individuals impacted
TJX Companies, Inc.	January 17, 2007	94 million
Visa, MasterCard, American Express	June 19, 2005	40 million
America Online	June 24, 2004	30 million
U.S. Department of Veterans Affairs	May 22, 2006	26 million
HM Revenue and Customs, TNT	November 20, 2007	25 million
T-Mobile, Deutsche Telekom	October 6, 2008	17 million
Archive Systems, Inc., Bank of New York Mellon Corporation	May 7, 2008	12 million
GS Caltex	September 6, 2008	11 million
Dai Nippon Printing Company	March 12, 2007	9 million
Certegy Check Services, Inc., Fidelity National Information Services	July 3, 2007	9 million
Best Western International, Inc.	August 23, 2008	8 million
TD Ameritrade	September 14, 2007	6 million

Source: Open Security Foundation's DataLossDB, http://datalossdb.org.

The number of incidents is alarming, as is the lack of initiative by some companies in informing the people whose data was stolen. Organizations are reluctant to announce data breaches due to the ensuing bad publicity and potential for lawsuits by angry customers. However, victims whose personal data was compromised during a data breach need to be informed so that they can take protective measures.

The cost to an organization that suffers a data breach can be quite high, by some estimates nearly $200 for each record lost. Nearly half the cost is typically due to lost business opportunity associated with the customers who've been lost due to the incident. Other costs include public-relations-related costs to manage the firm's reputation, and increased customer-support costs for information hotlines and credit monitoring services for victims.[33] Hannaford Brothers—a supermarket chain with 167 stores in five northeastern states and Florida—suffered a data breach in which credit and debit card data was captured illegally as cards were swiped at the checkout line. Within days of the breach becoming public knowledge, multiple class actions were filed against the company, alleging that it was negligent for failing to maintain adequate computer security. Hannaford is likely facing years of litigation; tens of millions of dollars in legal fees, settlement costs, and customer credit monitoring services; and reduction of sales revenue due to loss of customer good will.[34]

There is no federal law requiring that organizations reveal a data breach. The state of California passed a data security breach notification law in 2002. It was enacted when the state's payroll database was breached and victims were not notified for six weeks. The law requires that "the disclosure shall be in the most expedient time possible and without unreasonable delay, consistent with the legitimate needs of law enforcement.[35] More than half the states have now implemented similar legislation.

Purchase of Personal Data

There is a black market in personal data. Credit card numbers can be purchased in bulk quantity for as little as $.40 each, while the logon name and PIN necessary to access a bank account can be had for just $10.[36] A full set of identity information—including date of birth, address, Social Security number, and telephone number—sells for between $1 and $15.[37]

Phishing

As discussed in Chapter 3, phishing is an attempt to steal personal identity data by tricking users into entering information on a counterfeit Web site. Spoofed e-mails lead consumers to counterfeit Web sites designed to trick them into divulging personal data. Users have learned through sad experience that simply accessing a phishing Web site can trigger an automatic and transparent download of malware known as spyware to a computer.

Spyware

Spyware is keystroke-logging software downloaded to users' computers without the knowledge or consent of the user. It is often marketed as a spouse monitor, child monitor, or surveillance tool. Spyware creates a record of the keystrokes entered on the computer, enabling the capture of account usernames, passwords, credit card numbers, and other sensitive information. The spy can view the Web sites visited as well as transcripts of chat logs. Spyware operates even if the infected computer is not connected to the Internet, continuing to record each keystroke until the next time the user connects to the Internet. Then, the data captured by the spyware is e-mailed directly to the spy or is posted to a Web site where the spy can view it. Spyware frequently employs sophisticated methods to avoid detection by popular software packages that are specifically designed to combat it. Consumers' fear of spyware has become so widespread that many people now delete e-mail from unknown sources without even opening the messages. This trend is seriously damaging the effectiveness of e-mail as a means for legitimate companies to communicate with customers.

In 2007, the FBI planted spyware on the computer of a 15-year-old student in an attempt to identify him as the person responsible for sending numerous bomb threats to his high school. The FBI first obtained a warrant to allow the agency to install a program called the Computer and Internet Protocol Address Verifier on the student's computer. The software recorded the IP addresses, dates, and times of each communication sent from the student's computer. The student was sentenced to 90 days in juvenile detention and fined $8,852.[38]

Congress passed the Identity Theft and Assumption Deterrence Act of 1998 (full text at *www.ftc.gov/os/statutes/itada/itadact.htm*) to fight identity fraud, making it a federal felony punishable by a prison sentence of three to 25 years. The act also assigns the FTC the task of helping victims restore their credit and erase the impact of the imposter. Although people have been convicted under this act, researchers estimate that fewer than one in 700 identity crimes leads to a conviction.[39]

Identity Theft Monitoring Services

There are numerous identity theft monitoring services, which offer a wide range of coverage. Basic monitoring services cost about $10 per month and provide protection by monitoring the three major credit reporting agencies (TransUnion, Equifax, and Experian) for anyone using your personal data to apply for a new credit card, cell phone, or loan. More expensive services monitor additional databases (e.g., financial institutions, utilities, and the DMV). Subscribers to these services receive a phone call or e-mail if suspicious activity is detected.

Consumer Profiling

Companies openly collect personal information about Internet users when they register at Web sites, complete surveys, fill out forms, or enter contests online. Many companies also obtain information about Web surfers through the use of **cookies**, text files that a Web site can download to visitors' hard drives so that it can identify visitors on subsequent visits. Companies also use tracking software to allow their Web sites to analyze browsing habits and deduce personal interests and preferences. The use of cookies and tracking software is controversial because companies can collect information about consumers without their explicit permission. Outside of the Web environment, marketing firms employ similarly controversial means to collect information about people and their buying habits. Each time a consumer uses a credit card, redeems frequent flyer points, fills out a warranty card, answers a phone survey, buys groceries using a store loyalty card, orders from a mail-order catalog, or registers a car with the DMV, the data is added to a storehouse of personal information about that consumer, which may be sold or shared with third parties. In many of these cases, consumers never explicitly consent to submitting their information to a marketing organization.

Aggregating Consumer Data

Marketing firms aggregate the information they gather about consumers to build databases that contain a huge amount of consumer data. They want to know as much as they can about consumers—who they are, what they like, how they behave, and what motivates them to buy. The marketing firms provide this data to companies so that they can tailor their products and services to individual consumer preferences. Advertisers use the data to more effectively target and attract customers to their messages. Ideally, this means that buyers should be able to shop more efficiently and find products that are well suited for them. Sellers should be better able to tailor their products and services to meet their customers' desires and to increase sales. However, concerns about how all this data is actually used is a major reason why many potential Web shoppers have not yet made online purchases.

Large-scale marketing organizations such as DoubleClick employ advertising networks to serve ads to thousands of Web sites. When someone clicks on an ad at a company's Web site, tracking information about the person is gathered and forwarded to DoubleClick, which stores it in a large database. This data includes a record of the ad on which the person clicked and what the person bought. A group of Web sites served by a single advertising network is called a collection of **affiliated Web sites**.

Collecting Data from Web Site Visits

Marketers use cookies to recognize return visitors to their sites and to store useful information about them. The goal is to provide customized service for each consumer. When someone visits a Web site, the site asks that person's computer if it can store a cookie on the hard drive. If the computer agrees, it is assigned a unique identifier, and a cookie with this identification number is placed on its hard drive. Cookies allow marketers to collect **clickstream data**—information gathered by monitoring a consumer's online activity. During a Web-surfing session, three types of data are gathered. First, as one browses the Web, "GET" data is collected. GET data reveals, for example, that the consumer visited an affiliated book site and requested information about the latest Dean Koontz book. Second, "POST" data is captured. POST data is entered into blank fields on an affiliated Web page when a consumer signs up for a service, such as the Travelocity service that sends an e-mail when airplane fares change for flights to favorite destinations. Third, the marketer monitors the consumer's surfing throughout any affiliated Web sites, keeping track of the information the user sought and viewed. Thus, as a person surfs the Web, a tremendous amount of data is generated for marketers and sellers.

After cookies have been stored on your computer, they work behind the scenes whenever you surf the Web, searching for information about you in network advertising databases. If a match is found, the information stored there about you can be used to tailor the ads and promotions presented as you browse Web sites. The marketer knows what ads have been viewed most recently and makes sure that they aren't shown again, unless the advertiser has decided to market using repetition. The marketer also tracks what sites are visited and uses the data to make educated guesses about the kinds of ads that would be most interesting to you.

There are four ways to limit or even stop the deposit of cookies on your hard drive: (1) adjust your browser settings so that your computer will not accept cookies; (2) manually delete cookies from your hard drive; (3) download and install a cookie-management program; or (4) use anonymous browsing programs that don't accept cookies. (For example, *www.anonymizer.com* offers anonymous surfing services; by switching on the Anonymizer privacy button or going through its Web site, you can hide your identity from nosy Web sites.) However, an increasing number of Web sites lock visitors out unless they allow cookies to be deposited on their hard drives.

Personalization Software

In addition to using cookies to track consumer data, online marketers use **personalization software** to optimize the number, frequency, and mixture of their ad placements, and to evaluate how visitors react to new ads. The goal is to turn first-time visitors to a site into paying customers and to facilitate greater cross-selling activities.

There are several types of personalization software. For example, rules-based personalization software uses business rules tied to customer-supplied preferences or online behavior to determine the most appropriate page views and product information to display when a user visits a Web site. For instance, if you use a Web site to book airline tickets to a popular vacation spot, rules-based software might ensure that you are shown ads for rental cars.

Collaborative filtering offers consumer recommendations based on the types of products purchased by other people with similar buying habits. For example, if you bought a book by Dean Koontz, a company might recommend Stephen King books to you, based on the fact that a significant percentage of other customers bought books by both authors.

Demographic filtering is another form of personalization software. It augments click-stream data and user-supplied data with demographic information associated with user zip codes to make product suggestions. Microsoft has captured age, sex, and location information for years through its various Web sites, including MSN and Hotmail. It has accumulated a vast database on tens of millions of people, each assigned a global user ID. Microsoft has also developed a technology based on this database that enables marketers to target one ad to men and another to women. Additional information such as age and location can be used as ad-selection criteria.

Yet another form of personalization software, contextual commerce, associates product promotions and other e-commerce offerings with specific content a user may receive in a news story online. For example, as you read a story about white-water rafting, you may be offered a deal on rafting gear or a promotion for a white-water rafting vacation in West Virginia. Instead of simply bombarding customers at every turn with standard sales promotions that result in tiny response rates, marketers are getting smarter about where and how they use personalization. They are also taking great care to measure whether personalization is paying off. The intended result is that effective personalization increases online sales and improves consumer relationships.

Online marketers cannot capture personal information, such as names, addresses, and Social Security numbers, unless people provide them. Without this information, companies can't contact individual Web surfers who visit their sites. Data gathered about a user's Web browsing through the use of cookies is anonymous, as long as the network advertiser doesn't link the data with personal information. However, if a Web site visitor volunteers personal information, a Web site operator can use it to find additional personal information that the visitor may not want to disclose. For example, a name and address can be used to find a corresponding phone number, which can then lead to obtaining even more personal data. All this information becomes extremely valuable to the Web site operator, who is trying to build a relationship with Web site visitors and turn them into customers. The operator can use this data to initiate contact or sell it to other organizations with which they have marketing agreements.

Consumer Data Privacy

Consumer data privacy has grown into a major marketing issue. Companies that can't protect or don't respect customer information often lose business and some become defendants in class action lawsuits stemming from privacy violations. For example, privacy groups spoke out vigorously to protest the proposed merger of Web ad server DoubleClick and database marketing company Abacus Direct. The groups were concerned that the information stored in cookies would be combined with data from mailing lists, thus revealing the Web users' identities. This would enable a network advertiser to identify and track the habits of unsuspecting consumers. Public outrage and the threat of lawsuits forced DoubleClick to back off this plan.

Opponents of consumer profiling are also concerned that personal data is being gathered and sold to other companies without the permission of consumers who provide the data. After the data has been collected, consumers have no way of knowing how it is used or who is using it. For example, when Toysmart.com went bankrupt in 2000, it planned to sell customer information from its Web site to the highest bidder, to earn cash to pay its employees and creditors. This data included names, addresses, and ages of customers and their children. TRUSTe had licensed Toysmart.com to put the TRUSTe privacy seal on its Web site, provided that Toysmart.com never divulged customer information to a third party. Because Toysmart.com was planning to violate that agreement, TRUSTe submitted a legal brief asking the bankruptcy court to withhold its approval for the proposed sale. TRUSTe officials also registered a complaint with the FTC, which launched an investigation and then filed suit to stop Toysmart.com from selling its customer list and related information, in violation of the privacy policy that appeared on the company's Web site. Finally, Walt Disney Company, which owned 60 percent of Toysmart.com, bought the list and "retired it"—both to protect customers' privacy and to put an end to the controversy.

One potential solution to consumer privacy concerns is a screening technology called the **Platform for Privacy Preferences (P3P)**, which helps shield users from sites that do not provide the level of privacy protection they desire. Instead of forcing users to find and read through the privacy policy for each site they visit, Web browser software using the P3P protocol downloads the privacy policy from each site, scans it, and notifies users if the policy does not match their preferences. (Of course, unethical marketers can post a privacy policy that does not accurately reflect how data is treated.) The World Wide Web Consortium—an international industry group whose members include Apple, Ericsson, and Microsoft—created P3P and is supporting its development.

Treating Consumer Data Responsibly

When dealing with consumer data, strong measures are required to avoid customer relationship problems. The most widely accepted approach to treating consumer data responsibly is for a company to adopt the Fair Information Practices and the 1980 OECD privacy guidelines. Under these guidelines, an organization collects only personal information that is necessary to deliver its product or service. The company ensures that the information is carefully protected and accessible only by those with a need to know, and that consumers can review their own data and make corrections. The company informs customers if it intends to use customer information for research or marketing, and it provides a means for them to opt out.

As a result of increased focus on the topic of data privacy, many companies recognize the need to establish corporate data privacy policies with an increasing number appointing executives to oversee their data privacy policies and initiatives. Some companies are appointing a **chief privacy officer (CPO)**. A CPO is a senior manager within an organization whose role is to ensure that the organization does not violate government regulations while reassuring customers that their privacy will be protected. In order for the CPO to be effective, the organization must give him or her the power to stop or modify major company marketing initiatives if necessary. The CPO's general duties include training employees about privacy; checking the company's privacy policies for potential risks; figuring out if gaps exist and, if so, how to fill them; and developing and managing a process for customer privacy disputes.

The CPO should be briefed on marketing programs, information systems, and databases involving the collection or dissemination of consumer data while these projects are still in the planning phase. The rationale for early involvement in such initiatives is to ensure that potential problems can be identified at the earliest stages, when it is easier and cheaper to fix them. Some organizations fail to address privacy issues early on, and it takes a negative experience to make them appoint a CPO. For example, U.S. Bancorp, a bank with more than $250 billion in assets as of early 2009, appointed a CPO, but only after spending $3 million to settle a lawsuit that accused the bank of selling confidential customer financial information to telemarketers. Table 4-5 provides useful guidance for ensuring that your organization treats consumer data responsibly. The preferred answer to each question is *yes*.

TABLE 4-5 Manager's checklist for treating consumer data responsibly

Question	Yes	No
Does your company have a written data privacy policy that is followed?		
Can consumers easily view your data privacy policy?		
Are consumers given an opportunity to opt in or opt out of your data policy?		
Do you collect only the personal information needed to deliver your product or service?		
Do you ensure that the information is carefully protected and accessible only by those with a need to know?		
Do you provide a process for consumers to review their own data and make corrections?		
Do you inform your customers if you intend to use their information for research or marketing and provide a means for them to opt out?		
Have you identified a person who has full responsibility for implementing your data policy and dealing with consumer data issues?		

Workplace Monitoring

As discussed in Chapter 2, many organizations have developed a policy on the use of IT in the workplace in order to protect against employee abuses that reduce worker productivity or expose the employer to harassment lawsuits. For example, an employee may sue his or her employer for creating an environment conducive to sexual harassment if other employees are viewing pornography online while at work and the organization takes no measures to stop such viewing. (E-mail containing crude jokes and cartoons or messages that discriminate against others based on sex, race, or national origin can also spawn lawsuits.) The institution and communication of an IT usage policy establishes boundaries of acceptable behavior and enables management to take action against violators. Table 4-6 summarizes the extent of workplace monitoring.

TABLE 4-6 Extent of workplace monitoring

Subject of workplace monitoring	Percent of employers that monitor workers	Percent of companies that have fired employees for abuse or violation of company policy
E-mail	43%	28%
Web surfing	66%	30%
Time spent on phone as well as phone numbers called	45%	6%

Source: American Management Association Press Room, "2007 Electronic Monitoring and Surveillance Survey," February 28, 2008, http://press.amanet.org/press-releases/177/2007-electronic-monitoring-surveillance-survey (accessed April 14, 2009).

The potential for decreased productivity, coupled with increased legal liabilities from computer users, have led employers to monitor workers to ensure that the corporate IT usage policy is followed. Many major U.S. firms find it necessary to record and review employee communications and activities on the job, including phone calls, e-mail, Internet connections, and computer files. Some are even videotaping employees on the job. In addition, some companies employ random drug testing and psychological testing. With few exceptions, these increasingly common (and many would say intrusive) practices are perfectly legal.

The Fourth Amendment to the Constitution protects citizens from unreasonable government searches and is often invoked to protect the privacy of government employees. Public-sector workers can appeal directly to the "reasonable expectation of privacy" standard established by the 1967 Supreme Court ruling in *Katz v. United States*.

However, the Fourth Amendment cannot be used to limit how a private employer treats its employees because such actions are not taken by the government. As a result, public-sector employees have far greater privacy rights than those in private industry. Although private-sector employees can seek legal protection against an invasive employer under various state statutes, the degree of protection varies widely by state. Furthermore, state privacy statutes tend to favor employers over employees. For example, to successfully sue an organization for violation of their privacy rights, employees must prove that they were in a work environment where they had a reasonable expectation of privacy. As a result, courts typically rule against employees who file privacy claims for being monitored while using company equipment. A private organization can defeat a privacy claim simply by proving that an employee had been given explicit notice that e-mail, Internet use, and files on company computers were not private and that their use might be monitored. When an employer engages in workplace monitoring, though, it must ensure that it treats all types of workers equally. For example, a company could get into legal trouble for punishing an hourly paid employee more seriously for visiting inappropriate Web sites than it punished a monthly paid employee.

Society is struggling to define the extent to which employers should be able to monitor the work-related activities of employees. On the one hand, employers want to be able to guarantee a work environment that is conducive to all workers, ensure a high level of worker productivity, and limit the costs of defending against privacy-violation lawsuits filed by disgruntled employees. On the other hand, privacy advocates want federal legislation that keeps employers from infringing on the privacy rights of employees. Such legislation would require prior notification to all employees of the existence and location of all electronic monitoring devices. Privacy advocates also want restrictions on the types of information collected and the extent to which an employer may use electronic monitoring. As a result, many privacy bills are being introduced and debated at the state and federal levels. As the laws governing employee privacy and monitoring continue to evolve, business managers must stay informed in order to avoid enforcing outdated usage policies. Organizations with global operations face an even bigger challenge because the legislative bodies of other countries also debate these issues.

Advanced Surveillance Technology

A number of advances in information technology—such as surveillance cameras, facial recognition software, and satellite-based systems that can pinpoint a person's physical location—provide exciting new data-gathering capabilities. However, these advances can also diminish individual privacy and complicate the issue of how much information should be captured about people's private lives.

Advocates of advanced surveillance technology argue that people have no legitimate expectation of privacy in a public place. Critics raise concerns about the use of surveillance to secretly store images of people, creating a new potential for abuse, such as intimidation of political dissenters or blackmail of people caught with the "wrong" person or in the "wrong" place. Critics also raise the possibility that such technology may not identify people accurately.

Camera Surveillance

London has one of the world's largest public surveillance systems—the average person there might be photographed by 300 cameras in the course of a day.[40] A number of U.S. cities plan to expand their surveillance systems accordingly. Chicago, which has the largest public video surveillance system in the United States, is expanding its 2,000-camera network and encouraging businesses to provide the city with live feeds from their surveillance cameras.[41] As part of its ongoing effort to protect the nation's largest transit system, New York's Metropolitan Transportation Authority awarded a $212 million contract in 2005 to place 1,000 video cameras and 3,000 motion sensors in its subways, commuter railroads, bridges, tunnels, and transit hubs, including Grand Central Station (shown in Figure 4-3).[42]

FIGURE 4-3   Commuters at Grand Central Station

A *smart surveillance system*, which singles out people who are acting suspiciously, is under development in Australia. In a smart surveillance system, computers learn what "normal" behavior is and then look for patterns of behavior outside the norm. When the system detects unusual behavior, it alerts authorities so that they can take preemptive action.

Facial Recognition Software

There have been numerous experiments done using facial recognition software to help identify criminal suspects, with mixed results. The Rampart Division of the Los Angeles Police Department tested such software over a two-month period in 2004, resulting in 19 arrests and the exoneration of one man the officers had suspected of being someone else. Officials in Tampa, Florida, stopped using it in 2003 because it didn't result in any arrests. And at Boston's Logan International Airport in 2002, two systems failed to identify 96 people who had volunteered to help test it, but correctly identified 153 other volunteers.

Polar Rose is a small start-up firm in Sweden whose mission is "to enhance the user experience of digital media by adding meaning to photos and video by recognizing people in them."[43] A browser plug-in enables users to tag and name people in photos appearing anywhere on the Web. The software uses the tagged photos to construct a 3D image of the person that can be used to recognize that person in any public photo on the Web. Google has developed a similar face recognition technology called Picasa. Such software can be used to perform an exhaustive search of the Web and find photos of people to identify where they have been, whom they were with, and what they were doing.

GPS Chips

From automobiles to cell phones, Global Positioning System (GPS) chips are being placed in many devices to precisely locate users. The FCC has asked cell phone companies to implement methods for locating users so that police, fire, and medical personnel can be accurately dispatched to assist 911 callers. Similar location-tracking technology is also available for personal digital assistants, laptop computers, trucks, and boats. Parents can place one of these chips in their teenager's car, then use software to track the car's whereabouts.

Banks, retailers, and airlines are eager to gain real-time access to consumer location data, and have already devised a number of new services they want to provide—sending digital coupons for stores that particular consumers are near, providing the location of the nearest ATM, and updating travelers on flight and hotel information. Airlines are considering the use of wireless devices both to enable passengers to check in for flights when they are close to the gate, and to monitor when each person passes through the gate.

Businesses claim that they will respect the privacy of wireless users and allow them to opt in or opt out of marketing programs that are based on their location data. Wireless spamming is a distinct possibility—a user might continuously receive wireless ads, notices for local restaurants, and shopping advice while walking down the street. Another concern is that the data could be used to track people down at any time or to figure out where they were at some particular instant. The potential to reveal one's location when using a cell phone might cause some people to reconsider using one in the future.

Summary

- The use of information technology in business requires balancing the needs of those who use the information that is collected against the rights and desires of the people whose information is being used. A combination of approaches—new laws, technical solutions, and privacy policies—is required to balance the scales.

- The Fourth Amendment reads: "The right of the people to be secure in their persons, houses, papers, and effects, against unreasonable searches and seizures, shall not be violated, and no Warrants shall issue, but upon probable cause, supported by Oath or affirmation, and particularly describing the place to be searched, and the persons or things to be seized." The courts have ruled that without a reasonable expectation of privacy, there is no privacy right to protect.

- Few laws provide privacy protection from private industry.

- There is no single, overarching national data privacy policy.

- There are a number of federal laws that provide protection for personal financial data, including the Fair Credit Reporting Act (regulates operations of credit-reporting bureaus) and the Gramm-Leach-Bliley Act (established guidelines for the collection and disclosure of personal financial information; required financial institutions to document their data security plan; and encouraged institutions to implement safeguards against pretexting).

- The Health Insurance Portability and Accountability Act defined numerous standards to improve the portability and continuity of health insurance coverage; reduce fraud, waste, and abuse in health insurance care and healthcare delivery; and simplify the administration of health insurance.

- The Children's Online Privacy Protection Act requires Web sites that cater to children to offer comprehensive privacy policies, notify parents or guardians about their data collection practices, and receive parental consent before collecting any personal information from children under the age of 13.

- The Communications Act of 1934 established the Federal Communications Commission and gave it responsibility for regulating all non-federal-government use of radio, television, and interstate telecommunications as well as all international communications that originate or terminate in the United States.

- Title III of the Omnibus Crime Control and Safe Streets Act regulates the interception of wire (telephone) and oral communications.

- The Foreign Intelligence Surveillance Act (FISA) describes procedures for the electronic surveillance and collection of foreign intelligence information between foreign powers and agents of foreign powers.

- The Foreign Intelligence Surveillance Amendments Act amended FISA procedures for gathering foreign intelligence and implemented legal protections for electronic communications service providers who had previously provided consumer data to the NSA and CIA.

- The Electronic Communications Privacy Act (ECPA) deals with the protection of communications while in transit from sender to receiver; the protection of communications held in electronic storage; and the prohibition of devices to record dialing, routing, addressing, and signaling information without a search warrant.

- The Communications Assistance for Law Enforcement Act (CALEA) requires the telecommunications industry to build tools into its products that federal investigators can use—after gaining a court order—to eavesdrop on conversations and intercept electronic communications.

- The USA PATRIOT Act modified 15 existing statutes and gave sweeping new powers both to domestic law enforcement and to international intelligence agencies, including increasing the ability of law enforcement agencies to eavesdrop on telephone communication, intercept e-mail messages, and search medical, financial, and other records; the act also eased restrictions on foreign intelligence gathering in the United States.

- Various organizations have defined guidelines to protect the transborder data flow of personal data. The Organisation for Economic Co-operation and Development created the Fair Information Practices—privacy guidelines that are often held up as the model for organizations to adopt for the ethical treatment of consumer data.

- The European Union Data Protection Directive requires member countries to ensure that data transferred to non–European Union countries is protected. It also bars the export of data to countries that do not have data privacy protection standards comparable to those of the European Union.

- The United States relies on self-regulation overseen by the Department of Commerce and organizations such as the Better Business Bureau Online (BBB*OnLine*) and TRUSTe.

- The Freedom of Information Act (FOIA) grants citizens the right to access certain information and records of the federal government upon request.

- The Privacy Act prohibits U.S. government agencies from concealing the existence of any personal data record-keeping system.

- Identity theft occurs when someone steals key pieces of personal information to impersonate a person; it is the fastest growing form of fraud in the United States.

- Identity thieves often create data breaches, purchase personal data, employ phishing, and install spyware to capture personal data.

- Companies use many different methods to collect personal data about visitors to their Web sites, including depositing cookies on visitors' hard drives and capturing click-stream, GET, and POST data.

- Marketers use personalization software to optimize the number, frequency, and mixture of their ad placements.

- Consumer data privacy has become a major marketing issue—companies that cannot protect or do not respect customer information have lost business and have become defendants in class actions stemming from privacy violations.

- One approach to treating consumer data responsibly is to adopt the Fair Information Practices; some companies also appoint a chief privacy officer.

- Many organizations have developed an IT usage policy to protect against employee abuses that can reduce worker productivity and expose employers to harassment lawsuits.

Self-Assessment Questions

The answers to the Self-Assessment Questions can be found in Appendix G.

1. The U.S. Supreme Court has ruled that the concept of privacy is protected by the _____.

2. _____ is the combination of communications privacy and data privacy.

3. Legislation that protects people from data privacy abuses by private industry is almost non-existent. True or False?

4. An act designed to promote accuracy, fairness, and privacy of information in the files of credit-reporting companies is the:

    a. Gramm-Leach-Bliley Act

    b. Fair Credit Reporting Act

    c. HIPAA

    d. USA PATRIOT Act

5. If someone refuses to give an institution the right to collect and share personal data about oneself, he or she is said to _____.

6. According to the Children's Online Privacy Protection Act, a Web site that caters to children must:

    a. offer comprehensive privacy policies

    b. notify parents or guardians about its data-collection practices

    c. receive parental consent before collecting any personal information from preteens

    d. all of the above

7. The _____ established the Federal Communications Commission and made it responsible for regulating all non–federal government use of radio and TV broadcasting and all interstate telecommunications.

8. _____ *v. United States* is a famous court ruling that helped form the basis for the requirement that there be a reasonable expectation of privacy for the Fourth Amendment to apply.

9. The Electronic Communications Privacy Act deals with

    a. the protection of communications while in transit from sender to receiver

    b. the protection of communications held in electronic storage

    c. the prohibition of devices to record dialing, routing, addressing, and signaling information without a search warrant

    d. all of the above

10. Which of the following identifies the numbers dialed for outgoing calls?

    a. pen register

    b. wiretap

    c. trap and trace

    d. all of the above

11. _____ gave sweeping new powers to law enforcement agencies to search telephone, e-mail, medical, financial, and other records; it also eased restrictions on foreign intelligence gathering in the United States.

12. Under the USA PATRIOT Act, the FBI can issue a(n) _____ to compel banks, Internet service providers, and credit reporting companies to turn over information about their customers without a court order simply on the basis that the information is needed for an ongoing investigation.

13. The European philosophy of addressing privacy concerns employs strict government regulation, including enforcement by a set of commissioners; it differs greatly from the U.S. philosophy of having no federal privacy policy. True or False?

14. Which of the following is *not* a technique frequently employed by identity thieves?

    a. hacking databases

    b. spyware

    c. phishing

    d. trap and trace

15. _____ is used by marketers to optimize the number, frequency, and mixture of their ad placements.

16. Over 25 percent of employers have fired workers for violating or abusing their corporate e-mail policy. True or False?

Discussion Questions

1. Provide a brief historical perspective on the right to privacy. Can you identify key laws and legal rulings that provide the basis for the right to privacy?

2. Prepare a set of arguments that would support the contention that the USA PATRIOT Act was overreaching in both its scope and its approach. Then prepare a set of arguments that support the USA PATRIOT Act as the proper and appropriate way to protect the United States from further terrorist acts.

3. Compare the philosophies of the European Union and the United States regarding data privacy. Which approach do you think is better? Why?

4. The American Civil Liberties Union (ACLU) has challenged the use of National Security Letters by the FBI in court several times. These lawsuits are in various stages of hearings and appeals. In one lawsuit, *Doe v. Holder*, the Court of the Southern District of New York and, upon appeal, the Second Circuit Court of Appeals ruled that the NSL gag provision—which prohibits NSL recipients from informing anyone, even the person who is the subject of the

NSL request, that the government has secretly requested his or her records—violates the First Amendment. Do research to document the status of this lawsuit.

5. What benefits can consumer profiling provide to you as a consumer? Do these benefits outweigh the loss of your privacy?

6. How much effort should a Web site operator be required to take to prevent preteen visitors who lie about their age from visiting the operator's adult-oriented Web site? Should Web site operators be prosecuted under COPPA if preteens provide false information in order to gain access to such sites?

7. A FOIA exemption prevents disclosure of records if it would invade someone's personal privacy. Develop a hypothetical example in which a person's privacy interests are clearly outweighed by competing public interests. Develop another hypothetical example in which a person's privacy interests are not outweighed by competing public interests.

8. Summarize the Fourth Amendment to the U.S. Constitution. Does it apply to the actions of private industry?

9. Do you feel that information systems to fight terrorism should be developed and used even if they infringe on privacy rights or violate the Privacy Act of 1974 and other such statutes?

10. Research the Web to find the latest developments on the Platform for Privacy Preferences (P3P). Write a short report summarizing your findings.

11. Why do employers monitor workers? Do you think they should be able to do so? Why or why not?

12. Do you think that law enforcement agencies should be able to use advanced surveillance technology, such as surveillance cameras combined with facial recognition software? Why or why not?

What Would You Do?

1. Your friend is considering using an online service to identify people with compatible personalities and attractive physical features who would be interesting to date. Your friend must first submit some basic personal information, then complete a five-page personality survey, and finally provide several recent photos. Would you advise your friend to do this? Why or why not?

2. You have 15 years of experience in sales and marketing with three different organizations. You currently hold a middle management position and have been approached by a headhunter to make a move to a company seeking a chief privacy officer. The headhunter claims the position would represent a good move for you in your career path leading to an eventual vice president of marketing position. You are not sure about the exact responsibilities of the CPO position or what authority the person in this position would have at this particular company. How would you go about evaluating this opportunity to determine if it actually is a good career move or a dead end?

3. As the information systems manager for a small manufacturing plant, you are responsible for all aspects of the use of information technology. A new inventory control system is being implemented to track the quantity and movement of all finished products stored in a local warehouse. Each time a forklift operator moves a case of product, he or she must first scan

the UPC code on the case. The product information is captured, as is the day, time, and forklift operator identification number. This data is transmitted over a LAN to the inventory control computer, which then displays information about the case and where it should be placed in the warehouse.

The warehouse manager is excited about using case movement data to monitor worker productivity. He will be able to tell how many cases per shift each operator moves, and he plans to use this data to provide performance feedback that could result in pay increases or termination. He has asked you if there are any potential problems with using the data in this manner, and if so, what should be done to avoid them. How would you respond?

4. You are a writer for a tabloid magazine and want to get some headline-grabbing news about the stars of a popular TV show. You decide to file a separate Freedom of Information Act request for each of the show's stars with the FBI. Would you consider this an ethical approach to getting the information you want? Do you think that the FBI would honor your request?

5. You have been asked to help develop a company policy on what should be done in the event of a data breach, such as unauthorized access to your firm's customer database containing some 1.5 million records. What sort of process would you use to develop such a policy? What resources would you call on?

6. You are a new brand manager for a product line of Coach purses. You are considering the purchase of customer data from a company that sells a large variety of women's products online. In addition to name, mailing address, and e-mail address of customers, the data provides an approximate estimate of customers' annual income based on the zip code in which they live, census data, and highest level of education achieved. You could use the estimate of annual income to identify likely purchasers of your high-end purses, and use e-mail addresses to send e-mails announcing the new product line and touting its many features. List the advantages and disadvantages of such a marketing strategy. Would you recommend this means of promotion in this instance? Why or why not?

7. Your company is rolling out a training program to ensure that everyone is familiar with the company's IT usage policy. As a member of the Human Resources organization, you have been asked to develop a key piece of the training relating to why this policy is needed. What kind of concerns can you expect your audience to raise? How can you deal with this anticipated resistance to the policy?

8. You are the CPO of a midsized manufacturing company, with sales of more than $250 million per year and almost $50 million from online sales. You have been challenged by the vice president of sales to change the company's Web site data privacy policy from an opt-in policy to an opt-out policy and to allow the sale of customer data to other companies. The vice president has estimated that this change would bring in at least $5 million per year in added revenue with little additional expense. How would you respond to this request?

Cases

1. Is Your Passport Secure?

In August 2007, the United States began issuing electronic passports to its citizens. These e-passports were identical to regular U.S. passports save for the addition of a radio frequency identification chip embedded in the back cover. A **Radio Frequency Identification (RFID) chip** listens for a radio query and responds by transmitting its own unique ID code.

The federal government pushed for the adoption of e-passports in order to automate identity verification, speed up immigration inspections, and increase border protection. Yet the information technology security industry has continuously raised concerns about e-passports, suggesting that they might actually increase security risks and identity theft.

Following the terrorist attacks of September 11, 2001, the administration of George W. Bush pushed through numerous measures designed to increase the security of U.S. citizens at home and abroad. As the administration and the newly established Department of Homeland Security (DHS) began considering how to better secure all U.S. ports of entry, many began to advocate for a transition to electronic identifications both for U.S. citizens and for citizens of the 27 countries with which the United States shares a visa waiver agreement. The federal government felt that e-passports would cut down on the number of fake passports with which international criminals and terrorists could gain access to the United States and other countries. In 2002, Congress passed the Enhanced Border Security and Visa Entry Reform Act, which mandated that an e-passport program be in place by 2004.

Privacy groups and the IT security community immediately raised concerns about the initiative. The chips, they argued, could be used for tracking and surveillance, identity theft, and identity cloning. In addition, criminals or terrorists could potentially read the chips in order to identify and target U.S. citizens from among a larger group of people.[44] These concerns were based on the federal government's decision to employ a radio frequency contactless chip in the e-passports. RFID was designed as a means of identifying and tracking objects from a distance utilizing a scanner and a chip embedded into the object. RFID technology is widely used by libraries, museums, airports, ranches, and toll-road systems. Corporations use RFID to track inventory and manage their supply chains. Data stored on RFID chips are considerably easier to read and intercept than data imprinted onto contact chips. RFID chips are also currently in use on enhanced driver's licenses (EDL) and passport cards, which are used to facilitate border crossings between the United States and both Canada and Mexico. (Passport cards cannot be used for international air travel.) The chips on passport cards contain only a unique identifier that is used to query a secure government database to pull up a photo and personal information, which the border inspectors can use to verify the identity of individuals.

Despite the concerns expressed, the Government Printing Office began testing e-passports in 2004.[45] The DHS began searching for ways to make the RFID chip more secure, but this research would take time. The State Department asked Congress to extend the e-passport deadline twice, and the DHS launched a second round of testing in early 2006.[46] The DHS and the State Department were ready to meet the congressional deadline set for October 2006.

Then in August 2006, German IT security consultant Lukas Grunwald arrived at DEFCON—an annual convention of hackers—and demonstrated how he could clone the RFID chip in his German e-passport using a laptop, a $200 RFID reader equipped with an antenna, and an inexpensive smart card writer. The cloned chip could then be pasted onto a forged passport.[47] The demonstration made headlines around the world.

U.S. government officials responded by arguing that Grunwald didn't understand the measures that the State Department had taken to safeguard U.S. e-passports. To prevent skimming—the reading of information off the chip—the government had reduced the distance from which the chip can be read from 20 feet to a few inches. In addition, the government placed metallic material on the cover of the passports to block any attempt to read the passport until it is opened. The government also adopted a protocol it calls Basic Access Control, which requires that the RFID reader prove to the chip that it is authorized to read the chip before the chip will allow the reader to access its data.[48] Furthermore, government officials responded by explaining that the clerk receiving the e-passport could compare the biometric data on the chip with that in the hardcopy of the passport as well as with the individual carrying the passport.[49]

At this same DEFCON convention, however, another group of individuals from the mobile security company Flexilis distributed a video showing how the RFID chip in U.S. passports could be used to identify U.S. citizens among a crowd. The security professionals reasoned that e-passports could easily open up half an inch or more if placed in a handbag. In this case, the metallic covering would no longer protect the data stored on the chip. The video showed that if the e-passport was slightly open, an RFID reader could read an identifier on the chip that would be unique to U.S. passports. Terrorists or criminals could then use this information to target U.S. citizens. In the video, the RFID reader triggers an explosion that harms the dummy carrying the passport.[50]

The federal government responded to this concern by explaining that although each chip presents a unique identifier (UID) to the reader, every time the e-passport is scanned the chip presents a new UID. Since this UID is randomly generated, it cannot be associated with the previous UID presented and hence individuals cannot be tracked.[51]

On January 30, 2009, IT security researcher Chris Paget posted a video on YouTube showing how he skimmed several passport cards and enhanced driver's licenses within half an hour while driving through San Francisco at about 30 miles an hour. He made adjustments to an RFID scanner that he purchased on eBay for $250 so that he could scan the cards from a distance of 20 feet. Paget read the card's unique identifier, which could be used to track individuals. This data could potentially be correlated with name and address information retrieved with another type of RFID reader, such as one used to read credit cards, to verify the identity of the person being tracked.[52] As Paget's demonstration made the news, not all reporters differentiated between e-passports and passport cards and EDL cards.

RFID experts were quick to point out that the RF contactless chips in e-passports are not the same as EPC Gen 2 RFID tags used in the passport and EDL cards that Paget was able to scan. These RFID tags do not have any true means of encrypting or authenticating data.[53] As a result, many organizations have criticized the State Department's decision to use EPC Gen 2 RFID tags in its passport cards as well as the decisions of state governments—such as Arizona and Washington—to issue these tags in its EDL cards.

The State Department has issued over 700,000 passport cards, which are primarily used by people who frequently cross the northern and southern borders. The State Department justifies its use of these RFID tags by arguing that they contain no personal data and that the cards use state-of-the-art security measures, such as laser engraving, to prevent counterfeiting. The State Department also issues a protective sleeve that should prevent RFID readers from skimming data from the cards. It is clear, however, that the cards that Paget was able to scan did not possess the sleeve or that the sleeve did not work as intended.[54]

Passport cards and EDL cards, however, are still optional. U.S. citizens can choose to use the more secure e-passports. However, when applying for a U.S. passport today, an individual cannot elect to obtain an old-fashioned, chipless passport. The U.S. government refuses to issue passports without the RFID contactless chip. If the chip becomes damaged, however, the passport holder is not required to obtain a new passport. As a result, the now well-known German IT security researcher and hacker Lukas Grunwald has a recommendation for those with privacy concerns: stick the e-passport in the microwave for a few seconds and destroy the chip.

Discussion Questions

1. What type of security breaches could occur with e-passports?
2. Why are passport cards less secure than e-passports? What types of breaches could occur with passport and EDL cards?
3. What measures do you think federal and state governments should take to protect the privacy of individuals when issuing these electronic identity cards?

2. How Secure Is Our Healthcare Data?

The American Recovery and Reinvestment Act (ARRA) of 2009 funneled billions of dollars into the development of health information technology, promoting the use of electronic health record (EHR) systems. EHR systems store medical records, allow healthcare professionals to access medical records remotely, and are intended to reduce medical errors and make the healthcare system more efficient. Only 17 percent of U.S. physicians and less than 10 percent of U.S. hospitals have even a basic EHR system.

ARRA was advanced as part of a wider effort to stimulate job creation. The initiative may well create jobs; however, by encouraging the swift adoption of EHR systems, will the initiative improve or endanger patient privacy? Security breaches into medical records are already commonplace. Some breaches result in serious consequences:

- In December 2006, federal prosecutor Susan Harrison discovered that thousands of dollars had disappeared from her checking account. Two former employees of the Virginia Mason Medical Center in Seattle had snuck into the center at night using fake IDs. They accessed medical records and stole the personal information of 42 individuals, including Harrison's husband, and used it to withdraw money from their bank accounts.
- Susan Pugh White, a diabetes education nurse at Randolph Hospital in North Carolina, stole the personal information of 12 patients from their medical records and used that information to run up credit card bills in their names. She bought $60,000 worth of merchandise, including a large-screen television and a lawn mower. She was sentenced to nearly three years in prison in June 2006.
- Between 2006 and 2007, Richard Yaw Adjei stole the identities of over 400 patients treated in Midwestern hospitals who used a Delaware-based billing company. Adjei had paid a claims processor at the company to pass him the names, birth dates, Social Security numbers, and medical information of these individuals. He then used this data to submit 163 tax returns and consequently obtained hundreds of thousands of dollars in tax rebates. Adjei was arrested and sentenced to six years and three months in jail.

As healthcare organizations rush to implement new EHR systems, will these types of breaches become more commonplace? Will new security threats emerge?

In 2008, the Healthcare Information and Management Systems Society (HIMSS) reported that between 2006 and 2007, over 1.5 million patient records were subject to security breaches in hospitals alone.[55] However, not all breaches are committed by individuals with malicious intent. Some commit security breaches out of curiosity—such as the 39 employees of New York City's hospital system who peeked at the records of a seven-year-old girl whose death had made her a tabloid sensation.[56] Others commit breaches out of carelessness. Television news investigators in Houston found computer printouts, prescription labels, and pill bottles with more than 200 patients' personal information in unlocked dumpsters outside 24 local pharmacies.[57]

But while an identity thief digging through a dumpster might recover personal information on 10, 20, maybe 100 individuals, a thief that steals a laptop or an electronic storage device has access to thousands of medical records. In 2006 in Ohio, California, Pennsylvania, Michigan, New York, Minnesota—in state after state, laptops containing medical records were stolen, most commonly from offices and vehicles of health provider employees. Sometimes these records were password protected and encrypted; sometimes they were not. These data breaches affect hundreds of thousands of people.[58] And although healthcare institutions often notify individuals when their EHRs have been compromised, it is the individuals who bear the responsibility for protecting their own financial assets and credit once their file has been breached.

Title II of the Health Insurance Portability and Accountability Act (HIPAA), commonly called the Administrative Simplification Provisions, mandates the creation of standards to protect the confidentiality of electronic transactions within the healthcare system. The HIPAA Security Rule safeguards the confidentiality of electronic patient information. In implementing the Security Rule, the U.S. Department of Health and Human Services (HHS) regulated enforcement tasks to the Centers for Medicare and Medicaid Services (CMS). CMS has the authority to interpret the provisions, conduct compliance reviews, and impose monetary penalties on organizations that do not comply with HIPAA security regulations.

Although CMS was awarded these far-reaching authorities, for the first decade after HIPAA was enacted, the agency chose not to conduct compliance reviews. Instead, it relied entirely on individuals filing complaints against institutions as a means to ensure compliance. By January 31, 2009, CMS had received a total of 1,044 complaints. Not a single monetary penalty had been imposed.[59]

HHS's Office of Inspector General (OIG) launched an investigation into CMS enforcement of the HIPAA Security Rule. It began auditing hospitals in 2006 to determine whether CMS reliance on complaints was sufficient to promote compliance with the HIPAA privacy provisions. OIG first inspected Piedmont Hospital in Atlanta, Georgia, where it evaluated access to patient data, the monitoring of information system activities, and security violations. As a result of the OIG inspection, some healthcare institutions began to take a much closer look at how they could ensure compliance.

In 2007, a number of large healthcare corporations—such as CVS Caremark, Johnson & Johnson, and Philips Healthcare—and technology giant Cisco Systems established the Health Information Trust Alliance, an organization dedicated to developing a common security framework for electronic medical data. A HITRUST survey in 2008 found that 85 percent of health information technology executives supported such an effort.[60] Already organizations like HIMSS engaged in research and professional development to support EHR security. The common security framework went beyond this to provide shared standards and procedures for access control, human resource security, physical and environmental security, information security incident management, and many other areas directly or indirectly related to privacy breaches.

By 2008, OIG released its report, finding that CMS reliance on complaints was an ineffective means to ensure HIPAA compliance. Although CMS officially refuted OIG's conclusions, it agreed to begin compliance reviews. In January 2008, CMS conducted its first review at a hospital about which it had received a complaint regarding a lost laptop. As a result of the review, the hospital developed a plan that included the implementation of procedures requiring laptops to be secured to the workstation where they are used, the training of individuals working with mobile hardware and electronic storage devices, and 24-hour surveillance systems. The hospital was to make sure employees did not have greater access to EHRs than their jobs required. In addition, the hospital instituted measures that assured that the IT department would be notified when system users were terminated so that they could no longer access the EHRs.[61]

The question remains, however, as to whether these efforts can forestall a major proliferation in security breaches as the United States rushes to implement EHR technology in the wake of ARRA. In addition to providing monetary incentives for healthcare organizations to adopt electronic health records, ARRA expands the privacy provisions under HIPAA. Specifically, it extends the Security Rule to business associates of healthcare institutions. In the past, security breaches have been committed not only by these institutions but also by outside contractors that they hire to achieve their IT objectives. ARRA also requires healthcare institutions to notify individuals who are affected by security breaches, provides for increased fines for noncompliance, and authorizes state attorneys general to prosecute institutions violating HIPAA regulations.

The clear intent of these provisions is to increase enforcement of HIPAA, but what will be the actual outcome of these new policies? The development of procedures, standards, and technology is left to the private sector, which clearly recognizes that there are major challenges ahead and has begun to organize to meet these challenges. But will the private sector be able to produce secure solutions in time? Will suitable technology become accessible and affordable to smaller and rural healthcare institutions that are targeted by ARRA? Will CMS and newly authorized government bodies institute effective enforcement policies? These issues will determine in large part how secure health data will be in the coming years.

Discussion Questions

1. What type of security breaches of medical records are common today?
2. What measures are being taken by the government and private industry to safeguard EHRs?
3. How do you think the implementation of ARRA will affect the privacy of our healthcare and personal data? What breaches do you foresee? How can they be forestalled?

3. Is Google Watching You?

In January 2009, a nine-year-old girl from Massachusetts was kidnapped by her grandmother. To find the girl, a local police officer and deputy fire chief called AT&T to obtain the GPS coordinates of the girl's cell phone. These coordinates can identify a phone's location within about 900 feet. The officers had a problem, though. The coordinates showed that the girl was in Virginia. They couldn't just drive to the general area and have a look around. So, instead, they entered the coordinates into Google Maps and used the Google service Street View. Street View provides a 360-degree panoramic view of any location (into which Google has sent a camera-loaded vehicle) from the street. Sure enough, the officers saw a long building with a red roof that looked like a hotel. The agents zoomed in on Street View until they could read the street signs. Then they searched Google to get the hotel's address and called the Virginia State Police. Half an hour later, the girl was recovered.[62]

Reporters celebrated the officers' ingenuity and Google technology. But not everyone is so pleased with Google Street View. In 2008, a couple from Pennsylvania sued Google for posting photos of their home. They claimed that Google had violated their privacy and devalued their property, which they had bought in part because their house could not be seen from the road. The couple claimed that the Google Street View vehicle must have entered their private driveway to take the photos. Google argued that Street View provides an easy method for homeowners and others to request the removal of photos from the application. The company had incorporated this removal feature into the Help section for good reason. After its 2007 launch, users combing through the mapping software came across a number of inappropriate shots, including one of a woman in her underwear. Google removed these photos after being notified. In 2009, a U.S. magistrate judge ruled against the couple.[63]

Some critics have also argued that Google Street View could be used by pedophiles to find schools, parks, and homes where children live and play. Others respond that pedophiles can obtain this information from many other sources.[64]

Street View is only one of a large number of applications that the company offers. Generating $21.8 billion in revenue just 10 years after its founding in 1998, Google operates the most popular search engine in the world.[65] Almost all of its revenue derives from its online advertising system embedded in the free products and services it offers to end users. Google offers free e-mail, Web page creation and publishing, blogging, Internet messaging, Web photo albums, online calendars, video sharing, document sharing, and many other products. It provides Web page translation, as well as searches for scholarly literature, patents, financial news, movies, programming code, and lots of other specialized information. Its mapping tools include Google Earth, through which users can view satellite photos taken of any location on earth, and Google Sky, through which users can view celestial bodies.

With this large assortment of innovative offerings, it is not surprising that Google inspires a host of privacy concerns. First and perhaps foremost is the concern generated by the wealth of personal information Google is able to assemble using a user's login identity. To use Google products, users must first create a Google login through which they provide an e-mail address, or in the case of Google e-mail service, a first and last name. Users must also agree to terms of service for each product that allow Google to collect and keep information on the user in its database. Google then tracks what the user does—what words the user enters in the search engine, what sites the user visits, and the time and date of these events. Google uses this information to fine-tune advertising to the user's needs and preferences. In this way, Google can target their clients' advertisements to a very specific demographic (age, location, profession, hobbies) within a population.[66] For example, if a user enters "brakes" as a search term, Google might display advertisements from automotive repair companies within the user's geographic area. Although Google collects search information for advertising purposes, it also stores this information in a Google database.

In order to retrieve historical data on the user or gain information about a user's identity before he or she has logged in, Google places cookies on the user's computer. One frequent complaint against Google is that it was the first company to extend the expiration of its cookies far into the future. The company gave its cookies an expiration date of 2038, hoping to track user preferences over time.[67] After privacy advocates sharply criticized this decision, Google decided to set its cookies to automatically expire after two years. This concession sounded good until people realized that any time a user visits a Google site, the cookie is automatically renewed for two more years.[68] Google defenders point out that users can delete the cookie at any time, manually or automatically whenever they close their browser. Google critics argue that many users aren't aware of this functionality.

New protests were raised when Google released its new Web browser, Chrome. When a user conducts a search on Google, the company usually collects the user's IP address, the cookie identifier, and the search term.[69] Chrome makes it possible for Google to store not only the search term but any alphanumeric combination entered into the location bar, where users typically enter the URL of the Web site they want to visit. Chrome stores this information whether or not the user presses the Enter key after typing in the URL.[70] Some privacy advocates were incensed. Others, however, pointed out that Chrome had added an incognito mode. This feature allows the user to search the Web without having the Web pages and downloaded files stored in the computer's download or browsing history. Although this feature keeps browsing information private from other individuals who might access a user's computer, it does nothing to prevent Google from storing that information within its database.

One additional fear is that Google will share the information it collects with other sources. Some critics fear what could happen if the government gained access to Google's database. Would this usher in a totalitarian era in which Big Brother (the government) watches and controls every step citizens take? In fact, in 2005, the Department of Justice (DOJ) did attempt to gain access to two months' worth of Google's data as part of its attempt to fight child pornography on the Web. Unlike other Internet service providers who had cooperated with the federal government, Google refused. The corporation argued that Google had pledged to keep its users' personal information private and that to hand over data would violate the trust of the users. Second, Google contended that in handing over this data, the company would be forced to reveal trade secrets regarding its search technology. Finally, Google questioned whether the reason behind the DOJ's demand was justifiable and in fact lawful—based on the Electronic Communications Privacy Act.[71] In March 2006, the judge ruled largely in Google's favor, allowing the DOJ access to part of Google's index of sites, but preventing it from accessing search-term data.[72]

Other privacy concerns have arisen over the malfunctioning of Google's applications. For example, on February 24, 2009, an employee of a Dutch marketing firm reported a bug related to Google's document sharing application, Google Docs.[73] When one of his coworkers tried to share a specific document with one additional person, the application shared the document not only with that person but with all the other people with whom the employee had ever shared a document. The result was a significant breach of confidentiality. Google sent out a warning note to users affected by the bug and fixed it.[74] However, technology reporters regarded the incident as a major privacy blunder.

There can be no question that privacy questions abound concerning Google and its plethora of online applications. Some argue that Google—with its "do no harm" motto—has done well with privacy relative to other companies. However, there is one point everyone can agree on: the best way for users to prevent an online breach of their privacy is to learn more about the online products and services they use.

Discussion Questions

1. How does Google's business model use personal data?

2. What do you think are the major privacy concerns raised by Google's business model and applications?

3. Do you think Google has taken adequate measures to protect its users' privacy? Explain your answer.

End Notes

1 Benjamin Franklin, Pennsylvania Assembly: Reply to the Governor, November 11, 1755.

2 "IRS Combats Its In-House Snoops," *Wall Street Journal*, December 19, 2007.

3 Kevin Poulsen, "Five IRS Employees Charged with Snooping on Tax Returns," *Wired*, May 13, 2008.

4 Treasury Inspector General for Tax Administration, "The Internal Revenue Service Deployed Two of Its Most Important Modernized Systems with Known Security Vulnerabilities," September 24, 2008, www.treas.gov/tigta/auditreports/2008reports/200820163fr.pdf.

5 Jaikumar Vijayan, "Two New IRS Systems Have Major Security Weaknesses, Federal Report Says," *Computerworld*, October 17, 2008, www.computerworld.com/action/article.do?command=viewArticleBasic&articleId=9117447.

6 Treasury Inspector General for Tax Administration, "The Internal Revenue Service Deployed the Modernized e-File System With Known Security Vulnerabilities," December 30, 2008, www.treas.gov/tigta/auditreports/2009reports/200920026fr.pdf.

7 Tomasz Zukowski and Irwin Brow, "Examining the Influence of Demographic Factors on Internet Users Information Privacy Concerns," ACM International Conference Proceedings Series, vol. 226, 2007.

8 Privacy Protection Study Commission, *Personal Privacy in an Information Society: The Report of the Privacy Protection Study Commission*, transmitted to President Jimmy Carter on July 12, 1977, http://aspe.hhs.gov/datacncl/1977privacy/toc.htm.

9 *Olmstead v. United States*, 227 U.S. 438 (1928), http://supreme.justia.com/us/277/438/case.html.

10 Roger Clarke, "Introduction to Dataveillance and Information Privacy, and Definitions of Terms," August 15, 1997, www.rogerclarke.com/DV/Intro.html#Priv.

11 David Leonhardt, "Washington's Invisible Hand," *New York Times*, September 28, 2008.

12 George V. Hulme, "Protecting Privacy," *InformationWeek*, April 16, 2001.

13 Penelope Patsuris, "Eli Lilly Exposes Prozac.com Subscribers," *Forbes*, July 5, 2001, www.forbes.com/2001/07/05/0705lilly.html.

14 "Teenagers 'Spend an Average of 31 Hours Online,'" *Telegraph*, February 13, 2009, www.telegraph.co.uk/scienceandtechnology/technology/4574792/Teenagers-spend-an-average-of-31-hours-online.html.

15 Federal Trade Commission, *UMG Recordings, Inc. to Pay $400,000, Bonzi Software, Inc. to Pay $75,000 to Settle COPPA Civil Penalty Charges*, September 13, 2006, www.ftc.gov/opa/2004/02/bonziumg.shtm.

16 "FTC Fines Xanga for Violating Kids' Privacy," MSNBC.com, September 7, 2006, www.msnbc.msn.com/id/14718350/.

17 Federal Trade Commission, "Frequently Asked Questions about the Children's Online Privacy Protection Rule," www.ftc.gov/privacy/coppafaqs.shtm, October 7, 2008.

18 U.S. Code 18 (1986), §§ 2510–22, www.it.ojp.gov/default.aspx?area=privacy&page=1284#contentTop.

19 *Katz v. United States*, 389 U.S. 347 (1967), http://supreme.justia.com/us/389/347.

20 U.S. Courts, Federal Judiciary, *Wiretaps Up 20 Percent in 2007*, April 30, 2008, www.uscourts. gov/Press_Releases/2008/wiretap.cfm.

21 U.S. Code: Title 50, Chapter 36, Subchapter I, Section 1802, www.law.cornell.edu/uscode/ uscode50/usc_sec_50_00001802----000-.html.

22 Andrew Harris, "Spy Agency Sought U.S. Call Records Before 9/11," June 30, 2006, Bloomberg.com.

23 American Civil Liberties Union, "The Foreign Intelligence Surveillance Act," www.aclu.org/ safefree/spying/fisa.html.

24 American Civil Liberties Union, "National Security Letters," www.aclu.org/safefree/national securityletters/index.html.

25 BBB*OnLine, www.bbb.org/online.*

26 Saul Hansell, "Will the Profit Motive Undermine the Trust in Truste?," *New York Times*, July 15, 2008.

27 William Douglas, "Freedom of Information Act Turns 40," *Seattle Times*, March 11, 2007.

28 Allan Robert Adler, "Step-by-Step Guide to Using the Freedom of Information Act," www.skepticfiles.org/aclu/foia.htm.

29 Bonneville Power Administration, "Procedures for Requesting a (FOIA)," www.bpa.gov/ corporate/public_affairs/FOIA/Procedures.cfm.

30 Allan Robert Adler, "Step-by-Step Guide to Using the Freedom of Information Act," www.skepticfiles.org/aclu/foia.htm.

31 Javelin Strategy & Research, "Latest Javelin Research Shows Identity Fraud Increased 22 Percent, Affecting Nearly Ten Million Americans: But Consumer Costs Fell Sharply by 31 Percent," February 9, 2009, www.javelinstrategy.com/2009/02/09/latest-javelin-research-shows-identity-fraud-increased-22-percent-affecting-nearly-ten-million-americans-but-consumer-costs-fell-sharply-by-31-percent.

32 Mathew Schwartz, "ID Theft Monitoring Services: What You Need to Know," *InformationWeek*, May 9, 2008.

33 Thomas Claburn, "The Cost of Data Loss Rises," *InformationWeek*, November 28, 2007.

34 "Hannaford Bros. Faces Class Action over Data Breach," ConsumerAffairs.com, March 21, 2008, www.consumeraffairs.com/news04/2008/03/hannaford_data3.html.

35 "HIPAA Compliance Strategies," AISHealth.com, February 25, 2008, www.aishealth.com.

36 Mathew Schwartz, "ID Theft Monitoring Services: What You Need to Know," *InformationWeek*, May 9, 2008.

37 Jacob Leibenluft, "Credit Card Numbers for Sale," *Slate*, April 24, 2008.

38 Gregg Keizer, "FBI Planted Spyware on Teen's PC to Trace Bomb Threats," *Computerworld*, July 19, 2007.

39 Steven Levy and Brad Stone, "Grand Theft Identity," *Newsweek*, July 4, 2005.

40 Mike Corning, "U.S. Cities Focus on Spy Cameras," *Chicago Tribune*, August 10, 2005.

41 Mike Corning, "U.S. Cities Focus on Spy Cameras," *Chicago Tribune*, August 10, 2005.

42 Sewell Chan and Shadi Rahimi, "M.T.A. to Keep Electronic Eye on the Subway," *New York Times*, August 24, 2005.

43 Polar Rose, www.crunchbase.com/company/polarrose.

44 Robert Lemos, "Privacy Groups Assail Future Passport Technology," *SecurityFocus*, April 14, 2004, www.securityfocus.com/news/10908.

45 Claire Swedberg, "U.S. Tests E-Passports," *RFID Journal*, November 2, 2004, www.rfidjournal.com/article/view/1218/1/1.

46 Wilson P. Dizard III, "E-passport's First Deployment," Government Computer News, October 6, 2006, http://gcn.com/articles/2006/10/06/epassports-first-deployment.aspx.

47 Joris Evers, "Researchers: E-passports Pose Security Risk," ZDNet, August 5, 2006, http://news.zdnet.com/2100-1009_22-149117.html.

48 The U.S. Electronic Passport Frequently Asked Questions, U.S. State Department, http://travel.state.gov/passport/eppt/eppt_2788.html.

49 Brian Robinson, "Alliance: E-passports Secure," FCW.com, August 9, 2006, www.mrtdanalysis.org/press/English/FCW.pdf.

50 *RFID Passport Shield Failure Demo—Flexilis*, YouTube, August 2, 2006, www.youtube.com/watch?v=-XXaqraF7pl.

51 The U.S. Electronic Passport Frequently Asked Questions, U.S. State Department, http://travel.state.gov/passport/eppt/eppt_2788.html.

52 *Cloning Passport Card RFIDs in Bulk for under $250*, YouTube, January 30, 2009, www.youtube.com/watch?v=9isKnDiJNPk.

53 Smart Card Alliance, "Smart Card Alliance Clarifies Technology Usage in U.S. Electronic Passports versus Passport Cards and EDLs," February 6, 2009, www.smartcardalliance.org/articles/2009/02/06/smart-card-alliance-clarifies-technology-usage-in-u-s-electronic-passports-versus-passport-cards-and-edls.

54 U.S. Department of State, "The U.S. Passport Card Is Now in Production," http://travel.state.gov/passport/ppt_card/ppt_card_3926.html.

55 "2008 HIMSS Analytics Report: Security of Patient Data," April 2008, www.mmc.com/knowledgecenter/Kroll_HIMSS_Study_April2008.pdf.

56 Milt Freudenheim and Robert Pear, "Health Hazard: Computers Spilling Your History," *New York Times*, December 3, 2006.

57 "Pharmacies Dump Medical Information in Trash," KPRC Local 2 (Houston), November 15, 2006.

58 Health Privacy Stories, Health Privacy Project, www.healthprivacy.org/usr_doc/Privacystories.pdf.

59 "OCR Receives 700 HIPAA Complaints in January 2009," Health Information Privacy/Security Alerts, HIPAA Enforcement Statistics for March 2009, www.melamedia.com/HIPAA.Stats.home.html.

60 George Hulme, "The Security and Privacy of Healthcare Data," *InformationWeek*, August 20, 2008, www.informationweek.com/blog/main/archives/2008/08/the_security_an.html.

61 "HIPAA Compliance Review Information and Examples," Centers for Medicare and Medicaid Services, HHS, www.cms.hhs.gov/Enforcement/09_HIPAAComplianceReviewInformationand Examples.asp.

62 "Google Search Finds Missing Child," BBC News, January 9, 2009, http://news.bbc.co.uk/2/hi/technology/7820984.stm.

63 Ben Leach, "Couple Who Sued Google over Street View Photos of Home Lose Privacy Case," *Telegraph*, February 19, 2009, www.telegraph.co.uk/scienceandtechnology/technology/google/4695714/Couple-who-sued-Google-over-Street-View-photos-of-home-lose-privacy-case.html.

64 Sharon Fisher, "Eek! Bryan Fischer Calls Google Street View a Tool for Pedophiles," New West Community Blogs, October 14, 2008, www.newwest.net/topic/article/eek_bryan_fischer_calls_google_street_view_a_tool_for_pedophiles/C564/L564/.

65 Google Inc., *BusinessWeek*, http://investing.businessweek.com/research/stocks/earnings/earnings.asp?symbol=GOOG.O (accessed March 13, 2009).

66 Jennifer Laycock, "Google AdWords Opens Up Demographic Bidding," *Search Engine Guide*, March 25, 2008.

67 Walaika Haskins, "Google Puts 2-Year Expiration Date on Cookies," *TechNewsWorld*, July 17, 2007, www.technewsworld.com/story/58362.html?wlc=1237383339.

68 Walaika Haskins, "Google Puts 2-Year Expiration Date on Cookies," *TechNewsWorld*, July 17, 2007, www.technewsworld.com/story/58362.html?wlc=1237383339.

69 Chris Soghoian, "Debunking Google's Log Anonymization Propaganda," *CNET*, September 11, 2008, http://news.cnet.com/8301-13739_3-10038963-46.html?tag=mncol;title.

70 Ina Fried, "Google's Omnibox Could Be Pandora's Box," *CNET*, September 3, 2008, http://news.cnet.com/8301-13860_3-10031661-56.html.

71 Nicole Wong, "Response to the DoJ Motion," The Official Google Blog, February 17, 2006, http://googleblog.blogspot.com/2006/02/response-to-doj-motion.html.

72 Anne Broache, "Judge: Google Must Give Feds Limited Access to Records," *CNET*, March 18, 2006, http://news.cnet.com/Judge-Google-must-give-feds-limited-access-to-records/2100-1028_3-6051257.html.

73 Richard de Vries, "Question: BUG Adding a Contributor to Multiple Documents Gives Everyone Acces [sic]," Google Docs Help, February 24, 2009, www.google.com/support/forum/p/Google+Docs/thread?tid=534196d50ef662e9&hl=en.

74 Jason Kincaid, "Google Privacy Blunder Shares Your Docs without Permission," *TechCrunch*, March 7, 2009, www.techcrunch.com/2009/03/07/huge-google-privacy-blunder-shares-your-docs-without-permission.

CHAPTER 5

FREEDOM OF EXPRESSION

VIGNETTE

Sexting

Over the years, the courts have made many decisions regarding what kinds of communication are obscene and whether or not they are protected by our First Amendment rights to freedom of expression. The line between what is legal and what is not has been in constant flux. Now comes a new test for our society: sexting. **Sexting**—sending sexual messages, nude or seminude photos, or sexually explicit videos over a cell phone—is a fast-growing trend among teens and young adults. According to a survey by the National Campaign to Prevent Teen and Unplanned Pregnancy, one in five teenagers have engaged in sexting, including 22 percent of teen girls, 18 percent of teen boys, and 11 percent of young teen girls aged 13 to 16.[1] Increasingly, teens are suffering the consequences of this new fad.

Jessie Logan was a good kid—lively, artistic, and fun. But in her last year at Sycamore High School, she made a terrible mistake. She used her cell phone to take nude photos of herself and then sent them to her boyfriend. After the couple broke up, Jessie's ex-boyfriend forwarded the photos to several other teenage girls; eventually, the pictures were sent to hundreds of teens in the Cincinnati area. Classmates, and even kids she did not know, started teasing her. They called her a tramp and worse; some even threw things at her. Instead of attending classes, she began sleeping in her car in the school parking lot or hiding in the bathroom, skipping classes to avoid further embarrassment. She stopped interacting with her friends, and her grades dropped. Finally, in the summer of 2008, Jessie was so full of despair that she hung herself.[2]

Jessie's mother, Cynthia Logan, wondered why school officials and authorities didn't do more to help Jessie. The school superintendent, Adrienne James, said that the sexting problem had been addressed at a parents' meeting. No other action could be taken, James said, because some of the students involved attended other school districts and because Jessie had taken the photos at home—not at school. A school resource officer said that he reprimanded students who harassed Jessie. The officer also spoke to a prosecutor who told him that nothing could be done because Jessie was 18 years old.[3]

But were Jessie's ex-boyfriend's actions legal? Certainly, if they involved exposing underage children to the nude photos, the ex-boyfriend would have been violating laws protecting children from exploitation and pornography. To date, however, the ex-boyfriend and other students who distributed the photos have not been arrested or charged with any crime.

Authorities, however, are starting to come down hard on teens under the age of 18 who engage in sexting:

- Seventeen-year-old Alex Phillips of La Crosse, Wisconsin, received nude photos from his 16-year-old girlfriend. In 2008, he posted the pictures on MySpace with obscene captions as a means of "venting" after the breakup. When police asked Phillips to remove the photos from the Web site, he refused. Police charged Phillips with possession of child pornography, sexual exploitation of a child, and defamation. These charges were eventually dropped, and he was charged with causing mental harm to a child.[4] As part of a plea bargain, Phillips was eventually sentenced to three years of probation and 100 hours of community service.[5]

- In Middletown, Ohio, a 13-year-old boy was arrested after a photo of an eighth-grade girl involved in sexual activity was found on his cell phone by school officials. He had shared the photo with other students at a skating party.[6]

- Phillip Albert sent nude pictures of his 16-year-old ex-girlfriend to 70 people, including her parents and grandparents, after she taunted him. The 18-year-old was sentenced to five years of probation and will be registered as a sex offender until he reaches the age of 43.[7]

Teens are also increasingly being charged for merely exchanging nude photos of themselves over their cell phones. A group of Pennsylvania teenagers may be charged with disseminating and possessing child pornography after three 13-year-old girls sent nude or seminude images of themselves to three 16- and 17-year-old boys. Two Ohio teenagers, four Alabama middle-school students, and many teens in other states have been arrested on similar charges for taking and sharing nude or seminude photos of themselves.

Across the nation, lawmakers and citizens have begun to debate whether teenagers involved in sexting should be charged with such serious child-porn related offenses. Some people argue that applying child pornography laws to teens is too harsh—that the purpose of these laws is to protect children, not prosecute them. In Utah, the legislature recently changed sexting from a felony to a misdemeanor.[8] Authorities in Montgomery County, Ohio, have created a program for teens arrested for sexting. This program will prevent some first-time offenders from being registered as sex offenders—a designation that can stay with them for up to 20 years.[9] In other instances, authorities have not prosecuted teens under child pornography laws, but brought them up on lesser charges.

The crackdown is part of a greater effort to send a message: sexting can be dangerous. The practice can have serious consequences, which those who engage in it may not initially be aware of—creating a situation that can get out of hand very easily. In 24 percent of the 2,100 cases of child pornography that the National Center for Missing and Exploited Children has handled, the children took the photos themselves.[10] Sometimes these photos are used for purposes other than just distribution. In Wisconsin, a teenage boy enticed other boys at his school to sext him by pretending to be a girl. He then blackmailed seven of them into performing sexual favors for him in exchange for not distributing their photos.[11]

When photos depict a person who is over 18, the recipient often has the right to share these photos with others who are over 18—even when sharing is not the "right" thing to do. In these cases, there's no legal recourse for anyone to take to stop that person. In the meantime, underage participants—those who have the least life experience to help them make good decisions—are being held accountable. Until we learn how to deal with this new phenomenon appropriately in homes, in schools, and in the courts, many teens and young adults will likely continue to get hurt.

Questions to Consider

1. Does sexting represent a form of expression that is protected by the First Amendment?
2. What can be done to protect people from the dangers of sexting while still safeguarding our First Amendment rights?

LEARNING OBJECTIVES

As you read this chapter, consider the following questions:

1. What is the basis for the protection of freedom of expression in the United States, and what types of expression are not protected under the law?
2. What are some key federal laws that affect online freedom of expression, and how do they impact organizations?
3. What important freedom of expression issues relate to the use of information technology?

FIRST AMENDMENT RIGHTS

The Internet enables a worldwide exchange of news, ideas, opinions, rumors, and information. Its broad accessibility, open discussions, and anonymity make the Internet a remarkable communications medium. It provides an easy and inexpensive way for a speaker to send a message to a large audience, potentially thousands of people worldwide. In addition, given the right e-mail addresses, a speaker can aim a message with laser accuracy at a select subset of powerful and influential people.

People must often make ethical decisions about how to use such incredible freedom and power. Organizations and governments have attempted to establish policies and laws to help guide people, as well as to protect their own interests. Businesses, in particular, have sought to conserve corporate network capacity, avoid legal liability, and improve worker productivity by limiting the non-business use of IT resources.

The right to freedom of expression is one of the most important rights for free people everywhere. The First Amendment to the U.S. Constitution (shown in Figure 5-1) was adopted to guarantee this right and others. Over the years, a number of federal, state, and local laws have been found unconstitutional because they violated one of the tenets of this amendment.

FIGURE 5-1 The U.S. Constitution

The First Amendment reads:

Congress shall make no law respecting an establishment of religion, or prohibiting the free exercise thereof; or abridging the freedom of speech, or of the press; or the right of the people peaceably to assemble, and to petition the government for a redress of grievances.

In other words, the First Amendment protects Americans' rights to freedom of religion and freedom of expression. This amendment has been interpreted by the Supreme Court as applying to the entire federal government, even though it only expressly refers to Congress.

Numerous court decisions have broadened the definition of speech to include nonverbal, visual, and symbolic forms of expression, such as flag burning, dance movements, and hand gestures. Sometimes the speech at issue is unpopular or highly offensive to a majority of people; however, the Bill of Rights provides protection for minority views. The Supreme Court has also ruled that the First Amendment protects the right to speak anonymously as part of the guarantee of free speech.

However, the Supreme Court has held that the following types of speech are not protected by the First Amendment and may be forbidden by the government: perjury, fraud, defamation, obscene speech, incitement of panic, incitement to crime, "fighting words," and sedition (incitement of discontent or rebellion against a government). Two of these types of speech—obscene speech and defamation—are particularly relevant to information technology.

Obscene Speech

Miller v. California is the 1973 Supreme Court case that established a test to determine if material is obscene and therefore not protected by the First Amendment. Marvin Miller, after conducting a mass mailing campaign to advertise the sale of adult material, was convicted of violating a California statute prohibiting the distribution of obscene material. Some unwilling recipients of Miller's brochures complained to the police, initiating the legal proceedings. Although the brochures contained some descriptive printed material, they primarily consisted of pictures and drawings explicitly depicting men and women engaged in sexual activity. In ruling against Miller, the Supreme Court determined that speech can be considered obscene and not protected under the First Amendment based on the following three questions:

1. Would the average person, applying contemporary community standards, find that the work, taken as a whole, appeals to the prurient interest?
2. Does the work depict or describe, in a patently offensive way, sexual conduct specifically defined by the applicable state law?
3. Does the work, taken as a whole, lack serious literary, artistic, political, or scientific value?

These three tests have become the U.S. standard for determining whether something is obscene. The requirement that a work be assessed by its impact on an average adult in a community has raised many questions:

- Who is an average adult?
- What are contemporary community standards?
- What is a community? (This question is particularly relevant in cases in which potentially obscene material is displayed worldwide on the Web.)

Defamation

The right to freedom of expression is restricted when the expressions, whether spoken or written, are untrue and cause harm to another person. Making either an oral or a written statement of alleged fact that is false and that harms another person is **defamation**. The harm is often of a financial nature, in that it reduces a person's ability to earn a living, work in a profession, or run for an elected office. An oral defamatory statement is **slander**, and a written defamatory statement is **libel**. Because defamation is defined as an untrue statement of fact, truth is an absolute defense against a charge of defamation. Although people have the right to express opinions, they must exercise care in their online communications to avoid possible charges of defamation. Organizations must also be on their guard and prepared to take action in the event of libelous attacks against them.

FREEDOM OF EXPRESSION: KEY ISSUES

Information technology has provided amazing new ways to communicate with people around the world, but with these new methods come new responsibilities and new ethical dilemmas. This section discusses a number of key issues related to freedom of expression, including controlling access to information on the Internet, anonymity on the Internet, defamation and hate speech, corporate blogging, and pornography.

Controlling Access to Information on the Internet

Although there are clear and convincing arguments to support freedom of speech online, the issue is complicated by the ease with which children can access the Internet. Even some advocates of free speech acknowledge the need to restrict children's Internet access, but it is difficult to restrict their access without also restricting adults' access. In attempts to address this issue, the U.S. government has passed laws, and software manufacturers have invented special software to block access to objectionable material. The following sections summarize these approaches.

Communications Decency Act (CDA) (1996)

The Telecommunications Act became law in 1996. Its purpose was to allow freer competition among phone, cable, and TV companies. Embedded in the Telecommunications Act was the **Communications Decency Act (CDA)**, aimed at protecting children from pornography. The CDA imposed $250,000 fines and prison terms of up to two years for the transmission of "indecent" material over the Internet.

In February 1996, the American Civil Liberties Union and 18 other organizations filed a lawsuit challenging the criminalization of so-called indecency on the Web under the CDA. After a three-judge federal panel ruled unanimously that the law unconstitutionally restricted free speech, the government appealed to the Supreme Court (shown in Figure 5-2) in a case that became known as *Reno v. ACLU*. Examples of indecency identified as potentially criminal by government witnesses included Web postings of a photo of actress Demi Moore, naked and pregnant on the cover of *Vanity Fair*, and any online use of the infamous "seven dirty words." In addition to the ACLU, the plaintiffs of the original suit included Planned Parenthood, Stop Prisoner Rape, Human Rights Watch, and the Critical Path AIDS Project. Many of these organizations feared that much of their online material could be classified as indecent, because examples cited by government witnesses included speech about abortion, prisoner rape, safe-sex practices, and other sexually related topics. The plaintiffs argued that much of this information was important to both minors and adults.

The problem with the CDA was its broad language and vague definition of *indecency*, a standard that was left to individual communities to determine. In June 1997, the Supreme Court ruled the law unconstitutional and declared that the Internet must be afforded the highest protection available under the First Amendment.[12] The Supreme Court said in its ruling that "the interest in encouraging freedom of expression in a democratic society outweighs any theoretical but unproven benefit of censorship." The ruling also said that "the growth of the Internet has been and continues to be phenomenal. As a matter of constitutional tradition, and in the absence of evidence to the contrary, we presume government regulation of the content of speech is more likely to interfere with the free exchange of ideas than to encourage it."[13] The ruling essentially applied the same free-speech protections to communication over the Internet as exists for print communication.

FIGURE 5-2 U.S. Supreme Court building

If the CDA had been judged constitutional, it would have opened all aspects of online content to legal scrutiny. Many current Web sites would probably either not exist or look much different today had the law not been overturned. Web sites that might have been deemed indecent under the CDA would be operating under an extreme risk of liability.

Child Online Protection Act (COPA) (1998)

In October 1998, the **Child Online Protection Act (COPA)** was signed into law. (This act is not to be confused with the Children's Online Privacy Protection Act [COPPA], discussed in Chapter 4.) The law states that "whoever knowingly and with knowledge of the character of the material, in interstate or foreign commerce by means of the World Wide Web, makes any communication for commercial purposes that is available to any minor and that includes any material that is harmful to minors shall be fined not more than $50,000, imprisoned not more than 6 months, or both." (Subsequent sections of the act allow for penalties of up to $150,000 for each day of violation.)[14]

The law became a significant battleground for proponents of free speech. Not only could it affect sellers of explicit material online and their potential customers, but it could ultimately set standards for Internet free speech. Supporters of COPA (primarily the Department of Justice) argued that the act protected children from online pornography while preserving the rights of adults. However, privacy advocacy groups—such as the Electronic Privacy Information Center, the ACLU, and the Electronic Frontier Foundation—claimed that the language was overly vague and limited the ability of adults to access material protected under the First Amendment.

Following a temporary injunction as well as numerous hearings and appeals, the Supreme Court ruled in June 2004 that there would be "a potential for extraordinary harm and a serious chill upon protected speech" if the law went into effect.[15] The ruling made it clear that COPA was unconstitutional and could not be used to shelter children from online pornography.

Internet Filtering

An **Internet filter** is software that can be used to block access to certain Web sites that contain material deemed inappropriate or offensive. The best Internet filters use a combination of URL, keyword, and dynamic content filtering. With URL filtering, a particular URL or domain name is identified as belonging to an objectionable site, and the user is not allowed access to it. Keyword filtering uses keywords or phrases—such as *sex*, *Satan*, and *gambling*—to block Web sites. With dynamic content filtering, each Web site's content is evaluated immediately before it is displayed, using such techniques as object analysis and image recognition.

Organizations may direct their network administrators to install filters on employees' computers to prevent them from viewing sites that contain pornography or other objectionable material. Employees who are unwillingly exposed to such material would have a strong case for sexual harassment. The use of filters can also ensure that employees do not waste their time viewing non-business-related Web sites. According to TopTenREVIEWS, the top rated Internet filters for home users for 2009 include Net Nanny 6.0 (shown in Figure 5-3), Safe Eyes, CYBERsitter, WiseChoice.NET, and CyberPatrol.[16]

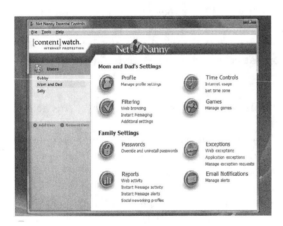

FIGURE 5-3   Screen shot from Net Nanny 6.0

Another popular Internet filter, ADL HateFilter, can be downloaded from the Anti-Defamation League's Web site (*www.adl.org*). If users try to access a blocked site—one that advocates bigotry, hatred, or violence toward groups on the basis of their ethnicity, race, religion, or sexual orientation—HateFilter's redirect feature offers to link them to related ADL educational materials.

A filtering tool designed for use by Web site owners is available through ICRA (formerly the Internet Content Rating Association), which is part of the nonprofit Family Online Safety Institute (FOSI), whose members as of January 2009 included such Internet industry leaders as AOL Europe, BellSouth, British Telecom, IBM, Google, Microsoft, MySpace,

Verizon, and Yahoo!. FOSI's mission is to enable the public to make informed decisions about electronic media through the open and objective labeling of content. Its goals are to protect children from potentially harmful material while safeguarding free speech online.

In the ICRA rating system, Web authors fill out an online questionnaire to describe the content of their site. The questionnaire covers such broad topics as the presence of chat rooms or other user-generated content, the language used, nudity and sexual content, the depiction of violence, alcohol and drugs, gambling, and suicide. Within each broad category, the Web author is asked whether specific items or features are present on the site. Based on the author's responses, ICRA generates a content label (a short piece of computer code) that the author adds to the site. This label conforms to an Internet industry standard known as the Platform for Internet Content Selection (PICS). Internet users can then set their browsers to allow or disallow access to Web sites based on the information declared in the content label and their own preferences.

Note that ICRA does not rate Web content—the content providers do. However, relying on Web site authors to do their own ratings has its weaknesses. For one, many hate sites and sexually explicit sites don't have ICRA ratings. Thus, these sites won't be blocked unless a browser is set to block *all* unrated sites, which would make Web surfing a virtually useless activity in that it would block many acceptable sites, as well. Site labeling also depends on the honesty with which Web site authors rate themselves. If authors lie when completing the ICRA questionnaire, their site could receive a content label that doesn't accurately reflect the site's content. For these reasons, site labeling is at best a complement to other filtering techniques.

Another approach to restricting access to Web sites is to subscribe to an Internet service provider (ISP) that performs the blocking. The blocking occurs through the ISP's server rather than via software loaded onto each user's computer. One ISP, ClearSail/Family.NET, prevents access to known Web sites that address such topics as bomb making, gambling, hacking, hate, illegal drugs, pornography, profanity, Satan, and suicide. ClearSail employees search the Web daily to uncover new sites to add to ClearSail's block list. The ISP blocks specific URLs and known pornographic hosting services, as well as other sites based on certain keywords. ClearSail's filtering blocks millions of Web pages. Newsgroups are also blocked because of the potential for pornography within them.

Children's Internet Protection Act (CIPA) (2000)

In another attempt to protect children from accessing pornography and other explicit material online, Congress passed the **Children's Internet Protection Act (CIPA)** in 2000. The act required federally financed schools and libraries to use some form of technological protection (such as an Internet filter) to block computer access to obscene material, pornography, and anything else considered harmful to minors. Congress did not specifically define which content or Web sites should be forbidden or which measures should be used—these decisions were left to individual school districts and library systems. Any school or library that failed to comply with the law would no longer be eligible to receive federal money through the E-Rate program, which provides funding to help pay for the cost of Internet connections. The following points summarize CIPA:

- Under CIPA, schools and libraries subject to CIPA will not receive the discounts offered by the E-Rate program unless they certify that they have certain Internet safety measures in place. These include measures to block or filter pictures that (1) are obscene, (2) contain child pornography, or (3) are harmful to minors (for computers that are used by minors).

- Schools subject to CIPA are required to adopt a policy to monitor the online activities of minors.
- Schools and libraries subject to CIPA are required to adopt a policy addressing (1) access by minors to inappropriate matter online; (2) the safety and security of minors when using e-mail, chat rooms, and other forms of direct electronic communications; (3) unauthorized access, including hacking and other unlawful activities by minors online; (4) unauthorized disclosure, use, and dissemination of personal information regarding minors; and (5) restricting minors' access to materials harmful to them. CIPA does not require the tracking of Internet use by minors or adults.[17]

Opponents of the law were concerned that it transferred power over education to private software companies who develop the Internet filters and define which sites to block. Furthermore, opponents felt that the motives of these companies were unclear—for example, some filtering companies track students' Web-surfing activities and sell the data to market research firms. Opponents also pointed out that some versions of these filters were ineffective, blocking access to legitimate sites and allowing access to objectionable ones. Yet another objection was that penalties associated with the act could cause schools and libraries to lose federal funds from the E-Rate program, which is intended to help bridge the digital divide between rich and poor, urban and rural. Loss of federal funds would lead to a less capable version of the Internet for students at poorer schools, which have the fewest alternatives to federal aid.

CIPA's proponents contended that shielding children from drugs, hate, pornography, and other topics was a sufficient reason to justify filters. They argued that Internet filters are highly flexible and customizable, and that critics exaggerated the limitations. Proponents pointed out that schools and libraries could elect not to implement a children's Internet protection program; they just wouldn't receive federal money for Internet access.

Many school districts have implemented programs consistent with CIPA. Acceptance of an Internet filtering system is more meaningful if the system and its rationale are first discussed with parents, students, teachers, and administrators. Then the program can be refined, taking into account everyone's feedback. An essential element of a successful program is to require that students, parents, and employees sign an agreement outlining the school district's acceptable-use policies for accessing the Internet. Controlling Internet access via a central district-wide network rather than letting each school set up its own filtering system reduces administrative effort and ensures consistency. Procedures must be defined to block new objectionable sites as well as remove blocks from Web sites that should be accessible.

Implementing CIPA in libraries is much more difficult because a library's services are open to people of all ages, including adults who have First Amendment rights to access a broader range of online materials than are allowed under CIPA. One county library was sued for filtering, while another was sued for not filtering enough. At least one federal court has ruled that a local library board may not require the use of filtering software on all library computers connected to the Internet. A possible compromise for public libraries with multiple computers would be to allow unrestricted Internet use for adults but to provide computers with only limited access for children.

The ACLU filed a suit to challenge CIPA. In May 2002, a three-judge panel in eastern Pennsylvania held that "we are constrained to conclude that the library plaintiffs must prevail in their contention that CIPA requires them to violate the First Amendment rights of their patrons, and accordingly is facially invalid" under the First Amendment. The ruling instructed the government not to enforce the act. This ruling, however, was reversed in June 2003 by the U.S. Supreme Court in *United States v. American Library Association*. The Supreme Court, in a 6-3 decision, held that public libraries must purchase filtering software and comply with all portions of CIPA.[18]

Rather than deal with all the technical and legal complications, some librarians say they wish they could simply focus on training students and adults to use the Internet safely and wisely.

Anonymity on the Internet

Anonymous expression is the expression of opinions by people who do not reveal their identity. The freedom to express an opinion without fear of reprisal is an important right of a democratic society. Anonymity is even more important in countries that don't allow free speech. However, in the wrong hands, anonymous communication can be used as a tool to commit illegal or unethical activities.

Anonymous political expression played an important role in the early formation of the United States. Before and during the American Revolution, patriots who dissented against British rule often used anonymous pamphlets and leaflets to express their opinions. England had a variety of laws designed to restrict anonymous political commentary, and people found guilty of breaking these laws were subject to harsh punishment—from whippings to hangings. A famous case in 1735 involved a printer named John Zenger, who was prosecuted for seditious libel because he wouldn't reveal the names of anonymous authors whose writings he published. The authors were critical of the governor of New York. The British were outraged when the jurors refused to convict Zenger, in what is considered a defining moment in the history of freedom of the press.

Other democracy supporters often authored their writings anonymously or under pseudonyms. For example, Thomas Paine was an influential writer, philosopher, and statesman of the Revolutionary War era. He published a pamphlet called *Common Sense*, in which he criticized the British monarchy and urged the colonies to become independent by establishing a republican government of their own. Published anonymously in 1776, the pamphlet sold more than 500,000 copies when the population of the colonies was estimated to have been less than 4 million; it provided a stimulus to produce the Declaration of Independence six months later.

Despite the importance of anonymity in early America, it took nearly 200 years for the Supreme Court to render rulings that addressed anonymity as an aspect of the Bill of Rights. One of the first rulings was in the 1958 case of *National Association for the Advancement of Colored People (NAACP) v. Alabama*, in which the court ruled that the NAACP did not have to turn over its membership list to the state of Alabama. The court believed that members could be subjected to threats and retaliation if the list were disclosed, and that disclosure would restrict a member's right to freely associate, in violation of the First Amendment.

Another landmark anonymity case involved a sailor threatened with discharge from the U.S. Navy because of information obtained from America Online (AOL). In 1998, following a tip, a Navy investigator asked America Online to identify the sailor, who used a pseudonym to post information in an online personal profile that suggested he might be gay. Thus, he could be discharged under the military's "don't ask, don't tell" policy on homosexuality. America Online admitted that its representative violated company policy by providing the information. A federal judge ruled that the Navy had overstepped its authority in investigating the sailor's sexual orientation and had also violated the Electronic Communications Privacy Act, which limits how government agencies can seek information from e-mail or other online data. The sailor received undisclosed monetary damages from America Online and, in a separate agreement, was allowed to retire from the Navy with full pension and benefits.[19]

Maintaining anonymity on the Internet is important to some computer users. They might be seeking help in an online support group, reporting defects about a manufacturer's goods or services, taking part in frank discussions of sensitive topics, expressing a minority or antigovernment opinion in a hostile political environment, or participating in chat rooms. Other Internet users would like to ban Web anonymity because they think that its use increases the risks of defamation, fraud, libel, and the exploitation of children.

When data is sent over the Internet, a computer's IP address (a numeric identification for a computer attached to the Internet) is logged by the ISP. The IP address can be used to identify the sender of an e-mail or an online posting. Internet users who want to remain anonymous can send e-mail to an **anonymous remailer** service, which uses a computer program to strip the originating IP number from the message. It then forwards the message to its intended recipient—an individual, a chat room, or a newsgroup—with either no IP address or a fictitious one. This ensures that no header information can identify the author. Some remailers route messages through multiple remailers to provide a virtually untraceable level of anonymity.

The use of a remailer keeps communications anonymous; what is communicated, and whether it is ethical or legal, is up to the sender. The use of remailers by people committing unethical or even illegal acts in some states or countries has spurred controversy. Remailers are frequently used to send pornography, to illegally post copyrighted material to Usenet newsgroups, and to send unsolicited advertising to broad audiences (spamming). An organization's IT department can set up a firewall to prohibit employees from accessing remailers or to send a warning message each time an employee communicates with a remailer.

John Doe Lawsuits

Businesses must protect against both the public expression of opinions that might hurt their reputations and the public sharing of company confidential information. When anonymous employees reveal harmful information online, the potential for broad dissemination is enormous, and it can require great effort to identify the people involved and stop them.

An aggrieved party can file a **John Doe lawsuit** against a defendant whose identity is temporarily unknown because he or she is communicating anonymously or using a pseudonym. Once the John Doe lawsuit is filed, the plaintiff can request court permission to issue subpoenas to command a person to appear under penalty. If the court grants permission, the plaintiff can serve subpoenas on any third party—such as an Internet service provider or a Web site hosting firm—that may have information about the true identity of the defendant. When, and if, the identity becomes known, the complaint is modified to show the correct name(s) of the defendant(s).

A company may file a John Doe lawsuit because it is upset by anonymous e-mail messages that criticize the company or reveal company secrets. For example, Raytheon filed a lawsuit in 1999 for $25,000 in damages against 21 John Does for allegedly revealing on a Yahoo! message board company financial results along with other information that the company claimed hurt its reputation. Raytheon received a court order to subpoena Yahoo! and several ISPs for the identity of the 21 unnamed defendants. Eventually, Raytheon traced the identities of all 21 people who posted the alleged company secrets. Four employees voluntarily left the company, and others received counseling about sharing confidential company information.[20]

America Online, EarthLink, NetZero, Verizon Online, and other ISPs receive more than a thousand subpoenas per year directing them to reveal the identity of John Does. Free-speech advocates argue that if someone charges libel, the anonymity of the Web poster should be preserved until the libel is proved. Otherwise, the subpoena power can be used to silence anonymous, critical speech.

Proponents of such lawsuits point out that most John Doe cases are based on serious allegations of wrongdoing, such as libel or disclosure of confidential information. For example, stock price manipulators can use chat rooms to affect the share price of stocks—especially those of very small companies that have just a few outstanding shares. In addition, competitors of an organization might try to create the feeling that the organization is a miserable place to work, which could discourage job candidates from applying, investors from buying stock, or consumers from buying company products. Proponents of John Doe lawsuits argue that perpetrators should not be able to hide behind anonymity to avoid responsibility for their actions.

Anonymity on the Internet is not guaranteed. By filing a lawsuit, companies gain immediate subpoena power, and many message board hosts release information as soon as it is requested, often without notifying the poster. Everyone who posts comments in a public place on the Web should consider the consequences if their identities were to be exposed. Furthermore, everyone who reads anonymous postings online should think twice about believing what they read.

The California State Court in *Pre-Paid Legal v. Sturtz et al.* set another legal precedent that refined the criteria that the courts apply when deciding whether or not to approve subpoenas requesting the identity of anonymous Web posters. The case involved a subpoena issued by Pre-Paid Legal Services (PPLS), which requested the identity of eight anonymous posters on Yahoo!'s Pre-Paid message board. Attorneys for PPLS argued that it needed the posters' identities to determine whether they were subject to a voluntary injunction that prevented former sales associates from revealing PPLS's trade secrets.

The Electronic Frontier Foundation (EFF) represented two of the John Does whose identities were subpoenaed. EFF attorneys argued that the message board postings cited by PPLS revealed no company secrets but were merely disparaging the company and its treatment of sales associates. They argued further that requiring the John Does to reveal their identities would let the company punish them for speaking out and set a dangerous precedent that would discourage other Internet users from voicing criticism. Without proper safeguards on John Doe subpoenas, a company could use the courts to uncover its critics.

EFF attorneys urged the court to apply the four-part test adopted by the federal courts in *Doe v. 2TheMart.com, Inc.* to determine whether a subpoena for the identity of the Web posters should be upheld. In that case, the federal court ruled that a subpoena should be enforced only when the following occurs:

- The subpoena was issued in good faith and not for any improper purpose.
- The information sought related to a core claim or defense.
- The identifying information was directly and materially relevant to that claim or defense.
- Adequate information was unavailable from any other source.

In August 2001, a judge in Santa Clara County Superior Court invalidated the subpoena to Yahoo! requesting the posters' identities. He ruled that the messages were not obvious violations of the injunctions invoked by PPLS, and that the First Amendment protection of anonymous speech outweighed PPLS's interest in learning the identity of the speakers.

Defamation and Hate Speech

In the United States, speech that is merely annoying, critical, demeaning, or offensive enjoys protection under the First Amendment. Legal recourse is possible only when hate speech turns into clear threats and intimidation against specific citizens. Persistent or malicious harassment aimed at a specific person can be prosecuted under the law, but general, broad statements expressing hatred of an ethnic, racial, or religious group cannot. A threatening private message sent over the Internet to a person, a public message displayed on a Web site describing intent to commit acts of hate-motivated violence at specific individuals, and libel directed at a particular person are all actions that can be prosecuted.

Although ISPs do not have the resources to prescreen content (and they do not assume any responsibility for content that is provided by others), many ISPs do reserve the right to remove content that, in their judgment, does not meet their standards. The speed at which content may be removed depends on how quickly such content is called to the attention of the ISP, how egregious the content is, and the general availability of ISP resources to handle such issues.

For example, AOL has documented a set of standards it calls the AOL Community Guidelines. AOL states clearly that it is not responsible for any failure or delay in removing material that violates these standards. To become an AOL subscriber, you must agree to follow these guidelines and acknowledge that AOL has the right to enforce them in its sole discretion. Thus, if you or anyone using your account violates the AOL Community Guidelines, AOL may take action against your account, ranging from issuing a warning to terminating your account.

The AOL Community Guidelines state the following in regard to the use of hate speech:

> "Do not use hate speech. Hate speech is unacceptable, and AOL reserves the right to take appropriate action against any account using the service to post, transmit, promote, distribute or facilitate distribution of content intended to victimize, harass, degrade or intimidate an individual or group on the basis of age, disability, ethnicity, gender, race, religion or sexual orientation."[21]

Because such prohibitions are included in the service contracts between a private ISP and its subscribers and do not involve the federal government, they do not violate the subscribers' First Amendment rights. Of course, ISP subscribers who lose an account for violating the ISP's regulations may resume their hate speech by simply opening a new account with some other, more permissive ISP.

Although they may implement a speech code, public schools and universities are legally considered agents of the government and therefore must follow the First Amendment's prohibition against speech restrictions based on content or viewpoint. Corporations, private schools, and private universities, on the other hand, are not part of state or federal government. As a result, they may prohibit students, instructors, and other employees from engaging in offensive speech using corporate-, school-, or university-owned computers, networks, or e-mail services.

Despite the protection of the First Amendment and the challenges posed by anonymous expression, there are instances of U.S. citizens being successfully sued or convicted of crimes relating to hate speech:

- A former student was sentenced to one year in prison for sending e-mail death threats to Asian American students at the University of California, Irvine. His e-mail was signed "Asian hater," and his letters stated that he would make it his career to find and kill every Asian himself.[22]

- A coalition of antiabortion groups was ordered to pay more than $100 million in damages after posting on a Web site information about doctors and clinic workers who perform abortions, including photos, home addresses, license plate numbers, and even the names of their spouses and children. Three of the doctors listed on the site were murdered, and others on the list were wounded. A jury found that the Web site provided information that resulted in a real threat of bodily harm and awarded damages. However, the U.S. Court of Appeals for the Ninth Circuit reversed the decision. The court ruled that the coalition made no statements mentioning violence and that publication of the personal information did not constitute a serious expression of intent to harm.[23]

- In 2002, Varian Medical Systems won a $775,000 verdict in a defamation and harassment lawsuit against two former employees who posted thousands of messages to a wide variety of online message boards accusing Varian managers of being homophobic and discriminating against pregnant women.[24]

Most other countries do not provide constitutional protection for hate speech. For example, promoting Nazi ideology is a crime in Germany, and denying the occurrence of the Holocaust is illegal in many European countries. Authorities in Britain, Canada, Denmark, France, and Germany have charged people for crimes involving hate speech on the Web.

Thousands of people faced the potential of criminal charges after posting hate messages and threats to the Facebook account of Brendan Sokaluk, who is accused of setting bushfires that killed 21 people in Victoria, Australia, in February 2009. "It's the cyber-world equivalent of angry mobs forming outside court, hurling abuse," said Michael Pearce, president of Liberty Victoria (one of Australia's leading civil liberties organizations). "There is a clear risk that these people are going to imperil a fair trial for the accused and also that they are in contempt of court."[25]

A U.S. citizen who posts material on the Web that is illegal in a foreign country can be prosecuted if he subjects himself to the jurisdiction of that country—for example, by visiting there. As long as the person remains in the United States, he is safe from prosecution, because U.S. laws do not allow a person to be extradited for engaging in an activity protected by the U.S. Constitution, even if the activity violates the criminal laws of another country.

Corporate Blogging

A growing number of organizations allow employees to create their own personal blogs relating to their employment. They see blogging as a new way to reach out to partners, customers, and other employees and to improve their corporate image. Under the best conditions, individual employees use their blogs to ask other employees for help with work-related problems, to share work-related information in a manner that invites conversation, or to invite others to refine or build on a new idea. However, most organizations are well aware that such blogs can also provide an outlet for uncensored commentary and interaction. Employees can use their blogs to criticize corporate policies and decisions. Employee blogging also involves the risk that employees might reveal company secrets or breach federal security disclosure laws.

Mark Jen, an associate product manager for Google, began a blog chronicling his experiences at Google after his first day on the job. He thought his entries might be of interest to his family and friends. Little did Mark realize that his blog would attract the attention of thousands of people curious about life inside Google. His entries candidly discussed his first day on the job, a global sales meeting, and Google's compensation package. His comments also included information about Google's future products and economic performance. Within a couple of days, Mark's audience had grown into the tens of thousands. The following week, Mark's blog was offline for a couple days. In his next posting, Mark revealed that he had been asked to take down sensitive information about the company. Then the entries stopped altogether, and rumors were rampant as the number of visitors to his blog approached 100,000 per day. A few weeks later, Mark finally checked back in to let his readers know that he had been fired. Within the blogosphere, Mark had become a cause célèbre, and Google's reputation suffered. The incident sent a shock wave through the IT industry, forcing companies to evaluate and establish their own blogging policies.[26]

Within a few months of Mark's dismissal, companies such as Sun, IBM, and Yahoo! began to formulate and publish employee blogging policies. Many guidelines suggested that employees use their common sense: Don't reveal company secrets. Don't insult people that you interact with for eight hours a day. Check with the Public Relations Department if you are not sure whether you should include specific information. Other guidelines provided suggestions that might improve the quality of a blog and make it more appealing to potential readers: Be interesting. Don't use pen names or ghost writers. Be authentic.

Pornography

Many people, including some free-speech advocates, believe that there is nothing illegal or wrong about purchasing adult pornographic material made by and for consenting adults. They argue that the First Amendment protects such material. On the other hand, most parents, educators, and other child advocates are concerned that children might be exposed to pornography. They are deeply troubled by its potential impact on children and fear that increasingly easy access to pornography encourages pedophiles and sexual predators.

Clearly, the Internet has been a boon to the pornography industry by providing fast, cheap, and convenient access to well over 4.2 million porn Web sites worldwide. Access via the Internet enables pornography consumers to avoid offending others or being embarrassed by others observing their purchases. There is no question that adult pornography on the Web is big business (with an estimated $4.9 billion in revenue in 2006) and generates a lot of traffic; it is estimated that there are 72 million visitors to pornographic Web sites monthly.[27]

Pornography purveyors are free to produce and publish whatever they want; however, if what they distribute or exhibit is judged obscene, they are subject to prosecution under the obscenity laws. U.S. organizations must be very careful when dealing with issues relating to pornography in the workplace. By providing computers, Internet access, and training in how to use those computers and the Internet, companies could be seen by the law as purveyors because they have enabled employees to store pornographic material and retrieve it on demand.

Many companies believe that they have a duty to stop the viewing of pornography in the workplace. As long as they can show that they took reasonable steps and determined actions to prevent it, they have a valid defense if they become the subject of a sexual harassment lawsuit. If it can be shown that a company made only a halfhearted attempt at stopping pornography, then its defense in court would be weak. Reasonable steps include establishing a computer usage policy that prohibits access to pornography sites, identifying those who violate the policy, and taking action against those users—no matter how embarrassing it is for the users or how harmful it might be for the company.

The Texas Workforce Commission has posted a sample Internet, e-mail, and computer usage policy at *www.texasworkforce.org/news/efte/internetpolicy.html*. Table 5-1 shows portions of the sample policy that pertain to the topics being discussed in this chapter.

TABLE 5-1 Portions of a sample Internet, e-mail, and computer usage policy

"Use of Company computers, networks, and Internet access is a privilege granted by management and may be revoked at any time for inappropriate conduct carried out on such systems, including, but not limited to:

- Violating the laws and regulations of the United States or any other nation or any state, city, province, or other local jurisdiction in any way;
- Engaging in unlawful or malicious activities;
- Using abusive, profane, threatening, racist, sexist, or otherwise objectionable language in either public or private messages;
- Sending, receiving, or accessing pornographic materials;
- Becoming involved in partisan politics;
- Maintaining, organizing, or participating in non-work-related Web logs ('blogs'), Web journals, 'chat rooms,' or private/personal/instant messaging."

Source: Texas Workforce Commission, www.texasworkforce.org/twcinfo/whatis.html.

A few companies take the opposite viewpoint—that they cannot be held liable if they don't know employees are viewing, downloading, and distributing pornography. Thus, they believe the best approach is to ignore the problem by never investigating it, thereby ensuring that they can claim that they never knew it was happening. Many people would consider such an approach unethical and would view management as shirking an important responsibility to provide a work environment free of sexual harassment. Employees unwillingly exposed to pornography would have a strong case for sexual harassment because they could claim that pornographic material was available in the workplace and that the company took inadequate measures to control the situation.

There are numerous federal laws addressing child pornography. Possession of child pornography is a federal offense punishable by up to five years in prison. The production and distribution of such materials carry harsher penalties; decades or even life in prison is not an unusual sentence. In addition to these federal statutes, all states have enacted laws against the production and distribution of child pornography, and all but a few states have outlawed the possession of child pornography. At least seven states have passed laws that require computer technicians who discover child pornography on clients' computers to report it to law enforcement officials.[28]

Controlling the Assault of Non-Solicited Pornography and Marketing (CAN-SPAM) Act (2003)

The **Controlling the Assault of Non-Solicited Pornography and Marketing (CAN-SPAM) Act** specifies requirements that commercial e-mailers must follow when sending out messages that advertise or promote a commercial product or service. The key requirements of the law include:

- The "From" and "To" fields in the e-mail, as well as the originating domain name and e-mail address, must be accurate and identify the person who initiated the e-mail.
- The subject line of the e-mail cannot mislead the recipient as to the contents or subject matter of the message.
- The e-mail must be identified as an advertisement and include a valid physical postal address for the sender.
- The e-mailer must provide a return e-mail address or some other Internet-based response procedure to enable the recipient to request no future e-mails, and the e-mailer must honor such requests to opt out.
- The e-mailer has 10 days to honor the opt-out request.
- Additional rules prohibit the harvesting of e-mail addresses from Web sites, using automated methods to register for multiple e-mail accounts, or relaying e-mail through another computer without the owner's permission.

Each violation of the provisions of the CAN-SPAM Act can result in a fine of up to $11,000. The Federal Trade Commission (FTC) is charged with enforcing the act. In addition, the FTC maintains a consumer complaint database relating to the law. Consumers can submit complaints online at *www.ftc.gov* or forward e-mail to the FTC at *spam@use.gov*.

There is considerable debate over whether the CAN-SPAM Act has helped control the growth of spam. After all, the act clearly defines the conditions under which the sending of spam is legal, and as long as mass e-mailers meet these requirements, they cannot be prosecuted. In addition, the FTC has done little to enforce the act. It is estimated that more than 100 billion spam messages are sent each year, yet the FTC has brought only 30 law-enforcement actions in the first five years of the act.[29] Some suggest that the act could be improved by penalizing the companies that use spam to advertise, as well as Internet service providers who support the spammers.

The CAN-SPAM Act can also be used in the fight against the dissemination of pornography. For example, three people were indicted by an Arizona grand jury in August 2005 for violating the CAN-SPAM Act by sending massive amounts of unsolicited e-mail advertising pornographic Web sites. The e-mails allegedly contained pornographic images, so the grand jury also leveled felony obscenity charges for transmission of hard-core pornography. The defendants face multiple counts of spamming and criminal conspiracy, which carry a maximum sentence of five years each. The number of people who received spam from the operation is estimated to be in the tens of millions.

Table 5-2 is a manager's checklist for dealing with issues of freedom of expression in the workplace. In each case, the preferred answer is *yes*.

TABLE 5-2 Manager's checklist for handling freedom of expression issues in the workplace

Question	Yes	No
Do you have a written data privacy policy that is followed?		
Does your corporate IT usage policy discuss the need to conserve corporate network capacity, avoid legal liability, and improve worker productivity by limiting the non-business use of information resources?		
Did the developers of your policy consider the need to limit employee access to non-business-related Web sites (for example, Internet filters, firewall configurations, or the use of an ISP that blocks access to such sites)?		
Does your corporate IT usage policy discuss the inappropriate use of anonymous remailers?		
Has your corporate firewall been set to detect the use of anonymous remailers?		
Has your company (in cooperation with legal counsel) formed a policy on the use of John Doe lawsuits to identify the authors of libelous, anonymous e-mail?		
Does your corporate IT usage policy make it clear that defamation and hate speech have no place in the business setting?		
Does your corporate IT usage policy prohibit the viewing or sending of pornography?		
Does your policy communicate if employee e-mail is regularly monitored for defamatory, hateful, and pornographic material?		
Does your corporate IT usage policy tell employees what to do if they receive hate mail or pornography?		

Summary

- The First Amendment protects Americans' rights to freedom of religion and freedom of expression. The Supreme Court has ruled that the First Amendment also protects the right to speak anonymously.

- Obscene speech, defamation, incitement of panic, incitement to crime, "fighting words," and sedition are not protected by the First Amendment and may be forbidden by the government.

- The Internet enables a worldwide exchange of news, ideas, opinions, rumors, and information. Its broad accessibility, open discussions, and anonymity make it a powerful communications medium. People must often make ethical decisions about how to use such remarkable freedom and power.

- Organizations and governments have attempted to establish policies and laws to help guide Internet use as well as protect their own interests. Businesses, in particular, have sought to conserve corporate network capacity, avoid legal liability, and improve worker productivity by limiting the non-business use of IT resources.

- Although there are clear and convincing arguments to support freedom of speech on the Internet, the issue is complicated by the ease with which children can use the Internet to gain access to material that many parents and others feel is inappropriate. The conundrum is that it is difficult to restrict children's Internet access without also restricting adults' access.

- The U.S. government has passed several laws to attempt to address this issue, including the Communications Decency Act (aimed at protecting children from online pornography) and the Child Online Protection Act (prohibited making harmful material available to minors via the Internet). Both were ruled unconstitutional.

- Software manufacturers have developed Web filters, which are designed to block access to objectionable material through a combination of URL, keyword, and dynamic content filtering.

- The Children's Internet Protection Act requires federally financed schools and libraries to use filters to block computer access to any material considered harmful to minors.

- Maintaining anonymity on the Internet is important to some computer users. Such users sometimes use an anonymous remailer service, which strips the originating IP number from the message and then forwards the message to its intended recipient.

- Many businesses monitor the Web for the public expression of opinions that might hurt their reputations. They also try to guard against the public sharing of company confidential information.

- Organizations may file a John Doe lawsuit to enable them to gain subpoena power in an effort to learn the identity of anonymous Internet users who have caused some form of harm through their postings.

- In the United States, speech that is merely annoying, critical, demeaning, or offensive enjoys protection under the First Amendment. Legal recourse is possible only when hate speech turns into clear threats and intimidation against specific citizens.

- Some ISPs have voluntarily agreed to prohibit their subscribers from sending hate messages using their services. Because such prohibitions can be included in the service contracts between a private ISP and its subscribers and do not involve the federal government, they do not violate subscribers' First Amendment rights.

- Numerous organizations allow employees to create their own personal blogs relating to their employment as a means to reach out to partners, customers, and other employees and to improve their corporate image.

- Organizations are advised to formulate and publish employee blogging policies to avoid potential negative consequences from employee criticism of corporate policies and decisions, or the revelation of company secrets or confidential financial data.

- Many adults and free-speech advocates believe that there is nothing illegal or wrong about purchasing adult pornographic material made by and for consenting adults. However, organizations must be very careful when dealing with pornography in the workplace. As long as companies can show that they were taking reasonable steps to prevent pornography, they have a valid defense if they are subject to a sexual harassment lawsuit.

- Reasonable steps include establishing a computer usage policy that prohibits access to pornography sites, identifying those who violate the policy, and taking action against those users—no matter how embarrassing it is for the users or how harmful it might be for the company.

- The CAN-SPAM Act specifies requirements that commercial e-mailers must follow in sending out messages that advertise a commercial product or service. The CAN-SPAM Act can also be used in the fight against the dissemination of pornography.

Self-Assessment Questions

The answers to the Self-Assessment Questions can be found in Appendix G.

1. The most basic legal guarantee to the right of freedom of expression in the United States is contained in the:
   a. Bill of Rights
   b. Fourth Amendment
   c. First Amendment
   d. U.S. Constitution

2. True or False? The right to freedom of expression has been broadened by the Supreme Court to include nonverbal, visual, and symbolic forms of expression.

3. An important Supreme Court case that established a test to determine if material is obscene and therefore not protected speech was _____.

4. A written statement that is false and that harms another person is called:

    a. a lie

    b. slander

    c. libel

    d. freedom of expression

5. The Communications Decency Act, which was passed in 1996 and was aimed at protecting children from online pornography, was eventually ruled unconstitutional in the _____ v. _____ lawsuit.

6. True or False? The Child Online Protection Act prohibited dissemination of harmful material to minors and was ruled unconstitutional.

7. The U.S. Supreme Court in *United States v. American Library Association* ruled that public libraries must install Internet filtering software to comply with all portions of:

    a. the Children's Internet Protection Act

    b. the Child Online Protection Act

    c. the Communications Decency Act

    d. the Internet Filtering Act

8. The best Internet filters rely on the use of:

    a. URL filtering

    b. keyword filtering

    c. dynamic content filtering

    d. all of the above

9. True or False? Anonymous expression, or the expression of opinions by people who do not reveal their identities, has been found to be unconstitutional.

10. A lawsuit in which the true identity of the defendant is temporarily unknown is called a _____.

11. The California State Court in *Pre-Paid Legal v. Sturtz et al.* set a precedent that courts apply when deciding whether to approve subpoenas requesting the identity of _____.

12. True or False? In the United States, speech that is merely annoying, critical, demeaning, or offensive enjoys protection under the First Amendment. Legal recourse is possible only when hate speech turns into clear threats and intimidation against specific citizens.

13. Which of the following statements about Internet pornography is true?

    a. The First Amendment is often used to protect distributors of adult pornography over the Internet.

    b. There are fewer than 60,000 Web sex sites.

    c. About one in six regular Internet users visits a Web sex site at least once per month.

    d. In contrast to adult pornography, few federal laws address child pornography.

14. The _____ Act specifies requirements that commercial e-mailers must follow in sending out messages that advertise or promote a commercial product or service.

Discussion Questions

1. Outline a scenario in which you might be acting ethically but might still want to remain anonymous while using the Internet. How might someone learn your identity even if you attempt to remain anonymous?

2. Why has the federal government had such difficulty enacting legislation to protect children from viewing "inappropriate material" online?

3. How did the Children's Internet Protection Act escape from being ruled unconstitutional? Talk to your local librarian and find out if the library has implemented filtering. If so, is it experiencing any problems enforcing the use of filters? Write a short paragraph summarizing your findings.

4. What can an ISP do to limit the distribution of hate e-mail? Why would such actions not be considered a violation of the subscriber's First Amendment rights?

5. Do research on one of the popular Internet filtering software tools. Write a brief discussion of how it operates, as well as its strengths and weaknesses.

6. Go to the FOSI Web site at *www.fosi.org*. After exploring this site, briefly describe how its ICRA rating process works. What are the advantages and disadvantages of this system?

7. What is a John Doe lawsuit? Do you think that a corporation should be allowed to use a subpoena to identify a John Doe before proving that the person has defamed the company? Why or why not?

8. Do research on the Web to locate an anonymous remailer. Find out what is required to sign up for this service and what fees are involved.

9. Look carefully at the e-mail you receive over the next few days. Are any of the e-mails advertisements for a commercial product or service that violate the CAN-SPAM Act? If so, what might you do to stop receiving such e-mail in the future?

10. Do you think further efforts to limit the dissemination of pornography on the Internet are appropriate? Why or why not?

11. Do research to find out if your school or employer has a set of policies that cover the use of corporate blogging. Do you consider these policies to be overly restrictive?

12. Do research to find out if your school or employer has a set of policies that cover the sending of hate e-mail. Do you consider these policies to be effective?

What Would You Do?

1. A coworker confides to you that he is going to begin sending e-mails to your employer's corporate blog site, which serves as a suggestion box. He plans to use an anonymous remailer and sign the messages "Anonymous." Your friend is afraid of retribution from superiors but wishes to call attention to instances of racial and sexual discrimination observed during his five years as an employee with the firm. What would you say to your friend?

2. A friend contacts you about joining his company, Anonymous Remailers Anonymous. He would like you to lead the technical staff and offers you a 20 percent increase in salary and

benefits over your current position. Your initial project would be to increase protection for users of the company's anonymous remailer service. After discussing the opportunity with your friend, you suspect that some of the firm's customers are criminal types and purveyors of pornography and hate mail. Although your friend cannot be sure, he admits it is possible that hackers and terrorists may use his firm's services. Would you accept the generous job offer? Why or why not?

3. Your 15-year-old nephew exclaims "Oh wow!" and proceeds to tell you about a sext message he just received from his 14-year-old girlfriend of three months. What would you say to your nephew?

4. You are a member of human resources and are working with a committee to complete your company's computer usage policy. What advice would you offer the committee regarding how to address online pornography? Would you suggest that the policy be laissez-faire, or would you recommend that it require strict enforcement of tough corporate guidelines? Why?

5. You are a member of your company's computer support group and have just helped a user upgrade his computer. As you run tests after making the upgrade, you are surprised to find that the user has disabled the Internet filter software that is supposed to be standard on all corporate computers. What would you do?

6. Imagine that you receive a hate e-mail at your school or job. What would you do? Does your school or workplace have a policy that covers such issues?

7. You are the chairperson of the board of directors of your county's public library system. The library is making plans to install Internet filtering software so that it will conform to the Children's Internet Protection Act and be eligible for federal funding. Outline a plan to implement this new program that appropriately involves patrons and employees of the library and ensures that the use of the filters will be accepted.

8. You are representing the IT organization in a meeting with representatives from Legal, Human Resources, and Finance to discuss the advisability of allowing employees to start their own corporate blogs. It is your turn to speak. What would you say?

Cases

1. The Online Reputation War

For years, Rahodeb had been attacking organic supermarket chain Wild Oats on Yahoo!'s Finance message board. "OATS has lost their way and no longer has a sense of mission," Rahodeb charged. "They are floundering around hoping to find a viable strategy that may stop their erosion." Rahodeb also attacked Wild Oats CEO Perry Odak: "Perhaps the OATS Board will wake up and dump Odak and bring in a visionary and highly competent C.E.O." When a few people on the message board suggested that Rahodeb might secretly be John Mackey, CEO of Whole Foods, Wild Oats's larger rival, Rahodeb denied the charge emphatically.[30]

Rahodeb also made more serious allegations against Wild Oats. In 2005, he predicted that the company would be sold when its stock plummeted or when the company went bankrupt.[31] Finally, in 2007, after Whole Foods announced that it was purchasing Wild Oats for $565 million, the news broke that Rahodeb was in fact John Mackey. The Federal Trade Commission (FTC) was intent on stopping the purchase under antitrust laws, and it felt that Mackey might have been trying to drive Wild Oats's stock prices down so that Whole Foods could purchase it more cheaply. Mackey insisted that he had posted those messages only because he had fun doing so.[32]

The FTC filed suit to obtain an injunction against Whole Foods' acquisition of Wild Oats. The acquisition initially went through after a U.S. district court failed to grant the FTC's request for an injunction to prevent the merger. However, in 2008, a U.S. court of appeals reversed the district court decision, and in 2009, Whole Foods agreed to sell the Wild Oats stores, the brand, and other assets it had acquired in the purchase.[33]

Posts to online message boards, blogs, user-generated news sites, and consumer sites, such as those made by Mackey, can destroy a company's reputation. Some posts are valid but harmful. In 2004, a person on an online bicycle forum explained how to open an expensive Kryptonite bicycle lock in seconds. The company had to replace 400,000 locks and redesign its product. Other posts have no basis in fact. In 2008, a citizen journalist posted a story on CNN's iReport announcing that Apple CEO Steve Jobs had suffered a heart attack. Its stock plummeted before Apple had time to deny the story.[34]

These examples are just a small part of an online war being waged daily over corporate reputations. With the social media boom, including an estimated 150,000 million blogs, the scope of this war is steadily increasing. In addition to blogs created by individuals, those provided by consumer Web sites also play a formidable role. Complaints.com, for example, founded its Web site in 1998 with the object of giving greater voice to consumer concerns. Users can file complaints about poor customer service, bad products, or scams. Today, the site lists company profiles of over 20 million businesses and has a feature whereby businesses can be contacted via e-mail or voice mail once complaints have been made against them.[35] The site, however, does not moderate submissions or check the legitimacy of the complaints. Similar Web sites have emerged: Ripoff Report, Yelp, CarComplaints.com, and many others. Although these sites offer consumer advocacy and services, they also injure corporate reputation whether the complaint is legitimate or not.

In response to fears of attack, many businesses hire online reputation management (ORM) companies. ORM companies monitor online social media to identify threats to the reputations of their clients from dissatisfied customers and disgruntled employees. Then they use a number of methods to try to reduce this threat.

Frequently, ORM businesses encourage their clients to initiate online efforts to show their company's products and services in a positive light. These efforts might include enhancing the company's online news page, enabling Web site visitors to sign up for regular company news updates, or establishing RSS feeds. ORM managers may ask company officials to post comments to newsgroups, online forums, or even blogs that are critical of the company.

To counter wild rumors, such as the one about Steve Jobs' supposed heart attack, ORM companies monitor the Web closely, develop relationships with editors at online magazine and industry portals, and form crisis-management teams that can react quickly. They sometimes contact the publisher of the content and request that the material be removed. If the publisher refuses, companies sometimes threaten legal action. The content might violate the Digital Millennium Copyright Act, a law that forbids the distribution of copyrighted materials. The content might also be libelous. A business whose reputation has been marred can sue on these grounds. In some cases, the threat of legal action is enough to get the publisher to capitulate. In others, it is not.

One ORM business went far beyond the use of legal tactics to try to get negative content removed from a site. A manager associated with DefamationAction.com and ComplaintRemover.com sent Ed Magedson—the founder of Ripoff Report—letters requesting that he delete information from his site. Eventually, the letters became threatening: "No matter where you go, we will cause you a problem. Your life is in danger until you comply with our demands. This is your last warning." The ORM manager is purported to have harassed Magedson's business partners and the law firm representing him.[36] The ORM manager involved in this case is considered to be a hard-core spammer; he is one of a hundred or so individuals that the Spamhaus Project alleges is responsible for 80 percent of the spam on the Internet.[37] In 2007, an Arizona federal district judge found him guilty of libel and making death threats.

This case, however, is an exception. Most ORM businesses suggest that companies deal professionally with consumer concerns. They also propose online PR campaigns that promote a positive image of their clients. So, the main question is not whether ORM companies are using some unethical strategies but whether it is ethical to use any strategy—search engine optimization or any other means—to suppress online criticism. When you search newsgroups and blogs for information about computers, cell phones, restaurants, or vacation rentals, are you getting an honest account or are you getting information that has been planted by someone with an agenda? With the wide berth given to consumers and businesses by the First Amendment, both must think carefully not just about what is legal but about what is ethical.

Discussion Questions

1. Do you think it is ethical to pursue a strategy to counter online criticism and defend your organization's reputation? Why or why not?

2. What steps do you think businesses should take when promoting their products and services online? What tactics, if any, should businesses avoid?

3. How should IT professionals and others respond to people with hidden agendas who post negative comments about them on social media sites? What should online visitors do to detect misinformation posted in blogs?

2. Defamation Incident Impacts Lives of All Involved

AutoAdmit was launched in 2004 by owner and creator Jarret Cohen as an online forum for current and prospective law students. The site receives well over 10,000 posts a day and has an estimated 700,000 unique readers per month. Even though AutoAdmit touts itself as the world's most prestigious college discussion board, it has often drawn criticism for its offensive and defamatory content.

In March 2007, TV news broadcasts and newspapers revealed that two female Yale Law School students alleged that defamatory postings about them had appeared in AutoAdmit. The postings included lewd and derogatory comments, made false claims about the students' academic records, alleged that the students had performed sexual favors to gain admittance to Yale and to improve their grades, and made threats of violent attacks. One of the women alleged that the postings had cost her job offers. The reporting of the alleged harassment triggered a new and even more vicious wave of harassment against the two students. The postings went on for over two years.

In June 2007, the students filed a lawsuit against Anthony Ciolli—a third-year law student at the University of Pennsylvania and AutoAdmit's chief education director at the time of the incident—and 28 anonymous posters for violation of privacy, defamation, and infliction of undue emotional distress. The lawyers for the plaintiffs were able to obtain John Doe subpoenas and learn the identities of some of the anonymous posters through various Internet service providers, even though AutoAdmit lacked any sort of IP number logging mechanism. Four months after filing, the plaintiffs dropped Ciolli from the lawsuit because of laws that protect Web site administrators from being sued over content posted on their sites. Ciolli also claimed that he did not have direct control over the content and that he tried to help the women.[38]

The uncovered anonymous posters show a wide range of reaction. One stated: "I said something really stupid on the [word deleted] Internet. I typed for literally, like, 12 seconds, and it devastated my life." After this poster's identity was made public, he gave up plans to attend law school and instead enlisted in the military. Another anonymous poster was concerned that he would lose his job if his identity was revealed and threatened to seek help online to corroborate all of the awful things that were said about the two women in order to defend himself. A third defendant had no remorse, considered the lawsuit to be frivolous, and was unconcerned because his parents' homeowner's policy would cover the cost of his defense.[39]

As this incident unfolded, Anthony Ciolli posted an online letter in which he and Jarret Cohen identified themselves as AutoAdmit's administrators. In the posting, they defended AutoAdmit's "free, uninhibited exchange of ideas." Subsequently, in an apparent change of heart, Ciolli resigned from AutoAdmit. Shortly thereafter, the managing partner at a prestigious Boston law firm stated that the firm had recently learned about the AutoAdmit controversy and Ciolli's involvement. The firm decided to revoke an offer of employment to Ciolli. In a letter to Ciolli, the managing partner stated that the contents of AutoAdmit are "antiethical" and violate the "principles of congeniality and respect that members of the legal profession should observe in their dealings with other lawyers. We expect any lawyer affiliated with our firm, when presented with the kind of language exhibited on the message board, to reject it and disavow any affiliation with it. You, instead, facilitated the expression and publication of such language." He went on to say that Ciolli's subsequent resignation from the site was "too late to ameliorate our concerns."[40]

In his response to the managing partner, Ciolli said that he was "still in the process of assessing all the lessons to be learned from the incident [including] the importance of good judgment and proceeding with caution." He suggested that the law firm defer his start date by a year to "allow me time to develop a series of positive contributions to the legal community that would go a long way toward strengthening my reputation and allaying your concerns."[41]

The managing partner responded that "none of the information you provided me resolves the concerns I expressed in my letter regarding your past affiliation with the site," and that the firm was revoking the employment offered to him.

In March 2008, Ciolli launched his own lawsuit against the two Yale Law School students. He claims they wrongly included him as a defendant in their defamation suit brought in June 2007. Even though they dismissed him from the case four months later, he alleges his connection with the lawsuit caused him to lose a full-time position with a prestigious Boston law firm.[42]

Discussion Questions

1. Did the two law school students overreact in filing a lawsuit over these postings? What sort of penalties should be levied against the anonymous posters?

2. Did Anthony Ciolli act ethically in his response to the postings on AutoAdmit; in his apparent "damage control" actions with the Boston law firm; in filing a lawsuit against the two Yale Law School students?

3. Do you think that the Boston law firm was right in revoking its job offer to Ciolli? Why or why not?

3. The Electronic Frontier Foundation (EFF)

The Electronic Frontier Foundation (EFF) was formed in 1990 by John Perry Barlow—a lyricist for the Grateful Dead—and Mitch Kapor—a founder of the Lotus Development Corporation, which developed the Lotus 1-2-3 spreadsheet software. EFF is a nonprofit, international advocacy and legal organization based in the United States. Its goal is to protect fundamental civil liberties relating to the use of technology, including free speech, privacy, innovation, and consumer rights. It frequently undertakes court cases as an advocate of preserving individual rights.

EFF's mission includes educating the press, policy makers, and the general public about civil liberties. Its Web site (*www.eff.org*) provides an extensive collection of information on such issues as censorship, free expression, digital surveillance, encryption, and privacy. The Web site gets more than 100,000 hits a day, and many visitors are managers who are responsible for making decisions about how to act ethically and legally in applying information technology.

Most of EFF's annual operating budget comes from individual donations from concerned citizens. EFF also has relationships with many IT-related organizations that provide it with hardware, software, and IT services, including Hewlett-Packard, Sun Microsystems, and Symantec. Many of these sponsors hope to do a better job of minimizing the negative impacts of IT on society, help IT users become more responsible in fulfilling their role and responsibilities, and eventually build a more knowledgeable marketplace for their products. In addition to its supporters, however, EFF has developed many critics over the years for what some see as its bias against most forms of regulation.

Discussion Questions

1. Visit the EFF Web site at *www.eff.org* and develop a list of its current "hot" issues. Research one EFF issue that interests you, and write a brief paper summarizing EFF's position. Discuss whether you support this position and why.

2. What reasons might a firm give for joining and supporting EFF?

3. The vice president of public affairs for your midsized telecommunications equipment company has suggested that the firm donate $10,000 in equipment and services to EFF and become a corporate sponsor. The CFO has asked if you, the CIO, support this action. What would you say?

End Notes

[1] "Sex and Tech: Results from a Survey of Teens and Young Adults," National Campaign to Prevent Teen and Unplanned Pregnancy, www.thenationalcampaign.org/sextech/PDF/SexTech_Summary.pdf.

[2] Mike Celizic, "Her Teen Committed Suicide over 'Sexting,'" TodayShow.com, March 6, 2009, www.msnbc.msn.com/id/29546030.

[3] Sheree Paolello, "Mom Loses Daughter over 'Sexting,' Demands Accountability," WLWT.com, March 5, 2009, www.wlwt.com/news/18866515/detail.html.

[4] "Teen Nabbed for Naked MySpace Photos," The Smoking Gun, May 21, 2008, www.thesmokinggun.com/archive/years/2008/0521081myspace1.html.

5 "UPDATE: Teen Put Girl's Nude Pics on MySpace," nbc15.com, March 6, 2009, www.nbc15.com/state/headlines/38449989.html.

6 Jennifer Baker, "Sex Images Found on Boy's Phone," Cincinnati.com, March 20, 2009, http://news.cincinnati.com/article/20090319/NEWS0107/303190036.

7 Bianca Prieto, "Teens Learning There Are Consequences to 'Sexting,'" *Seattle Times*, March 11, 2009, http://seattletimes.nwsource.com/html/nationworld/2008845324_sexting12.html.

8 Wendy Koch, "Teens Caught 'Sexting' Face Porn Charges," *USA Today*, March 11, 2009, www.usatoday.com/tech/wireless/2009-03-11-sexting_N.htm.

9 Lou Grieco, "County Eases 'Sexting' Penalty," *Dayton Daily News*, March 5, 2009, www.daytondailynews.com/n/content/oh/story/news/local/2009/03/05/ddn030509sexting.html.

10 Elaine O'Connor, "Sexting Trouble," Canada.com, March 18, 2009, www.canada.com/news/SEXTING+TROUBLE/1402955/story.html.

11 "Boy Posing as Girl on Facebook Extorts Sex," CBS News, February 5, 2009, www.cbsnews.com/stories/2009/02/05/national/main4777194.shtml?source=RSSattr=HOME_4777194.

12 Courtney Macavinta, "The Supreme Court Today Rejected the Communications Decency Act," *CNET*, June 26, 1997, http://news.cnet.com/High-court-rejects-CDA/2009-1023_3-200957.html.

13 *Reno v. American Civil Liberties Union, et al.,* 521 U.S. 844 (1997), www.law.cornell.edu/supct/html/96-511.ZO.html.

14 Title XIV Child Online Protection Act, http://epic.org/free_speech/censorship/copa.html.

15 *Ashcroft v. American Civil Liberties Union, et al.,* 542 U.S. 656 (2004), www.law.cornell.edu/supct/html/03-218.ZS.html.

16 "2009 Internet Filter Software Product Comparisons," TopTenREVIEWS, www.internet-filter-review.toptenreviews.com.

17 Federal Communications Commission, "Children's Internet Protection Act: FCC Consumer Facts," www.fcc.gov/cgb/consumerfacts/cipa.html.

18 Northeastern Regional Information Center, "Children's Internet Protection Act," June 23, 2003, http://cipa.neric.org.

19 Frank Rich, "Journal; The 2 Tim McVeighs," *New York Times*, January 17, 1998, www.nytimes.com/1998/01/17/opinion/journal-the-2-tim-mcveighs.html?n=Top/Reference/Times%20Topics/Subjects/H/Homosexuality.

20 Dan Goodin, "Two 'John Does' Resign from Raytheon," *CNET*, April 6, 1999, http://news.cnet.com/Two-John-Does-resign-from-Raytheon/2100-1023_3-223888.html.

21 AOL, "AOL Member Community Guidelines," http://help.aol.com/help/viewContent.do?externalId=221226#.

22 Courtney Macavinta, "Prison Time for Email Threats," *CNET*, May 4, 1998, http://news.cnet.com/Prison-time-for-email-threats/2100-1023_3-210845.html.

23 Courtney Macavinta, "Abortion 'Hit List' Slammed in Court," *CNET*, February 2, 1999, http://news.cnet.com/2100-1023-221054.html.

24 Stephanie Armour, "Courts Frown on Online Bad-Mouthing, *USA Today*, January 7, 2002, www.usatoday.com/tech/news/2002/01/07/online-bad-mouthing.htm.

25 Selma Milovanovic, "Online Hate Mail Threat to Arson Case," *The Age*, February 17, 2009, www.theage.com.au/national/online-hate-mail-threat-to-arson-case-20090216-899r.html.

26 Evan Hansen, "Google Blogger: 'I was Terminated,'" *CNET*, February 11, 2005, http://news.cnet.com/Google-blogger-I-was-terminated/2100-1038_3-5572936.html.

27 Jerry Ropelato, "Internet Pornography Statistics," TopTenREVIEWS, http://internet-filter-review.toptenreviews.com/internet-pornography-statistics.html#anchor2.

28 National Conference of State Legislatures, "Child Pornography Reporting Requirements (ISPs and IT Workers)," December 31, 2008.

29 Scott Bradner, "The CAN-SPAM Act as a Warning," *Network World*, January 6, 2009, www.networkworld.com/columnists/2009/010609bradner.html.

30 Andrew Martin, "Whole Foods Executive Used Alias," *New York Times*, July 12, 2007, www.nytimes.com/2007/07/12/business/12foods.html.

31 "Whole Foods Is Hot, Wild Oats a Dud—So Said 'Rahodeb,'" *Wall Street Journal*, July 12, 2007, http://online.wsj.com/article/SB118418782959963745.html.

32 Andrew Martin, "Whole Foods Executive Used Alias," *New York Times*, July 12, 2007, www.nytimes.com/2007/07/12/business/12foods.html.

33 "Whole Foods, FTC Settle on Wild Oats," *CNNMoney.com*, March 6, 2009, http://money.cnn.com/2009/03/06/news/companies/wholefoods_ftc.reut.

34 Henry Blodget, "Apple Denies Steve Jobs Heart Attack Story: 'It Is Not True,'" *Business Insider*, October 3, 2008, www.businessinsider.com/2008/10/apple-s-steve-jobs-rushed-to-er-after-heart-attack-says-cnn-citizen-journalist.

35 "The All-New Complaints.com Puts the Power of Complaint Response in the Hands of Businesses and Web 2.0 Consumers," *e-releases*, March 11, 2009, www.ereleases.com/pr/allnew-complaintscom-puts-power-complaint-response-hands-businesses-web-20-consumers-16449.

36 Declan McCullagh, "Police Blotter: Dark Side of 'Reputation Defending' Service," *CNET*, June 29, 2007, http://news.cnet.com/Police-Blotter-Dark-side-of-reputation-defending-service/2100-1030_3-6194158.html.

37 Register of Known Spam Operations (ROKSO) Database, Spamhaus Project, www.spamhaus.org/rokso (accessed March 19, 2009).

38 Amir Efrati, "AutoAdmit Lawsuit Update: Ciolli Dropped," *Wall Street Journal*, November 9, 2007, http://blogs.wsj.com/law/2007/11/09/autoadmit-lawsuit-update-ciolli-dropped.

39 "AutoAdmit Update: What Happened to Those Commenters Who Thought They Were Anonymous?," Patterico's Pontifications, February 13, 2009, http://patterico.com/2009/02/13/audoadmit-update-what-happened-to-those-commenters-who-thought-they-were-anonymous.

40 Amir Efrati, "Law Firm Rescinds Offer to Ex-AutoAdmit Executive," May 3, 2007, *Wall Street Journal*, http://blogs.wsj.com/law/2007/05/03/law-firm-rescinds-offer-to-ex-autoadmit-director.

41 Amir Efrati, "Law Firm Rescinds Offer to Ex-AutoAdmit Executive," May 3, 2007, *Wall Street Journal*, http://blogs.wsj.com/law/2007/05/03/law-firm-rescinds-offer-to-ex-autoadmit-director.

42 Derek Tam, "Briefly: Former AutoAdmit Executive Continues Lawsuit," *Yale Daily News*, April 13, 2009, www.yaledailynews.com/articles/view/28661.

INTELLECTUAL PROPERTY

VIGNETTE

RIAA Fights Music Piracy

The Recording Industry Association of America (RIAA) is the trade group that represents the U.S. recording industry. Its members create, manufacture, and distribute 85 percent of the legitimate sound recordings produced and sold in the United States.[2]

Since 2003, the RIAA has taken legal action against more than 40,000 people, mostly for copyright infringement for illegally sharing or downloading music via the Internet.[3] Some of the lawsuits threatened fines of tens of thousands of dollars. Alleged violators who challenged the RIAA in court risked being assessed crippling fines; thus, most settled out of court.

The RIAA justifies its strong antipiracy efforts as a means of both protecting the ability of the recording industry to invest in new artists and new music, and giving legitimate online music sharing services a chance to be successful.[4] In addition, it claims that each year the recording industry loses about $5 billion and retailers lose about $1 billion worldwide from music piracy.[5] (Other sources cite different amounts, but still in the range of billions of dollars per year). According to the RIAA: "Just as we must hold accountable the businesses that encourage theft, [the] individuals who engage in illegal downloading must also know that there are consequences to their actions. If you violate the law and steal from record companies, musicians, songwriters, and everyone else involved in making music, you can be held accountable."[6]

In December 2008, the RIAA announced it was dropping its aggressive and highly controversial strategy of sending out pre-litigation letters asking alleged music pirates to stop their copyright infringement and pay a certain dollar amount or go to court. The RIAA is switching to a graduated response program:

- The RIAA will alert participating Internet service providers to what it believes to be illegal downloading activities by ISP customers.

- The ISP will either forward RIAA copyright infringement notices to its subscribers or notify its customers about the notices and ask them to cease and desist.

- The ISP can take a series of escalating sanctions against repeat offenders, ranging from slowing down the subscriber's network speed to terminating service.

- The RIAA reserves the right to sue flagrant copyright offenders.[7]

Questions to Consider

1. Is the RIAA's strong stand on copyright infringement helping or hurting the music recording industry?

2. Could an ISP's implementation and enforcement of the RIAA's multitier strategy have a negative impact on the ISP?

LEARNING OBJECTIVES

As you read this chapter, consider the following questions:

1. What does the term *intellectual property* encompass, and why are organizations so concerned about protecting intellectual property?

2. What are the strengths and limitations of using copyrights, patents, and trade secret laws to protect intellectual property?

3. What is plagiarism, and what can be done to combat it?

4. What is reverse engineering, and what issues are associated with applying it to create a look-alike of a competitor's software program?

5. What is open source code, and what is the fundamental premise behind its use?

6. What is the essential difference between competitive intelligence and industrial espionage, and how is competitive intelligence gathered?

7. What is cybersquatting, and what strategy should be used to protect an organization from it?

WHAT IS INTELLECTUAL PROPERTY?

Intellectual property is a term used to describe works of the mind—such as art, books, films, formulas, inventions, music, and processes—that are distinct, and owned or created by a single person or group. Intellectual property is protected through copyright, patent, and trade secret laws.

Copyright law protects authored works, such as art, books, film, and music; patent law protects inventions; and trade secret law helps safeguard information that is critical to an organization's success. Together, copyright, patent, and trade secret legislation forms a complex body of law that addresses the ownership of intellectual property. Such laws can also present potential ethical problems for IT companies and users—for example, some innovators believe that copyrights, patents, and trade secrets stifle creativity by making it harder to build on the ideas of others. Meanwhile, the owners of intellectual property want to control and receive compensation for the use of their intellectual property. Should the need for ongoing innovation or the rights of property owners govern how intellectual property is used?

Defining and controlling the appropriate level of access to intellectual property are complex tasks. For example, protecting computer software has proven to be difficult because it has not been well categorized under the law. Software has sometimes been treated as the expression of an idea, which can be protected under copyright law, but it has also been treated as a process for changing a computer's internal structure, making it eligible for protection under patent law. At one time, software was even judged to be a series of mental steps, making it inappropriate for ownership and ineligible for any form of protection.

COPYRIGHTS

Copyright and patent protection was established through the U.S. Constitution, Article I, section 8, clause 8, which specifies that Congress shall have the power "to promote the Progress of Science and useful Arts, by securing for limited Times to Authors and Inventors the exclusive Rights to their respective Writings and Discoveries."

A **copyright** is the exclusive right to distribute, display, perform, or reproduce an original work in copies or to prepare derivative works based on the work. Copyright protection is granted to the creators of "original works of authorship in any tangible medium of expression, now known or later developed, from which they can be perceived, reproduced, or otherwise communicated, either directly or with the aid of a machine or device."[8] The author may grant this exclusive right to others. As new forms of expression develop, they can be awarded copyright protection. For example, in the Copyright Act of 1976, audiovisual works were added, and computer programs were assigned to the literary works category.

Copyright infringement is a violation of the rights secured by the owner of a copyright. Infringement occurs when someone copies a substantial and material part of another's copyrighted work without permission. The courts have a wide range of discretion in awarding damages—from $200 for innocent infringement to $100,000 for willful infringement.

Copyright Term

Copyright law guarantees developers the rights to their works for a certain amount of time. Since 1960, the term of copyright has been extended 11 times from its original limit of 28 years. The Copyright Term Extension Act, also known as the Sonny Bono Copyright Term Extension Act, signed into law in 1998, established the following time limits:

- For works created after January 1, 1978, copyright protection endures for the life of the author plus 70 years.
- For works created but not published or registered before January 1, 1978, the term endures for the life of the author plus 70 years, but in no case expires earlier than December 31, 2004.
- For works created before 1978 that are still in their original or renewable term of copyright, the total term was extended to 95 years from the date the copyright was originally secured.[9]

These extensions were primarily championed by movie studios concerned about retaining rights to their early films. Opponents argued that lengthening the copyright period made it more difficult for artists to build on the work of others, thus stifling creativity and innovation. The Sonny Bono Copyright Term Extension Act was legally challenged by Eric Eldred, a bibliophile who wanted to put digitized editions of old books online. The case went all the way to the Supreme Court, which ruled the act constitutional in 2003.[10]

Eligible Works

The types of work that can be copyrighted include architecture, art, audiovisual works, choreography, drama, graphics, literature, motion pictures, music, pantomimes, pictures, sculptures, sound recordings, and other intellectual works, as described in Title 17 of the U.S. Code. To be eligible for a copyright, a work must fall within one of the preceding categories, and it must be original. Copyright law has proven to be extremely flexible in covering new technologies; thus, software, video games, multimedia works, and Web pages can all be protected. However, evaluating the originality of a work is not always a straightforward process, and disagreements over whether or not a work is original sometimes lead to litigation. For example, former Beatles member George Harrison was entangled for decades in litigation over similarities between his hit "My Sweet Lord," released in 1970, and "He's So Fine," composed by Ronald Mack and recorded by the Chiffons in 1962.[11]

Some works are not eligible for copyright protection, including those that have not been fixed in a tangible form of expression (such as an improvisational speech) and those that consist entirely of common information that contains no original authorship, such as a chart showing conversions between European and American units of measure.

Fair Use Doctrine

Copyright law tries to strike a balance between protecting an author's rights and enabling public access to copyrighted works. The fair use doctrine was developed over the years as courts worked to maintain that balance. The **fair use doctrine** allows portions of copyrighted materials to be used without permission under certain circumstances. Title 17, section 107, of the *U.S. Code* established four factors that courts should consider when deciding whether a particular use of copyrighted property is fair and can be allowed without penalty:

- The purpose and character of the use (such as commercial use or nonprofit, educational purposes)
- The nature of the copyrighted work
- The portion of the copyrighted work used in relation to the work as a whole
- The effect of the use on the value of the copyrighted work[12]

The concept that an idea cannot be copyrighted but the expression of an idea can be is key to understanding copyright protection. For example, an author cannot copy the exact words that someone else used to describe his feelings during a World War II battle, but he can convey the sense of horror that the other person expressed. Also, there is no copyright infringement if two parties independently develop a similar or even identical work. For example, if two writers happened to use the same phrase to describe a key historical figure, neither would be guilty of infringement. Of course, independent creation can be extremely difficult to prove or disprove.

Software Copyright Protection

The use of copyrights to protect computer software raises many complicated issues of interpretation. For example, a software manufacturer can observe the operation of a competitor's copyrighted program and then create a program that accomplishes the same result and performs in the same manner. To prove infringement, the copyright holder must show a striking resemblance between its software and the new software that could be explained only by copying. However, if the new software's manufacturer can establish that it developed the program on its own, without any knowledge of the existing program, there is no infringement. For example, two software manufacturers could conceivably develop separate programs for a simple game such as tic-tac-toe without infringing the other's copyright.

An area that holds the potential for software copyright infringement involves the sale of refurbished consumer computer supplies, such as toner and inkjet cartridges. One company that objected to the use of refurbished cartridges was Lexmark International, a manufacturer and supplier of printers and associated supplies. In 2002, Lexmark filed suit against Static Control Components (SCC), a producer of components used to make refurbished printer cartridges. The suit alleged that SCC's Smartek chips included Lexmark software in violation of copyright law. (The software is necessary to allow the refurbished toner cartridges to work with Lexmark's printers.) In February 2003, Lexmark was granted an injunction that prevented SCC from selling the chips until the case could be resolved at trial. However, the ruling was overturned in October 2003 by the U.S. Court of Appeals for the Sixth Circuit, which said that copyright law should not be used to inhibit interoperability between the products of rival vendors. The appeals court upheld its own decision in February 2005.[13]

The Prioritizing Resources and Organization for Intellectual Property (PRO-IP) Act of 2008

The Prioritizing Resources and Organization for Intellectual Property (PRO-IP) Act of 2008 increased trademark and copyright enforcement, and substantially increased penalties for infringement. For example, the penalty for infringement of a 10-song album was raised from $7,500 to $1.5 million. The law also created within the Justice Department the Office of the United States Intellectual Property Enforcement Representative, charged with creating and overseeing a Joint Strategic Plan against counterfeiting and privacy, and coordinating the efforts of the many government agencies that deal with these issues.[14] Tom Donohue, the president of the U.S. Chamber of Commerce, stated, "the PRO-IP Act sends the message to IP criminals everywhere that the U.S. will go the extra mile to protect American innovation."[15]

Meanwhile, opponents of the act proclaimed that "its penalties were far too harsh and that it didn't balance users' rights and concerns over those of major software, media and pharmaceutical companies."[16]

General Agreement on Tariffs and Trade (GATT)

The original General Agreement on Tariffs and Trade (GATT) was signed in 1947 by 150 countries. Since then, there have been eight rounds of negotiations addressing various trade issues. The Uruguay Round, completed in December 1993, resulted in a trade agreement among 117 countries. This agreement also created the World Trade Organization (WTO) in Geneva, Switzerland, to enforce compliance with the agreement. GATT includes a section covering copyrights called the Agreement on Trade-Related Aspects of Intellectual Property Rights (TRIPS), discussed in the following section. U.S. law was amended to be essentially consistent with GATT through both the Uruguay Round Agreements Act in 1994 and the Sonny Bono Copyright Term Extension Act in 1998. Despite GATT, copyright protection varies greatly from country to country, and extreme caution must be exercised on all international usage of any intellectual property.

The WTO and the WTO TRIPS Agreement (1994)

The World Trade Organization (WTO) deals with rules of international trade based on WTO agreements that are negotiated and signed by members of the world's trading nations. The WTO is headquartered in Geneva, Switzerland, and had 153 member nations as of July

2008. Its goal is to help producers of goods and services, exporters, and importers conduct their business.[17]

Many nations recognize that intellectual property has become increasingly important in world trade, yet the extent of protection and enforcement of intellectual property rights varies around the world. As a result, the WTO developed the **Agreement on Trade-Related Aspects of Intellectual Property Rights**, also known as the TRIPS Agreement, to establish minimum levels of protection that each government must provide to the intellectual property of all WTO members. This binding agreement requires member governments to ensure that intellectual property rights can be enforced under their laws and that penalties for infringement are tough enough to deter further violations. Table 6-1 provides a brief summary of copyright, patent, and trade secret protection under the TRIPS Agreement.

TABLE 6-1   Summary of the WTO TRIPS Agreement

Form of intellectual property	Key terms of agreement
Copyright	Computer programs are protected as literary works. Authors of computer programs and producers of sound recordings have the right to prohibit the commercial rental of their works to the public.
Patent	Patent protection is available for any invention—whether a product or process—in all fields of technology without discrimination, subject to the normal tests of novelty, inventiveness, and industrial applicability. It is also required that patents be available and patent rights enjoyable without discrimination as to the place of invention and whether products are imported or locally produced.
Trade secret	Trade secrets and other types of undisclosed information that have commercial value must be protected against breach of confidence and other acts that are contrary to honest commercial practices. However, reasonable steps must have been taken to keep the information secret.

Source: World Trade Organization, "Overview: The TRIPS Agreement," www.wto.org/english/tratop_e/trips_e/intel2_e.htm.

The World Intellectual Property Organization (WIPO) Copyright Treaty (1996)

The World Intellectual Property Organization (WIPO), headquartered in Geneva, Switzerland, is an agency of the United Nations established in 1967. WIPO is dedicated to developing "a balanced and accessible international intellectual property (IP) system, which rewards creativity, stimulates innovation and contributes to economic development while safeguarding the public interest."[18] It has 184 member nations and administers 24 international treaties. Since the 1990s, WIPO has strongly advocated for the interests of intellectual property owners.

The WIPO Copyright Treaty, adopted in 1996, provides additional copyright protections to address electronic media. The treaty ensures that computer programs are protected as literary works, and that the arrangement and selection of material in databases is also protected. It provides authors with control over the rental and distribution of their work, and prohibits circumvention of any technical measures put in place to protect the works. The WIPO Copyright Treaty is implemented in U.S. law through the Digital Millennium Copyright Act (DMCA), which is discussed in the following section.

The Digital Millennium Copyright Act (1998)

The Digital Millennium Copyright Act (DMCA) was signed into law in November 1998; it was written to bring U.S. law into compliance with the global copyright protection treaty from WIPO. The DMCA added new provisions, making it an offense to do the following:

- Circumvent a technical protection
- Develop and provide tools that allow others to access a technologically protected work
- Manufacture, import, provide, or traffic in tools that enable others to circumvent protection and copy a protected work

Violations of these provisions carry both civil and criminal penalties, including up to five years in prison, a fine of up to $500,000 for each offense, or both. Unlike traditional copyright law, the statute does not govern copying; instead, it focuses on the distribution of tools and software that can be used for copyright infringement as well as for legitimate non-infringing use. Although the DMCA explicitly outlaws technologies that can defeat copyright protection devices, it does permit reverse engineering for encryption, interoperability, and computer security research.

Several cases brought under the DMCA have dealt with the use of software to enable the copying of DVD movies. For example, motion picture companies supported the development and worldwide licensing of the Content Scramble System (CSS), which enables a DVD player (shown in Figure 6-1) or a computer drive to decrypt, unscramble, and play back motion pictures on DVDs, but not copy them. However, a software program called DeCSS can break the encryption code and enable users to copy DVDs. The posting of this software on the Web in January 2000 led to a lawsuit by major movie studios against its author. After a series of cases, courts finally ruled that the use of DeCSS violated the DMCA's anticircumvention provisions.

FIGURE 6-1   Several cases brought under the DMCA dealt with the use of software to enable the copying of DVD movies.

Opponents of DMCA say that it gives holders of intellectual property so much power that it actually restricts the free flow of information. For example, under DMCA, Internet service providers (ISPs) are required to remove access to Web sites that allegedly break copyright law—even before infringement has been proven. Companies that provide Internet access to music and videos face legal action and the failure of their businesses if they do not gain approval to publish content from the music and movie industries.

PATENTS

A **patent** is a grant of a property right issued by the United States Patent and Trademark Office (USPTO) to an inventor. A patent permits its owner to exclude the public from making, using, or selling a protected invention, and it allows for legal action against violators. Unlike a copyright, a patent prevents independent creation as well as copying. Even if someone else invents the same item independently and with no prior knowledge of the patent holder's invention, the second inventor is excluded from using the patented device without permission of the original patent holder. The rights of the patent are valid only in the United States and its territories and possessions. There were 157,774 patents granted in 2008, a slight increase over the 157,248 granted in 2007.[19]

The value of patents to a company cannot be underestimated. IBM obtained 4,186 patents in 2008, the 16th consecutive year it received more patents than any other company in the United States.[20] It has been estimated that IBM's licensing of patents and technologies generates several hundred million dollars in annual revenue.[21]

To obtain a U.S. patent, an application must be filed with the USPTO according to strict requirements. As part of the application, the USPTO searches the **prior art**—the existing body of knowledge that is available to a person of ordinary skill in the art—starting with patents and published material that have already been issued in the same area. The USPTO will not issue a patent for an invention whose professed improvements are already present in the prior art. Although the USPTO employs some 3,000 examiners to research the originality of each patent application, it still takes an average of 25 months to process one. Such delays can be costly for companies that want to bring patented products to market quickly.[22] As a result, in many cases, people who are trained in the patent process, rather than the inventors themselves, prepare patent applications.

The main body of law that governs patents is contained in Title 35 of the U.S. Code, which states that an invention must pass the following four tests to be eligible for a patent:

- It must fall into one of five statutory classes of items that can be patented: (1) processes, (2) machines, (3) manufactures (such as objects made by humans or machines), (4) compositions of matter (such as chemical compounds), and (5) new uses in any of the previous four classes.
- It must be useful.
- It must be novel.
- It must not be obvious to a person having ordinary skill in the same field.[23]

The U.S. Supreme Court has ruled that three classes of items cannot be patented: abstract ideas, laws of nature, and natural phenomena. Standing on its own, mathematical subject matter is also not entitled to patent protection. Thus, Pythagoras could not have patented his formula for the length of the hypotenuse of a right triangle ($c^2 = a^2 + b^2$).

Patent infringement, or the violation of the rights secured by the owner of a patent, occurs when someone makes unauthorized use of another's patent. Unlike copyright infringement, there is no specified limit to the monetary penalty if patent infringement is found. In fact, if a court determines that the infringement is intentional, it can award up to three times the amount of the damages claimed by the patent holder. The most common defense against patent infringement is a counterattack on the claim of infringement and the validity of the patent itself. Even if the patent is valid, the plaintiff must still prove that every element of a claim was infringed and that the infringement caused some sort of damage.

Software Patents

A software patent claims as its invention some feature or process embodied in instructions executed by a computer. Prior to 1981, the courts regularly turned down requests for such patents, giving the impression that software could not be patented. In the 1981 *Diamond v. Diehr* case, the Supreme Court granted a patent to Diehr, who had developed a process control computer and sensors to monitor the temperature inside a rubber mold. The USPTO interpreted the court's reasoning to mean that just because an invention used software did not mean that the invention could not be patented. Based on this ruling, courts have slowly broadened the scope of protection for software-related inventions.[24]

The creation of the U.S. Court of Appeals for the Federal Circuit in 1982 further improved the environment for the use of patents in software-related inventions. This court is charged with hearing all patent appeals and is generally viewed as providing stronger enforcement of patents and more effective punishment for willful infringement.

Since the early 1980s, the USPTO has granted as many as 20,000 software-related patents per year. Applications software, business software, expert systems, and system software have been patented, as well as such software processes as compilation routines, editing and control functions, and operating system techniques. Even electronic font types and icons have been patented.

Before obtaining a software patent, a developer should do a patent search, which can be lengthy and expensive. However, even a thorough search may not identify all potential infringements. The USPTO's classification system is complex, and software patents may be classified in several categories, making them difficult to find. If a patent search misses something, there is a risk of an expensive patent infringement lawsuit. The Software Patent Institute is building a database of information to document known patented software and assist the USPTO and others in researching prior art in the software arena.

Some software experts think that too many software patents are being granted, inhibiting new software development. For example, in September 1999, Amazon.com obtained a patent for "one-click shopping," based on the use of a shopping-cart purchase system for electronic commerce. In October 1999, Amazon.com sued Barnes & Noble for allegedly infringing this patent with its Express Lane feature. The filing of the suit prompted many complaints about the issuing of patents to business methods, which critics deride as overly broad and unoriginal concepts that do not merit patents. Some critics considered one-click shopping little more than a simple combination of existing Web technologies. Following preliminary court hearings and the discovery that others had used the one-click technology before Amazon.com even began business, Amazon.com and Barnes & Noble settled out of court in March 2002.

Software engineers rarely take the time to search patent databases for new inventions that could benefit their projects, partly because software patents are described in obscure language and partly because engineers risk paying triple damages for knowingly infringing one. As a result, many software patent infringements are for independent inventions—the same invention discovered by two different parties without knowledge of the other's work.

In an example of how the potential for infringement can arise due to independent inventions, Cygnus Systems alleged in late 2008 that Apple, Google, and Microsoft infringed a patent that Cygnus filed for in 2001. Cygnus says that the three firms violated its patent (which was finally cleared in March 2008) on the use of document-preview icons, or thumbnails. Cygnus alleges that Apple's iPhone, Safari Internet browser, and Mac OS X Leopard operating systems; Google's Chrome browser; and Microsoft's Vista OS operating system and Internet Explorer 8 all employ this technology.[25] Because this is such a commonly used technology, many more companies may be sued for patent infringement.

Software Cross-Licensing Agreements

Many large software companies have cross-licensing agreements in which each party agrees not to sue the other over patent infringements. For example, Microsoft is working to put in place 100 or more agreements with firms such as IBM, Sun Microsystems, SAP, Hewlett-Packard, Siemens, Cisco, Autodesk, Brother, Lexmark, Cadence, Pioneer, and Nikon by 2010.[26] This strategy to obtain the rights to technologies that it might use in its products provides a tremendous amount of development freedom to Microsoft without risk of expensive litigation.

Major IT firms usually have little interest in cross-licensing with smaller firms, so small businesses have no choice but to license patents if they use them. As a result, small businesses must pay an additional cost from which many larger companies are exempt. Furthermore, small businesses are generally unsuccessful in enforcing their patents against larger companies. Should a small business bring a patent infringement suit against a large firm, the larger firm can overwhelm the small business with multiple patent suits, whether they have merit or not. Considering that the average patent lawsuit costs $3 to $10 million and takes two to three years to litigate, a small firm often simply cannot afford to fight; instead, it usually settles and licenses its patents to the large company.[27]

IBM announced in 2009 that it would donate some 3,000 patents for free use by developers to help them innovate and build new hardware and software.[28] The announcement represented a major shift in IBM's intellectual property strategy and was meant to encourage other patent holders to donate their own intellectual property. It also placed IBM in direct opposition to Microsoft and other major software developers. IBM's strategy certainly differs from zealous patent defenders, such as major pharmaceutical and media companies, who are IBM customers. Hopefully for IBM, this difference in patent strategies will not hurt sales to these customers.

Defensive Publishing and Patent Trolls

Inventors sometimes employ a tactic called defensive publishing as an alternative to filing for patents. Under this approach, a company publishes a description of its innovation in a bulletin, conference paper, or trade journal, or on a Web site. Although this obviously provides competitors with access to the innovation, it also establishes the idea's legal existence as prior art. Therefore, competitors cannot patent the idea or charge licensing fees to other users of the technology or technique. This approach costs mere hundreds of dollars, requires no lawyers, and is fast.

A **patent troll** is a firm that acquires patents with no intention of manufacturing anything, instead licensing the patents to others. Intellectual Ventures is an example of such a firm; it has built a portfolio of more than 20,000 patents, most for IT-related technology. Google, Intel, eBay, NVIDIA, SAP, Sony, Microsoft, Nokia, and other IT firms invested money in Intellectual Ventures in exchange for licenses to patents in the portfolio. Some IT organizations pay large amounts of money for the right to use one or more of these patents, including Cisco Systems and Verizon Communications, who paid between $200 million and $400 million for patent licenses in 2008.[29]

Submarine Patents and Patent Farming

A **standard** is a definition that has been approved by a recognized standards organization or accepted as a de facto standard within a particular industry. Standards exist for communication protocols, programming languages, operating systems, data formats, and electrical interfaces. Standards are extremely useful because they enable hardware and software from different manufacturers to work together.

A technology, process, or principle that has been patented may be embedded—knowingly or unknowingly—within a standard. If so, the patent owner can either demand a royalty payment from any party that implements the standard or refuse to permit certain parties to use the patent, thus effectively blocking them from using the standard. A patented process or invention that is surreptitiously included within a standard without being made public until after the standard is broadly adopted is called a **submarine patent**. A devious patent holder might influence a standards organization to make use of its patented item without revealing the existence of the patent. Later, the patent holder might demand royalties from all parties that use the standard. This strategy is known as **patent farming**.

One possible example of a submarine patent used in patent farming could be U.S. Patent 5,838,906, which is owned by the University of California and licensed exclusively to a small software company called Eolas Technologies. The patent describes how a Web browser can use external applications. The University of California did not make the patent known for years and then sued Microsoft for use of the principle detailed in the patent. The university and Eolas received a $520 million award in August 2003 after a federal jury found that Microsoft's Internet Explorer browser infringed the patent. In November 2003, the patent office began a review of the patent based on a request from world-renowned Tim Berners-Lee, father of the World Wide Web and director of the World Wide Web Consortium. He argued that the 1998 patent should be invalidated because of the existence of prior art, or previous examples of the technology's use. In January 2004, a federal judge upheld the original decision, requiring Microsoft to pay $520 million on grounds that Internet Explorer infringed the patent. The judge rejected Microsoft's request for a new trial and ordered the payment of more than $45 million in interest.[30] In September 2005, the USPTO upheld the validity of the patent held by the University of California.[31]

Another possible example of patent farming involved Rambus, a designer and manufacturer of computer memory technology. Rambus allegedly influenced a standards organization to adopt its technology as part of an industry standard, without disclosing that it had a patent application in process. By 2000, Rambus was enmeshed in a series of lawsuits with many of the world's leading memory chip makers, including Infineon in Germany, Micron

Technology in the United States, and Hynix Semiconductor in South Korea. Rambus claimed that rival producers of dynamic RAM chips (shown in Figure 6-2) infringed its patents. Collectively, the penalties for patent infringement could have been worth hundreds of millions of dollars. Infineon and Rambus reached a settlement in March 2005 that required Infineon to pay Rambus $47 million for a global license to all existing and future Rambus patents for use in Infineon products until 2007.[32] In January 2009, a U.S. district court ruled that Rambus had destroyed documents pertinent to the Micron patent infringement lawsuit and barred it from enforcing patents against Micron.[33] In May 2009, the court ruled in favor of Rambus in its case against Hynix and ordered Hynix to raise money to pay Rambus $397 million for patent violations.[34]

FIGURE 6-2   Rambus was involved in litigation for years over memory chips.

TRADE SECRETS

In Chapter 2, a trade secret was defined as business information that represents something of economic value, has required effort or cost to develop, has some degree of uniqueness or novelty, is generally unknown to the public, and is kept confidential.

Trade secret protection begins by identifying all the information that must be protected—from undisclosed patent applications to market research and business plans—and developing a comprehensive strategy for keeping the information secure. Trade secret law protects only against the *misappropriation* of trade secrets. If competitors come up with the same idea on their own, it is not misappropriation; in other words, the law doesn't prevent someone from using the same idea if it was developed independently.

Trade secret law has several key advantages over the use of patents and copyrights in protecting companies from losing control of their intellectual property, as summarized in the following list:

- There are no time limitations on the protection of trade secrets, as there are with patents and copyrights.
- There is no need to file an application, make disclosures to any person or agency, or disclose a trade secret to outsiders to gain protection. (After the USPTO issues a patent, competitors can obtain a detailed description of it.)

- While patents can be ruled invalid by the courts, meaning that the affected inventions no longer have patent protection, this risk does not exist for trade secrets.
- No filing or application fees are required to protect a trade secret.

Because of these advantages, trade secret laws protect more technology worldwide than patent laws do.

Trade Secret Laws

Trade secret protection laws vary greatly from country to country. For example, the Philippines provides no legal protection for trade secrets. In some European countries, pharmaceuticals, methods of medical diagnosis and treatment, and information technology cannot be patented. Many Asian countries require foreign corporations operating there to transfer rights to their technology to locally controlled enterprises. (Coca-Cola reopened its operations in India in 1993 after halting sales for 16 years to protect the "secret formula" for its soft drink, even though India's vast population represented a huge potential market.) American businesses that seek to operate in foreign jurisdictions or enter international markets must take these differences into account.

Uniform Trade Secrets Act (UTSA)

The Uniform Trade Secrets Act (UTSA) was drafted in the 1970s to bring uniformity to all states in the area of trade secret law. The first state to enact the UTSA was Minnesota in 1981, followed by 39 more states and the District of Columbia. The UTSA defines a trade secret as "information, including a formula, pattern, compilation, program, device, method, technique, or process, that:

- Derives independent economic value, actual or potential, from not being generally known to, and not being readily ascertainable by, persons who can obtain economic value from its disclosure or use
- Is the subject of efforts that are reasonable under the circumstances to maintain its secrecy"[35]

Under these terms, computer hardware and software can qualify for trade secret protection by the UTSA.

The Economic Espionage Act (EEA) (1996)

The Economic Espionage Act (EEA) of 1996 imposes penalties of up to $10 million and 15 years in prison for the theft of trade secrets. Before the EEA, there was no specific criminal statute to help pursue economic espionage; the FBI was investigating nearly 800 such cases in 23 countries when the EEA was enacted.[36] Today, intellectual property loss from industrial and economic espionage costs U.S.-based businesses more than $300 billion annually, according to estimates by Guardsmark, a New York security services firm. As with the UTSA, information is considered a trade secret under the EEA only if companies take steps to protect it.

Employees and Trade Secrets

Employees are the greatest threat to the loss of company trade secrets—they might accidentally disclose trade secrets or steal them for monetary gain. Organizations must educate employees about the importance of maintaining the secrecy of corporate information. Trade secret information should be labeled clearly as confidential and should only be accessible by a limited number of people. Most organizations have strict policies regarding nondisclosure of corporate information.

Losing customer information to competitors is a growing concern in industries in which companies compete for many of the same clients. There are numerous cases of employees making unauthorized use of their employer's customer list. For example, the Ohio State Supreme Court upheld a verdict against a man who left a financial services firm and recruited former clients to start his own firm. His former employer sued him, even though the former employee had not stolen a client list. "This ruling says, it doesn't matter if the confidential list is on paper or in your memory if it qualifies as a trade secret," said Susan Guerette, a specialist in restrictive covenant and trade secret law at Pennsylvania-based Fisher & Phillips LLP.[37]

Legally, a customer list is not automatically considered a trade secret. If a company doesn't treat the list as valuable, confidential information internally, neither will the court. The courts must consider two main factors in making this determination. First, did the firm take prudent steps to keep the list secret? Second, did the firm expend money or effort to develop the customer list? The more the firm invested to build its customer list and the more that the list provides the firm with a competitive advantage, the more likely the courts are to accept the list as a trade secret.

Because organizations can risk losing trade secrets when key employees leave, they often try to prohibit employees from revealing secrets by adding **nondisclosure clauses** to employment contracts. Thus, departing employees cannot take copies of computer programs or reveal the details of software owned by the firm. Defining reasonable nondisclosure agreements can be difficult, as seen in the following example involving Apple. In addition to filing hundreds of patents on iPhone technology, the firm put into place a restrictive nondisclosure agreement to provide an extra layer of protection. Many iPhone developers complained bitterly about the tough restrictions, which prohibited them from talking about their coding work with anyone not on the project team and even prohibited them from talking about the restrictions themselves. Eventually, Apple admitted that its nondisclosure terms were overly restrictive and loosened them for iPhone software that was already released.[38]

Another option for preserving trade secrets is to have an experienced member of the Human Resources Department conduct an exit interview with each departing employee. A key step in the interview is to review a checklist that deals with confidentiality issues. At the end of the interview, the departing employee is asked to sign an acknowledgment of responsibility not to divulge any trade secrets.

Employers can also use noncompete agreements to protect intellectual property from being used by competitors when key employees leave. A **noncompete agreement** prohibits an employee from working for any competitors for a period of time, often one to two years. When courts are asked to settle disputes over noncompete agreements, they consider the reasonableness of the restriction and how the restriction protects the legitimate interests of the former employer. The courts also consider geographic area and the length of time of the restriction in relation to the pace of the industry.

The following is an example of a noncompete agreement:

> The employee agrees as a condition of employment that in the event of termination for any reason, he or she will not engage in a similar or competitive business

for a period of two years, nor will he or she contact or solicit any customer with whom Employer conducted business during his or her employment. This restrictive covenant shall be for a term of two years from termination, and shall encompass the geographic area within a 100-mile radius of Employer's place of business.

IBM sued Mark Papermaster, a microchip expert, for violating a noncompete agreement when he announced that he intended to leave the company to join Apple as its head of device hardware engineering. The lawsuit was settled when Papermaster agreed to report to IBM should he suspect that any breakthroughs he develops at Apple infringe on proprietary or confidential information he learned while working at IBM. Papermaster must also twice submit to IBM a written declaration that states he is not using confidential IBM material in his role at Apple.[39]

KEY INTELLECTUAL PROPERTY ISSUES

This section discusses several issues that apply to intellectual property and information technology, including plagiarism, reverse engineering, open source code, competitive intelligence, and cybersquatting.

Plagiarism

Plagiarism is the act of stealing someone's ideas or words and passing them off as one's own. The explosion of electronic content and the growth of the Web have made it easy to cut and paste paragraphs into term papers and other documents without proper citation or quotation marks. To compound the problem, hundreds of online "paper mills" enable users to download entire term papers. Although some sites post warnings that their services should be used for research purposes only, many users pay scant heed. As a result, plagiarism has become an issue from elementary schools to the highest levels of academia.

Plagiarism is also common outside academia. Popular literary authors, playwrights, musicians, journalists, and even software developers have been accused of it:

- Katie Couric gave a commentary about the joys of getting her first library card that plagiarized a column in the *Wall Street Journal*; the "commentary" was actually written by a network producer who was subsequently fired.[40]
- Reporter Jayson Blair resigned from the *New York Times* after he was accused of plagiarism and fabricating quotes and other information in news stories. Executive Editor Howell Raines and Managing Editor Gerald Boyd also resigned in the fallout from the scandal.[41]
- Science fiction writer Harlan Ellison successfully sued movie director James Cameron for taking key elements from two different episodes of the TV series *The Outer Limits*—written by Ellison—and using them in the 1984 classic movie *The Terminator*.[42]

Despite school codes of ethics that clearly define plagiarism and prescribe penalties ranging from no credit on a paper to expulsion, many students still do not understand what constitutes plagiarism. Some students believe that all electronic content is in the public domain, while other students knowingly commit plagiarism because they either feel pressure to achieve a high GPA or are too lazy or pressed for time to do original work.

Some instructors say that being familiar with a student's style of writing, grammar, and vocabulary enables them to determine if the student actually wrote a paper. In addition, plagiarism detection systems (see Table 6-2) allow teachers, corporations, law firms, and publishers to check for matching text in different documents as a means of identifying potential plagiarism. These systems work by checking submitted material against one or more of the following databases of electronic content:

- More than 5 billion pages of publicly accessible electronic content on the Internet
- Millions of works published in electronic form, including newspapers, magazines, journals, and electronic books
- A database of papers submitted to the plagiarism detection service from participating institutions

TABLE 6-2   Partial list of plagiarism detection services and software

Name of service	Web site	Provider
iThenticate	www.ithenticate.com	iParadigms
Turnitin	www.turnitin.com	iParadigms
SafeAssign	www.safeassign.com	Blackboard
Glatt Plagiarism Services	www.plagiarism.com	Glatt Plagiarism Services
EVE Plagiarism Detection	www.canexus.com/eve	CaNexus

Turnitin has been available since 1999 and is used in over 5,000 academic institutions around the world. It can check documents in 30 languages. iThenticate is available from the same company that created Turnitin but is designed to meet the needs of members of the information industry, such as publishers, research facilities, legal firms, government agencies, and financial institutions.[43]

Interestingly, four high school students brought a lawsuit against iParadigms, accusing the firm of copyright infringement. The basis of their lawsuit was that the firm's primary product, Turnitin, used archived student papers without their permission to assess the originality of newly submitted papers. However, both a district court and a court of appeals ruled that the use of student papers for purposes of plagiarism detection constitutes a fair use and is therefore not a copyright infringement. A U.S. court of appeals ruled that such use of student papers is "a highly transformative use that adds something new in purpose and character and does not harm the future marketability of the students' works."[44]

The following list shows some of the actions that schools can take to combat student plagiarism:

- Help students understand what constitutes plagiarism and why they need to properly cite sources.
- Show students how to document Web pages and materials from online databases.

- Schedule major writing assignments so that portions are due over the course of the term, thus reducing the likelihood that students will get into a time crunch.
- Make clear to students that instructors are aware of Internet paper mills.
- Ensure that instructors both educate students about plagiarism detection services and make them aware that they know how to use these services.
- Incorporate detection software and services into a comprehensive antiplagiarism program.

Reverse Engineering

Reverse engineering is the process of taking something apart in order to understand it, build a copy of it, or improve it. Reverse engineering was originally applied to computer hardware but is now commonly applied to software as well. Reverse engineering of software involves analyzing it to create a new representation of the system in a different form or at a higher level of abstraction. Often, reverse engineering begins by extracting design-stage details from program code. Design-stage details about an information system are more conceptual and less defined than the program code of the same system.

One frequent use of reverse engineering for software is to modify an application that ran on one vendor's database so that it can run on another's (for example, from Access to Oracle). Database management systems use their own programming language for application development. As a result, organizations that want to change database vendors are faced with rewriting existing applications using the new vendor's database programming language. The cost and length of time required for this redevelopment can deter an organization from changing vendors and deprive it of the possible benefits of converting to an improved database technology.

Using reverse engineering, a developer can use the code of the current database programming language to recover the design of the information system application. Next, code-generation tools can be used to take the design and produce code (forward engineer) in the new database programming language. This reverse-engineering and code-generating process greatly reduces the time and cost needed to migrate the organization's applications to the new database management system. No one challenges the right to use this process to convert applications that were developed in-house. After all, those applications were developed and are owned by the companies using them. It is quite another matter, though, to use this process on a purchased software application developed and licensed by outside parties. Most IT managers would consider this action unethical, because the software user does not actually own the right to the software. In addition, a number of intellectual property issues would be raised, depending on whether the software was licensed, copyrighted, or patented.

Other reverse-engineering issues involve tools called compilers and decompilers. A **compiler** is a language translator that converts computer program statements expressed in a source language (such as COBOL, Pascal, or C) into machine language (a series of binary codes of 0s and 1s) that the computer can execute. When a software manufacturer provides a customer with its software, it usually provides the software in machine-language form. Tools called reverse-engineering compilers, or **decompilers**, can read the machine language and produce the source code. For example, REC is a decompiler that reads an executable, machine-language file and produces a C-like representation of the code used to build the program.

Decompilers and other reverse-engineering techniques can be used to reveal a competitor's program code, which can then be used to develop a new program that either duplicates the original or interfaces with the program. Thus, reverse engineering provides a way to gain access to information that another organization may have copyrighted or classified as a trade secret.

The courts have ruled in favor of using reverse engineering to enable interoperability. In the early 1990s, video game maker Sega developed a computerized lock so that only Sega video cartridges would work on its entertainment systems. This essentially shut out competitors from making software for the Sega systems. *Sega Enterprises Ltd. v. Accolade, Inc.*, dealt with rival game maker Accolade's use of a decompiler to read the Sega software source code. With the code, Accolade could create new software that circumvented the lock and ran on Sega machines. An appeals court ultimately ruled that if someone lacks access to the unprotected elements of an original work and has a "legitimate reason" for gaining access to those elements, disassembly of a copyrighted work is considered to be a fair use under section 107 of the Copyright Act. The unprotected element in this case was the code necessary to enable software to interoperate with the Sega equipment. The court reasoned that to refuse someone the opportunity to create an interoperable product would allow existing manufacturers to monopolize the market, making it impossible for others to compete. This ruling had a major impact on the video game industry, allowing video game makers to create software that would run on multiple machines.[45]

Software license agreements increasingly forbid reverse engineering. As a result of the increased legislation affecting reverse engineering, some software developers are moving their reverse-engineering projects offshore to avoid U.S. rules.

The reverse engineering of copyrighted or patented hardware or software items done for the purposes of interoperability (for example, to support undocumented file formats or undocumented hardware peripherals) is often considered to be legal. However, copyright and patent owners often contest this in an attempt to stifle reverse engineering of their products for any reason.

The ethics of using reverse engineering are debated. Some argue that its use is fair if it enables a company to create software that interoperates with another company's software or hardware and provides a useful function. This is especially true if the software's creator refuses to cooperate by providing documentation to help create interoperable software. From the consumer's standpoint, such stifling of competition increases costs and reduces business options. Reverse engineering can also be a useful tool in detecting software bugs and security holes.

Others argue strongly against the use of reverse engineering, saying it can uncover software designs that someone else has developed at great cost and taken care to protect. This unfairly robs the creator of future earnings, opponents say, and without a payoff, what is the business incentive for software development?

Open Source Code

Historically, the makers of proprietary software have not made their source code available, but not all developers share that philosophy. **Open source code** is any program whose source code is made available for use or modification, as users or other developers see fit.

The basic premise behind open source code is that when many programmers can read, redistribute, and modify a program's code, the software improves. Programs with open source code can be adapted to meet new needs, and bugs can be rapidly identified and fixed. Open source code advocates believe that this process produces better software than the traditional closed model.

A considerable amount of open source code is available. A recent survey by Talend—an open source software provider—shows that the use of open source is growing, and that 35 percent of organizations use both proprietary and open source software. A common use of open source software is to move data from one application to another and to extract, transform, and load business data into large databases. Two frequently cited reasons for using open source software are that it provides a better solution to a specific business problem and that it costs less.[46] Open source software is used in applications developed for smartphones and other mobile devices, such as Apple's iPhone, Palm's Treo, and Research In Motion's BlackBerry. See Table 6-3 for a listing of commonly used open source software.

TABLE 6-3   Commonly used open source software

Open source software	Purpose
7-Zip	File compression
Ares Galaxy	Peer-to-peer file sharing
Audacity	Sound editing and special effects
Azureus	Peer-to-peer file sharing
Blender 3D	3D modeling and animation
eMule	Peer-to-peer file sharing
Eraser	Erase data completely
Firefox	Internet browser
OpenOffice	Word processing, spreadsheets, presentations, graphics, and databases
Video Dub	Video editing

Why would firms or individual developers create open source code if they do not receive money for it? Here are several reasons:

- Some people share code to earn respect for solving a common problem in an elegant way.
- Some people have used open source code that was developed by others and feel the need to pay back.
- A firm may be required to develop software as part of an agreement to address a client's problem. If the firm is paid for the employees' time spent to develop the software rather than for the software itself, it may decide to license the code as open source and use it either to promote the firm's expertise or as an incentive to attract other potential clients with a similar problem.

- A firm may develop open source code in the hope of earning software maintenance fees if end users need changes in the future.
- A firm may develop useful code but may be reluctant to license and market it, and so might donate the code to the general public.

There are various definitions of what constitutes open source code, each with its own idiosyncrasies. The GNU General Public License (GPL) was a precursor to the open source code defined by the Open Source Initiative (OSI). GNU is a computer operating system composed entirely of free software; its name is a recursive acronym for GNU's Not Unix. The GPL is intended to protect GNU software from being made proprietary, and it lists terms and conditions for copying, modifying, and distributing free software. The OSI is a nonprofit organization that advocates for open source and certifies open source licenses. Its certification mark, "OSI Certified," may be applied only to software distributed under an open source license that meets OSI criteria, as described at its Web site, *www.opensource.org*.

A software developer could attempt to make a program open source simply by putting it into the public domain with no copyright. This would allow people to share the program and their improvements, but it would also allow others to revise the original code and then distribute the resulting software as their own proprietary product. Users who received the program in the modified form would no longer have the freedoms associated with the original software. Use of an open source license avoids this scenario.

Bob Jacobsen is a physics professor at UC Berkeley and—through the JMRI project—the developer of model railroad software called LocoNet, for model railroaders who want to control their layouts from a computer. Jacobsen develops his software as open source and gives it away with the full source code. Matthew A. Katzer owns KAM Industries, a firm that sells model railroad software. Katzer claims that he filed a patent covering the transmission of model railroad commands between multiple devices in 2002. In August 2005, Katzer's firm began sending Jacobsen a monthly invoice for royalties for using his patent—initially $19 and then $29 for every copy of JMRI Jacobsen gave away. The total invoice was for over $200,000. In August 2006, Jacobsen filed a declaratory judgment in federal court to get a legal determination of his rights in this matter. His complaint alleged acts of unfair competition, libel, patent fraud, and misrepresentation against Katzer, and asked the court to issue an injunction against Katzer to prevent him from continuing his actions. One thing escalated to another, and in February 2009, Katzer sued Jacobsen for copyright infringement of software manuals and asked for damages in excess of $6 million. The software development community is watching to see how all this is settled as the case may set several legal precedents for open source software.[47]

Competitive Intelligence

Competitive intelligence is legally obtained information that is gathered to help a company gain an advantage over its rivals. For example, some companies have employees who monitor the public announcements of property transfers to detect any plant or store expansions of competitors. An effective competitive intelligence operation requires the continual gathering, analysis, and evaluation of data with controlled dissemination of useful information to decision makers. Competitive intelligence is often integrated into a company's strategic plan and decision making. Many companies, such as Eastman Kodak, Monsanto, and United

Technologies, have established formal competitive intelligence departments. Some companies have even employed former CIA analysts to assist them.

Competitive intelligence is not the same as **industrial espionage**, which employs illegal means to obtain business information not available to the general public. In the United States, industrial espionage is a serious crime that carries heavy penalties.

Almost all the data needed for competitive intelligence can be collected from examining published information or interviews, as outlined in the following list:

- 10-K or annual reports
- An SC 13D acquisition—a filing by shareholders who report owning more than 5 percent of common stock in a public company
- 10-Q or quarterly reports
- Press releases
- Promotional materials
- Web sites
- Analyses by the investment community, such as a Standard & Poor's stock report
- Dun & Bradstreet credit reports
- Interviews with suppliers, customers, and former employees
- Calls to competitors' customer service groups
- Articles in the trade press
- Environmental impact statements and other filings associated with a plant expansion or construction
- Patents

By coupling this data with analytical tools and industry expertise, an experienced analyst can make deductions that lead to significant information. According to Avinash Kaushik, self-described Analytics Evangelist for Google, "The Web is the best competitive intelligence tool in the world." Kaushik likens the failure to use such data to driving a car 90 miles an hour with the windshield painted black, then scraping off the paint and realizing "you're going 90 but everyone else is going 220 and you're going to die."[48]

Competitive intelligence gathering has become enough of a science that over two dozen colleges and universities offer courses or entire programs in this subject. Also, the Society of Competitive Intelligence Professionals (*www.scip.org*) offers ongoing training programs and conferences.

Without proper management safeguards, the process of gathering competitive intelligence can cross over to industrial espionage and dirty tricks (see Figure 6-3). One frequent trick is to enter a bar near a competitor's plant or headquarters, strike up a conversation, and ply people for information after their inhibitions have been weakened by alcohol.

FIGURE 6-3 Organizations must be careful that they use only legal means to obtain competitive business information.

Competitive intelligence analysts must avoid unethical or illegal actions, such as lying, misrepresentation, theft, bribery, or eavesdropping with illegal devices. See Table 6-4 for useful guidelines. The preferred answer to each question in the checklist is *yes*.

TABLE 6-4 A manager's checklist for running an ethical competitive intelligence operation

Question	Yes	No
Has the competitive intelligence organization developed a mission statement, objectives, goals, and a code of ethics?		
Has the company's legal department approved the mission statement, objectives, goals, and code of ethics?		
Do analysts understand the need to abide by their organization's code of ethics and corporate policies?		
Is there a rigorous training and certification process for analysts?		
Do analysts understand all applicable laws—domestic and international—including the Uniform Trade Secrets Act and the Economic Espionage Act, and do they understand the critical importance of abiding by them?		

(continued)

TABLE 6-4 A manager's checklist for running an ethical competitive intelligence operation (*continued*)

Question	Yes	No
Do analysts disclose their true identity as well as the name of their organization prior to any interviews?		
Do analysts understand that everything their firm learns about the competition must be obtained legally?		
Do analysts respect all requests for anonymity and confidentiality of information?		
Has the company's legal department approved the processes for gathering data?		
Do analysts provide honest recommendations and conclusions?		
Is the use of third parties to gather competitive intelligence carefully reviewed and managed?		

Failure to act prudently when gathering competitive intelligence can get analysts and companies into serious trouble. For example, Procter & Gamble (P&G) admitted publicly in 2001 that it unethically gained information about Unilever, its competitor in the multibillion-dollar hair-care business. Unilever markets brands such as Salon Selectives, Finesse, and Thermasilk, while P&G manufactures Pantene, Head & Shoulders, and Pert. Competitive intelligence managers at P&G hired a contractor, who in turn hired several subcontractors to spy on P&G's competitors. Unilever was the primary target.

In at least one instance, the espionage included going through dumpsters on public property outside Unilever corporate offices in Chicago. In addition, competitive intelligence operatives were alleged to have misrepresented themselves to Unilever employees, suggesting that they were market analysts. (P&G confirms the dumpster diving, but it denies that misrepresentation took place.) The operatives captured critical information about Unilever's brands, including new-product rollouts, selling prices, and operating margins.

When senior P&G officials discovered that the firm hired by the company was operating unethically, P&G immediately stopped the campaign and fired the three managers responsible for hiring the firm. P&G then did something unusual—it blew the whistle on itself, confessed to Unilever, returned stolen documents to Unilever, and started negotiations with them to set things straight. P&G's chairman of the board was personally involved in ensuring that none of the information obtained would ever be used in P&G business plans. Several weeks of high-level negotiations between P&G and Unilever executives led to a secret agreement between the two companies. P&G is believed to have paid tens of millions of dollars to Unilever. In addition, several hair-care product executives were transferred to other units within P&G.

Experts in competitive intelligence agree that the firm hired by P&G crossed the line of ethical business practices by sorting through Unilever's garbage. However, they also give P&G credit for going to Unilever quickly after it discovered the damage. Such prompt action was seen as the best approach. If no settlement had been reached, of course, Unilever could have taken P&G to court, where embarrassing details might have been revealed, causing more bad publicity for a company that is generally perceived as highly ethical. Unilever also stood to lose from a public trial. The trade secrets at the heart of the case may have been disclosed in depositions and other documents during a trial, which could have devalued proprietary data.[49]

Cybersquatting

A **trademark** is a logo, package design, phrase, sound, or word that enables a consumer to differentiate one company's products from another's. Consumers often cannot examine goods or services to determine their quality or source, so instead they rely on the labels attached to the products. Trademark law gives the trademark's owner the right to prevent others from using the same mark or a confusingly similar mark on a product's label.

The United States has a federal system that stores trademark information; merchants can consult this information to avoid adopting marks that have already been taken. Merchants seeking trademark protection apply to the USPTO if they are using the mark in interstate commerce or if they can demonstrate a true intent to do so. Trademarks can be renewed forever—as long as a mark is in use.

Companies that want to establish an online presence know that the best way to capitalize on the strengths of their brand names and trademarks is to make the names part of the domain names for their Web sites. When Web sites were first established, there was no procedure for validating the legitimacy of requests for Web site names, which were given out on a first-come, first-served basis. **Cybersquatters** registered domain names for famous trademarks or company names to which they had no connection, with the hope that the trademark's owner would eventually buy the domain name for a large sum of money.

The main tactic organizations use to circumvent cybersquatting is to protect a trademark by registering numerous domain names and variations as soon as the organization knows it wants to develop a Web presence (for example, UVXYZ.com, UVXYZ.org, and UVXYZ.info). In addition, trademark owners who rely on non-English-speaking customers should register their names in multilingual form. Registering additional domain names is far less expensive than attempting to force cybersquatters to change or abandon their domain names.

Other tactics can also help curb cybersquatting. For example, the Internet Corporation for Assigned Names and Numbers (ICANN) is a nonprofit corporation responsible for managing the Internet's domain name system. ICANN is in the process of adding seven new top-level domains (.aero, .biz, .coop, .info, .museum, .name, and .pro) to the system. Current trademark holders will be given time to assert rights to their trademarks in the new top-level domains before registrations are opened up to the general public. ICANN also has a Uniform Domain Name Dispute Resolution Policy, under which most types of trademark-based domain name disputes must be resolved by agreement, court action, or arbitration before a registrar will cancel, suspend, or transfer a domain name. The ICANN policy is designed to provide for the fast, relatively inexpensive arbitration of a trademark owner's complaint that a domain name was registered or used in bad faith.

The Anticybersquatting Consumer Protection Act (ACPA), enacted in 1999, allows trademark owners to challenge foreign cybersquatters who might otherwise be beyond the jurisdiction of U.S. courts. Also under this act, trademark holders can seek civil damages of up to $100,000 from cybersquatters that register their trade names or similar-sounding names as domain names. It also helps trademark owners challenge the registration of their trademark as a domain name even if the trademark owner has not created an actual Web site.

OnlineNIC was one of the very first domain registrars licensed by ICANN. During 2008, Verizon Communications, Microsoft, and Yahoo! each filed separate lawsuits against OnlineNIC because that firm registered hundreds of domain names identical or similar to their trademark names (e.g., *verizon-cellular.com*, *encarta.com*, and *yahoozone.com*). In December 2008, Verizon was awarded damages of $31.15 million. OnlineNIC was prohibited from registering any additional names containing Verizon trademarks, and it was ordered to transfer the disputed domain names to Verizon.[50]

Summary

- *Intellectual property* is a term used to describe works of the mind—such as art, books, films, formulas, inventions, music, and processes—that are distinct and owned or created by a single person or group.

- Copyrights, patents, trademarks, and trade secrets provide a complex body of law relating to the ownership of intellectual property, which represents a large and valuable asset to most companies. If these assets are not protected, other companies can copy or steal them, resulting in significant loss of revenue and competitive advantage.

- A copyright is the exclusive right to distribute, display, perform, or reproduce an original work in copies; prepare derivative works based on the work; and grant these exclusive rights to others.

- Copyright law has proven to be extremely flexible in covering new technologies, including software, video games, multimedia works, and Web pages. However, evaluating the originality of a work can be difficult and has led to litigation.

- Copyrights provide less protection for software than patents; software that produces the same result in a slightly different way may not infringe a copyright if no copying occurred.

- The fair use doctrine establishes four factors for courts to consider when deciding whether a particular use of copyrighted property is fair and can be allowed without penalty.

- The Digital Millennium Copyright Act makes it illegal to circumvent a technical protection or develop and provide tools that allow others to access a technologically protected work.

- The Prioritizing Resources and Organization for Intellectual Property (PRO-IP) Act of 2008 increased trademark and copyright enforcement; it also substantially increased penalties for infringement.

- The original General Agreement on Tariffs and Trade (GATT) was signed in 1947 by 150 countries. It created the World Trade Organization (WTO) in Geneva, Switzerland, to enforce compliance with the agreement. GATT includes a section covering copyrights called the Agreement on Trade-Related Aspects of Intellectual Property Rights (TRIPS).

- The WTO deals with rules of international trade based on WTO agreements that are negotiated and signed by members of the world's trading nations. Its goal is to help producers of goods and services, exporters, and importers conduct their business.

- The World Intellectual Property Organization (WIPO), headquartered in Geneva, Switzerland, is an agency of the United Nations established in 1967. It is dedicated to developing "a balanced and accessible international intellectual property (IP) system, which rewards creativity, stimulates innovation and contributes to economic development while safeguarding the public interest."

- A patent enables an inventor to sue people who manufacture, use, or sell the invention without permission while the patent is in force. A patent prevents copying as well as independent creation (which is allowable under copyright law).

- Unlike copyright infringement, for which monetary penalties are limited, if the court determines that a patent has been intentionally infringed it can award up to triple the amount of the damages claimed by the patent holder.

- To qualify as a trade secret, information must have economic value and must not be readily ascertainable. In addition, the trade secret's owner must have taken steps to maintain its secrecy. Trade secret laws do not prevent someone from using the same idea if it was developed independently or from analyzing an end product to figure out the trade secret behind it.

- Trade secret law has three key advantages over the use of patents and copyrights in protecting companies from losing control of their intellectual property: (1) there are no time limitations on the protection of trade secrets, unlike patents and copyrights; (2) there is no need to file any application or otherwise disclose a trade secret to outsiders to gain protection; and (3) there is no risk that a trade secret might be found invalid in court.

- To plagiarize is to steal someone's ideas or words and pass them off as one's own. Plagiarism detection systems enable people to check the originality of documents and manuscripts.

- Reverse engineering is the process of breaking something down in order to understand it, build a copy of it, or improve it. Reverse engineering was originally applied to computer hardware, but is now commonly applied to software.

- In some situations, reverse engineering might be considered unethical because it enables access to information that another organization may have copyrighted or classified as a trade secret.

- Recent court rulings and software license agreements that forbid reverse engineering, as well as restrictions in the DMCA, have made reverse engineering a riskier proposition in the United States.

- Open source code refers to any program whose source code is made available for use or modification, as users or other developers see fit. The basic premise behind open source code is that when many programmers can read, redistribute, and modify it, the software improves. Open source code can be adapted to meet new needs, and bugs can be rapidly identified and fixed.

- Competitive intelligence is legally obtained information that is gathered to help a company gain an advantage over its rivals. Competitive intelligence is not the same as industrial espionage, which employs illegal means to obtain business information that is not readily available to the general public. In the United States, industrial espionage is a serious crime that carries heavy penalties.

- Competitive intelligence analysts must take care to avoid unethical or illegal behavior, including lying, misrepresentation, theft, bribery, or eavesdropping with illegal devices.

- Cybersquatters register domain names for famous trademarks or company names to which they have no connection, with the hope that the trademark's owner will eventually buy the domain name for a large sum of money.

- The main tactic organizations use to circumvent cybersquatting is to protect a trademark by registering numerous domain names and variations as soon as they know they want to develop a Web presence.

Self-Assessment Questions

The answers to the Self-Assessment Questions can be found in Appendix G.

1. Which of the following is an example of intellectual property?
   a. a business process for the efficient handling of medical insurance claims
   b. a computer program
   c. the design for a computer wafer
   d. all of the above

2. The term of a copyright, originally 28 years, has been extended many times and now can be as long as the life of the author plus 70 years. True or False?

3. The _____ established four factors for courts to consider when deciding whether a particular use of copyrighted property is fair and can be allowed without penalty.

4. The _____ was signed into law in 1998 and was written in compliance with the global copyright protection treaty from WIPO; the law makes it illegal to circumvent a technical protection of copyrighted materials or to develop and provide tools that allow others to access a technologically protected work.

5. Not only does a _____ prevent copying, but it also prevents independent creation, unlike a copyright.

6. A _____ is a logo, package design, phrase, sound, or word that enables a consumer to differentiate one company's products from another's.

7. Many large software companies have cross-licensing agreements in which each agrees not to sue the other over _____.

8. A _____ is a form of protection for intellectual property that does not require any disclosures or filing of an application.
   a. copyright
   b. patent
   c. trade secret
   d. trademark

9. A customer list can be considered a trade secret if an organization treats the information as valuable and takes measures to safeguard it. True or False?

10. _____ established minimum levels of protection that each government must provide to the intellectual property of all WTO members.

11. Plagiarism is an issue only in academia. True or False?

12. The process of taking something apart in order to understand it, build a copy of it, or improve it is called _____.

13. Which of the following statements is true about open source code?:

    a. There is only one definition of open source code.

    b. Open source code advocates believe that the quality of open source code is on par with that of commercial software.

    c. A very limited amount of open source code is in use today.

    d. Putting source code into the public domain with no copyright is equivalent to creating open source code.

14. Almost all the data needed for competitive intelligence can be collected either from carefully examining published information or through interviews. True or False?

15. The main tactic used to circumvent cybersquatting is to register numerous domain name variations as soon as an organization thinks it might want to develop a Web presence. True or False?

Discussion Questions

1. Explain the concept that an idea cannot be copyrighted, but the expression of an idea can be, and why this distinction is a key to understanding copyright protection.

2. Briefly describe the provisions of the Digital Millennium Copyright Act. Explain why many consumers view the DMCA as overly protective.

3. What is a cross-licensing agreement? How do large software companies use them? Do you think their use is fair to small software development firms? Why or why not?

4. What is defensive publishing, and why might a company practice this tactic?

5. Discuss the ways in which a software manufacturer can protect the unauthorized use of its software. Which do you think is the best way for a software manufacturer to protect new software? Why?

6. Discuss the conditions under which a company's customer list can be considered a trade secret.

7. Identify and briefly discuss three key advantages that trade secret law has over the use of patents and copyrights in protecting intellectual property. Are there any drawbacks with the use of trade secrets to protect intellectual property?

8. What problems can arise in using nondisclosure and noncompete agreements to protect intellectual property?

9. What is the WTO, and what is the scope and intent of the TRIPS Agreement?

10. Outline an approach that a university might take to successfully combat plagiarism among its students.

11. Does the DMCA prohibit the use of reverse engineering? Explain your answer.

12. Why might an organization elect to use open source code instead of proprietary software?

13. How might a corporation use reverse engineering to convert to a new database management system? How might it use reverse engineering to uncover the trade secrets behind a competitor's software?

14. What recent developments have challenged the reverse engineering of software in the United States? Is this development good or bad for the software industry? Why?

15. Compare the key issues in the *Sega v. Accolade* and *Lexmark v. Static Control Components* reverse engineering lawsuits.

What Would You Do?

1. You have procrastinated too long and now your final paper for your junior English course is due in just five days—right in the middle of final exam week! The paper counts for half your grade for the term and would probably take you at least 20 hours to research and write. Your roommate, an English major with a 3.8 GPA, has suggested two options: (1) he will write an original paper for you for $100, or (2) he will show you two or three "paper mill" Web sites, from which you can download a paper for less than $35. You want to do the right thing, but writing the paper will take away from the time you have available to study for your final exam in three other courses. What would you do?

2. Your friend has e-mailed you the URL of a new Web site that permits free music downloading. The site maintains a central database of thousands of the most current and popular music tracks. You vaguely remember that the Recording Industry Association of America (RIAA) issued thousands of subpoenas to Internet service providers across the United States in an attempt to get the names of people downloading large volumes of music illegally. But this is legal, isn't it? You are simply downloading from a central database; you're not sharing your music with others. Are there any legal or ethical issues that would keep you from using this Web site?

3. Because of the amount of the expense, your company's CFO had to approve a $500,000 purchase order for hardware and software needed to upgrade the servers used to store data for the Product Development Department. Everyone in the department had expected an automatic approval, and they were disappointed when the purchase order request was turned down. Management said that the business benefits of the expenditure were not clear. Realizing that she needs to develop a more solid business case for the order, the vice president of product development has come to you for help. Can you help her identify arguments related to protecting intellectual property that might strengthen the business case for this expenditure?

4. You have been asked by the manager of software development to lead a small group of four or five software developers in an attempt to reengineer the latest release of the software by your leading competitor. The goal of the group is to identify features that could be implemented into the next few releases of your firm's software. You are told that the group would relocate from the United States to the island of Antigua, in the Caribbean Sea, to "reduce the risk of the group being distracted by the daily pressures associated with developing fixes and enhancements with the current software release." What questions would you have about this assignment? Would you consider taking this position?

5. You have been asked to lead your company's new competitive intelligence organization. What would you do to ensure that members of the new organization obey applicable laws and the company's own ethical policies?

6. You are the vice president for software development at a small, private firm. Sales of your firm's products have been strong, but you recently detected a patent infringement by one of your larger competitors. Your in-house legal staff has identified three options: (1) ignore the infringement out of fear that your larger competitor will file numerous countersuits; (2) threaten to file suit, but try to negotiate an out-of-court settlement for an amount of money that you feel your larger competitor would readily pay; or (3) point out the infringement and negotiate aggressively for a cross-licensing agreement with the competitor, which has numerous patents you had considered licensing. Which option would you pursue and why?

7. You are a human resources manager for a large software developer. You have the opportunity to hire three experienced members of a competitor's salesforce. Through informal means, you discover that all three signed nondisclosure agreements with their current employer. What issues might you face if the salespeople accept job offers from your firm? How could you minimize the potential problems? What additional steps would you take before authorizing your firm to extend a job offer?

8. You are beginning to feel very uncomfortable in your new position as a computer hardware salesperson for a firm that is the major competitor of your previous employer. Today, for the second time, someone has mentioned to you how valuable it would be to know what the marketing and new product development plans were of your ex-employer. You stated that you are unable to discuss such information under the nondisclosure contract signed with your former employer, but you know your response did not satisfy your new coworkers. You fear that the pressure to reveal information about the plans of your former company is only going to increase over the next few weeks. What do you do?

Cases

1. Intellectual Property and the War over Software Maintenance

In late 2006, Oracle Corporation noticed unusually heavy activity on its Customer Connection Web site—the password-protected site through which Oracle supplies customer support for its PeopleSoft and JD Edwards products. In November 2006, more than 10,000 copyrighted technical support items were downloaded by unauthorized users in what Oracle called "corporate theft on a grand scale."[51]

Oracle is one of the largest providers of business software in the world, with over 320,000 customers in over 145 countries using their applications. In 2005, the company purchased PeopleSoft (which had acquired JD Edwards & Company in 2003) after a federal judge ruled that the purchase would not violate antitrust laws. PeopleSoft offered top-notch human resource management, customer relations, finance, and enterprise performance systems. JD Edwards provided similar products, but for smaller companies that could not afford PeopleSoft. After the acquisition, Oracle announced that it planned to continue to support PeopleSoft and JD Edwards products and that it would eventually integrate them into Oracle's enterprise resource planning (ERP) product.

When Oracle made its acquisition, it gained not only the sales revenue generated by the PeopleSoft and JD Edwards products, but also the sizable maintenance contract revenue that came with those product lines. In return for a substantial annual fee, Oracle provides telephone and online support as well as access to instruction manuals, patches, code, and other documents through its Customer Connection Web site.

Oracle and other software vendors typically charge about 20 percent of the price of the original software package to maintain and provide support. So, if a package cost $1,000,000, a company would have to spend $200,000 each year to fix bugs that arose during and after installation or add enhancements released by the software developer. If the software provider continues to add value to the product each year, the client may not see these maintenance expenses as excessive. However, if the bugs are quickly ironed out and the software provider is doing little to maintain or improve the software package, a company might begin to look for other ways to obtain support for the product. Third-party vendors are one option, but the level of support they can provide depends on the level of access that the software provider is willing to offer these third parties. Not surprisingly, software providers are generally unwilling to cooperate with third parties that siphon off maintenance revenue. As a result, some business analysts argue that clients are held hostage by these software providers, forced into expensive contracts.[52]

By the fourth or fifth year of operation, a business may not need a great deal of support for its software systems. In this case, a third-party vendor that only charges 10 to 15 percent of the original cost of the software package may be a very good option. At least, that is what Honeywell, Lockheed Martin, Coors, and others thought when they switched their business to TomorrowNow Inc., a company formed by former PeopleSoft executives. TomorrowNow originally supported PeopleSoft products and then branched out to JD Edwards and other applications. At the time Oracle acquired PeopleSoft and the JD Edwards product line, TomorrowNow already had about 100 clients. As PeopleSoft clients became nervous about Oracle's plans for the software package, traffic on TomorrowNow's site jumped 300 percent, and TomorrowNow pulled some maintenance support clients away from Oracle.[53]

In 2005, just days after the acquisition, Oracle's chief competitor, SAP, bought TomorrowNow. Many analysts believe that SAP procured TomorrowNow to counter Oracle's acquisition of PeopleSoft. With this purchase, SAP could erode Oracle's maintenance contract revenue while encouraging PeopleSoft and JD Edwards customers to migrate to SAP's ERP system through its Safe Harbor migration program. The Safe Harbor initiative was aimed at clients who were already using a combination of software, such as PeopleSoft human resource and SAP financial systems. Clients would receive a 75 percent discount on SAP's ERP software, and TomorrowNow would continue to support PeopleSoft and JD Edwards products through the migration. Any clients who migrated to the SAP software would pay an annual maintenance fee based on the full value of the ERP system.[54] Although some analysts believed that TomorrowNow was too small to accommodate the potential client base that would migrate over, Oracle clearly saw this move as a threat.

Oracle filed a lawsuit on March 22, 2007, in U.S. district court alleging that SAP had violated the federal Computer Fraud and Abuse Act as well as California law through the actions of TomorrowNow. The complaint accused TomorrowNow of downloading thousands of documents from Oracle's Customer Connection Web site in late 2006 and early 2007. TomorrowNow purportedly downloaded copyrighted material containing software updates, patches, bug fixes, and instructions for PeopleSoft and JD Edwards products.

In late 2006, Oracle noticed that customers using expired and soon-to-be-expiring passwords were retrieving information and code from every library on the Web site with much greater frequency than usual. One of these "customers," who had previously averaged 20 downloads per month, averaged over 1,800 downloads per day over the course of four days. Although paying customers can access support materials for any PeopleSoft or JD Edwards product by entering a username and password, they have no legal rights to support documents for software products for which they have not purchased a maintenance contract. In this sweep of the Customer Connection Web site, many of these people downloaded resources unrelated to the systems they had paid Oracle to support. Oracle claims these "customers" had one thing in common: they were or would soon be clients of TomorrowNow. Moreover, Oracle claimed that the downloads all originated from an Internet protocol (IP address) in Bryan, Texas—the headquarters of TomorrowNow. Oracle believes that through its subsidiary, TomorrowNow, SAP downloaded an extensive collection of Oracle's copyrighted software and support materials onto SAP computers.[55]

Oracle Corporation claims that TomorrowNow was engaged in unethical business practices as far back as 2005. Oracle purports that TomorrowNow used "non-production" copies of People-Soft software to develop support solutions. Oracle alleges that SAP knew of these improprieties when it purchased TomorrowNow.[56] Oracle is seeking $1 billion in damages from SAP.[57]

SAP responded that TomorrowNow was acting on behalf of its clients who were authorized to make such downloads. SAP admitted, however, that in doing so, TomorrowNow also made some inappropriate downloads. In November 2007, SAP announced a change in management at TomorrowNow, and in July 2008, SAP decided to close the subsidiary. It helped TomorrowNow's customers transition back to Oracle or to third-party vendor Rimini Street for maintenance and support.[58]

Rimini Street is a third-party software maintenance company established by TomorrowNow cofounder Seth Ravin, a former PeopleSoft manager who left TomorrowNow three months after SAP acquired the company. In August 2007, when asked whether the Oracle lawsuit was scaring clients off third-party software support companies like Rimini Street, Ravin replied, "The lawsuit actually opened up business—because some people didn't even know there was a choice. A lot of people read the lawsuit and said, 'My God, Merck and Honeywell, and all these large companies are using third-party support.' And instead of saying, 'Wow, look what happened here!' The issue really is, 'Holy gee, am I the only guy paying full price?'"[59]

So, as two software giants wrestle, third-party vendors remain an option. The questions remain as to how and whether these businesses can open up the maintenance contract market to competition without violating federal law.

Discussion Questions

1. According to Oracle, how was TomorrowNow violating intellectual property law?
2. Why do you think Oracle sued SAP?
3. What do you think should be done, if anything, to open the maintenance contract market to third-party contractors?

2. Google Book Search Library Project

In 2005, Google announced the Google Book Search Library Project, a highly ambitious plan to scan and digitize books from various libraries, including the New York Public Library, as well as the libraries at Harvard University, Oxford University, Stanford University, and the University of Michigan. Google's goal is to "work with publishers and libraries to create a comprehensive, searchable, virtual card catalog of all books in all languages that helps users discover new books and publishers discover new readers."[60]

Because many of the books are protected under copyright law, Google needed a way to avoid problems with copyright infringement. Therefore, Google established a process requiring publishers and copyright holders to opt out of the program if they did not want their books to be searchable. Publishers and copyright holders were incensed and argued that they should control who can view and search their books. In October 2005, the Authors Guild and the Association of American Publishers (on behalf of McGraw-Hill, Simon & Schuster, John Wiley & Sons, Pearson Education, and the Penguin Group) filed suit against Google to stop the program. They argued that making a full copy of a copyright-protected book does not fit into the narrow exception to the law defined by fair use.

After more than two years of discussions, the parties negotiated a settlement in October 2008. The settlement did not resolve the legal dispute over whether Google's project is permissible as a fair use; however, it concluded the litigation, enabling the parties to avoid the cost and risk of a trial. The date of the hearing for the court to consider whether to grant final approval of the settlement is scheduled for October 2009 in the District Court for the Southern District of New York.[61]

The proposed settlement would give Google the right to display the books online and to profit from them by selling access to individual texts as well as selling subscriptions to its entire collection to libraries and other institutions. The resulting revenue would be shared among Google, authors, and publishers.[62] Readers would have several new ways to access books, both on a free and fee-based basis "via public and university libraries and through institutional subscriptions for academic, corporate, and government libraries and organizations."[63]

Google as well as many authors and publishers have defended the settlement, saying the project will benefit authors, publishers, and the public, and renew access to millions of out-of-print books.[64]

However, in a further complication, the Justice Department began an inquiry in April 2009 into the antitrust implications of the proposed settlement. Critics, including the Internet Archive and Consumer Watchdog, claim the settlement would unfairly give Google an exclusive license to profit from millions of books. In addition, "Google alone would have a license that covers millions of so-called orphan books, whose authors cannot be found or whose rights holders are unknown."[65]

Discussion Questions

1. Do you think that Google should have taken a different approach that would have allowed it to avoid litigation and a lengthy delay in implementing its Book Search Library Project? Please explain your answer.

2. As a potential user, are you in favor of or do you oppose the Book Search Library Project? Please explain your answer.

3. Do you think that the proposed settlement gives Google an unfair advantage to profit from creating an online service that allows people to access and search millions of books?

3. *Lotus v. Borland*

By the early 1990s, conflicting court decisions over the previous decade had confused software developers, especially regarding the "look and feel" of software. Must every software product have its own unique look and feel? Could a company improve its products by incorporating features that existed in competing products? Was it legal to build and market a clone—a product externally identical to another product? The *Lotus v. Borland* lawsuit and associated appeals lasted more than five years, and went all the way to the U.S. Supreme Court. The case set a precedent that clarified the limits of software copyright protection.

The first electronic spreadsheet on the market was Daniel Bricklin's VisiCalc, introduced in 1979. The program enabled users to easily prepare budgets, forecast profits, analyze investments, and summarize tax data. VisiCalc was a huge success, selling more than 100,000 copies in its first year on the market—at a time when only a few million people owned PCs. For the first time, users found an application compelling enough to make them want to buy a computer. VisiCalc and other early spreadsheet programs are credited with being catalysts for the PC revolution, but neither the software nor the concept of a spreadsheet program was patented or copyrighted.

Mitch Kapor and Jonathan Sachs founded Lotus Development Corporation in 1982. In 1983, Lotus released its flagship spreadsheet product, Lotus 1-2-3, which, like VisiCalc, enabled users to perform accounting functions on a computer. Lotus 1-2-3 was highly successful, generating sales of $53 million in its first year and $156 million the next year.[66]

Users manipulated and controlled the Lotus 1-2-3 program through a series of menu commands, such as Copy, File, Print, and Quit. Users could select these commands either by highlighting them on the screen or by typing their first letter. Lotus 1-2-3 also allowed users to write a macro—a sequence of commands activated by a single keystroke. The program could then recall and perform the designated commands automatically.

Borland Software Corporation was founded in 1983 with the objective of developing a program superior to Lotus 1-2-3. In 1987, Borland released its Quattro spreadsheet program to the public. Quattro gave users a choice regarding how they communicated with the spreadsheet program—by using menu commands designed by Borland, or by using the commands and command structure found in Lotus 1-2-3. The latter capability helped attract Lotus 1-2-3 users to the Borland software. To provide this dual functionality, Borland copied only the words and structure of Lotus's menu command hierarchy—not any of its computer code.

Lotus sued Borland in 1990, charging that the execution of commands in Quattro copied the look and feel of the Lotus 1-2-3 interface. In August 1992, a U.S. district court found that Borland included in its spreadsheet "a virtually identical copy of the entire Lotus 1-2-3 menu tree" and infringed the copyright of Lotus. The judge held that the menu command hierarchy—the selection of menu items and their arrangement in an inverted tree structure—was a protectable element of the program that Borland had infringed. The judge also ruled that Borland's "key reader"—a program feature that enabled Quattro to interpret and execute Lotus 1-2-3 macros—was infringing because it employed a table that reproduced the entire Lotus 1-2-3 menu command hierarchy, with the first letter of each command substituted for the full command name.[67]

The court subsequently entered an injunction against Borland that prohibited further sales or distribution of its spreadsheet products. In response, Borland appealed, and began shipping new versions of its spreadsheet that did not include the infringing features. The district court reaffirmed its decision in July 1993.

In March 1994, Borland decided to get out of the spreadsheet business, and sold its Quattro spreadsheet software to Novell Inc. One year later, the U.S. Court of Appeals for the First Circuit reversed the district court ruling that Borland had infringed the copyright of Lotus 1-2-3. The court found that the Lotus menu command hierarchy was a "method of operation" that was excluded from copyright protection because the command hierarchy "provides the means by which users control and operate Lotus 1-2-3." The ruling was welcomed by Borland, which would have found it extremely difficult to pay the estimated $100 million in damages sought by Lotus.[68]

The case was appealed to the U.S. Supreme Court, and in January 1996, five years after the suit began, the court affirmed the appeals court ruling for Borland. The case was significant for the software industry, which had been riddled with infringement lawsuits due to ambiguities in copyright law. The ruling made it clear that software copyrights could be successfully challenged, which further discouraged the use of copyright to protect software innovations. As a result, developers had to go through the more difficult and expensive patent process to protect their software products. These additional costs would be either absorbed by developers or passed on to users.[69]

Borland is still in business, providing products and services targeted to software developers, including a Borland version of the C++ programming language. Meanwhile, IBM purchased Lotus in 1995 for $3.5 billion. Lotus 1-2-3 is still widely used, although it is not as popular as the Microsoft Excel spreadsheet program.

Discussion Questions

1. Go to your school's computer lab or a PC software store and experiment with current versions of any two of the Quattro, Excel, or Lotus 1-2-3 spreadsheet programs. Write a brief paragraph summarizing the similarities and differences in the "look and feel" of these two programs.

2. The courts took several years to reverse their initial decision and rule in favor of Borland. What impact did this delay have on the software industry? How might things have been different if Borland had received an initial favorable ruling?

3. Assume that you are the manager of Borland's software development. With the benefit of hindsight, what different decisions would you have made about Quattro?

End Notes

[1] Stephen Shankland, "Sendmail May Turn Tools over to Open Source," *CNET*, April 6, 2006, http://news.cnet.com/Sendmail-may-turn-tools-over-to-open-source/2100-7344_3-6058499.html.

[2] RIAA, "Who We Are," www.riaa.com/aboutus.php.

[3] Sarah McBride, "Music File-Sharing Decision to Have Broad Impact," *Wall Street Journal*, August 15, 2008, http://online.wsj.com/article/SB121875652064642585.html.

[4] RIAA, "For Students Doing Reports," www.riaa.com/faq.php.

[5] RIAA, "Piracy: Online and on the Street," www.riaa.com/physicalpiracy.php.

[6] RIAA, "For Students Doing Reports," www.riaa.com/faq.php.

[7] Jaikumar Vijayan, "RIAA Shifts Gears on Music Piracy, Says It Won't File More Suits," *Computerworld*, December 19, 2008.

[8] U.S. Code, Title 17, § 102(a).

[9] United States Copyright Office, "Copyright Law of the United States of America and Related Laws Contained in Title 17 of the United States Code," www.copyright.gov/title17/92chap3.html.

[10] "Sonny Bono Copyright Term Extension Act," EconomicExpert.com, www.economicexpert.com/a/Sonny:Bono:Copyright:Term:Extension:Act.htm.

[11] UCLA Law and Columbia Law School Copyright Infringement Project, *Bright Tunes Music v. Harrisongs Music,* http://cip.law.ucla.edu/cases/case_brightharrisongs.html.

[12] U.S. Code, Title 17, § 107.

[13] James Niccolai, "Court Won't Block Low-Cost Cartridges," *PC World*, February 22, 2005, www.pcworld.com/article/119747/court_wont_block_lowcost_cartridges.html.

[14] K. C. Jones, "Groups Weigh in on Intellectual Property Protection Act," *InformationWeek*, October 14, 2008.

[15] "IP Enforcement Bill Becomes Law," *National Journal*, October 13, 2008, http://techdailydose.nationaljournal.com/2008/10/ip-enforcement-bill-becomes-la.php.

[16] Steven Schwankert, "Bush Enacts PRO-IP Anti-piracy Law," *PC World*, October 14, 2008.

[17] World Trade Organization, "What Is the WTO?," www.wto.org/english/thewto_e/whatis_e/whatis_e.htm.

[18] World Intellectual Property Organization, "What Is WIPO?," www.wipo.int/about-wipo/en/what_is_wipo.html.

[19] Dan Burger, "IBM Piles on the Patents, Promises to Publish Plenty," *IT Jungle*, January 19, 2009, www.itjungle.com/tfh/tfh011909-story03.html.

[20] Dan Burger, "IBM Piles on the Patents, Promises to Publish Plenty," *IT Jungle*, January 19, 2009, www.itjungle.com/tfh/tfh011909-story03.html.

[21] Joff Wild, "More on the IBM $1 Billion Patent Licensing Urban Legend," *Intellectual Asset Management*, March 27, 2008, www.iam-magazine.com/blog/Detail.aspx?g=9be3f156-79b1-49f4-abf1-9bee7e788501.

[22] "The Patent Filing Process," Essortment, www.essortment.com/all/patentfilingpr_rrgx.htm.

[23] U.S. Code, Title 35, § 103–4.

[24] *Diamond v. Diehr*, 450 U.S. 175 (1981), BitLaw, www.bitlaw.com/source/cases/patent/Diamond_v_Diehr.html.

[25] Viktoria Carrella, "Cygnus Sues Google, Apple, Microsoft; Recession the Problem?," *Associated Content,* December 31, 2008.

[26] Ina Fried, "Microsoft--License to Deal," *CNET*, November 8, 2004, http://news.cnet.com/Microsoft--license-to-deal/2100-1012_3-5440881.html.

[27] "Patent Litigation Costs: How Much Does It Cost to Protect a Patent?," www.inventionstatistics.com/Patent_Litigation_Costs.html.

[28] Dan Burger, "IBM Piles on the Patents, Promises to Publish Plenty," *IT Jungle*, January 19, 2009, www.itjungle.com/tfh/tfh011909-story03.html.

[29] Amol Sharma and Don Clark, "Tech Guru Riles the Industry by Seeking Huge Patent Fees," *Wall Street Journal*, September 17, 2008.

[30] Associated Press, "Federal Judge Upholds $520 Million Verdict Against Microsoft," *InformationWeek*, January 15, 2004, www.informationweek.com/news/windows/showArticle.jhtml?articleID=17301451.

[31] Ina Fried, "Microsoft Loses in Eolas Patent Ruling," *CNET*, September 29, 2005, http://news.cnet.com/Microsoft-loses-in-Eolas-patent-ruling/2100-1012_3-5885657.html.

[32] Dawn Kawamoto, "Rambus and Infineon Reach Settlement," *CNET*, March 21, 2005, http://news.cnet.com/Rambus-and-Infineon-reach-settlement/2100-1014_3-5628224.html.

[33] David Lawsky, "UPDATE 3-US Judge Rules Against Rambus in Micron Case," Reuters UK, January 10, 2009, http://uk.reuters.com/article/rbssTechMediaTelecomNews/idUKN0927671220090110.

[34] Dylan McGrath, "Rambus: Court Orders Hynix to Post Bond," *EE Times*, May 26, 2009.

[35] National Conference of Commissioners on Uniform State Laws, "Uniform Trade Secrets Act," http://nsi.org/Library/Espionage/usta.htm.

[36] Gerald J. Mossinghoff, J. Derek Mason, Ph.D., and David A. Oblon, "The Economic Espionage Act: A New Federal Regime of Trade Secret Protection," Oblon Spivak, www.oblon.com/media/index.php?id=59.

[37] Barbara Rose, "Non-Compete Clause Tying Hands of Employees," *Chicago Tribune*, February 25, 2008, http://archives.chicagotribune.com/2008/feb/25/business/chi-mon_space_0225feb25.

[38] Thomas Claburn, "Apple's Controversial iPhone Developer Agreement Published," *InformationWeek*, October 28, 2008, www.informationweek.com/news/personal_tech/iphone/showArticle.jhtml?articleID=211601121.

[39] Paul McDougall, "Papermaster Settlement: Apple Innovations May Require IBM's OK," *InformationWeek*, January 27, 2009, www.informationweek.com/news/showArticle.jhtml?articleID=212902972.

[40] Howard Kurtz, "'Katie's Notebook' Item Cribbed from *W.S. Journal*," *Washington Post*, April 11, 2007, www.washingtonpost.com/wp-dyn/content/article/2007/04/10/AR2007041001537.html.

[41] Jacques Steinberg, "Times's 2 Top Editors Resign after Furor on Writer's Fraud," *New York Times*, June 6, 2003, www.nytimes.com/2003/06/06/business/media/06PAPE.html.

[42] Pat Sierchio, "Feisty, Prolific SF Author Harlan Ellison Bares 'Sharp Teeth' in Bio-Pic," *Jewish Journal*, August 2, 2007, www.jewishjournal.com/arts/article/feisty_prolific_sf_author_harlan_ellison_bares_sharp_teeth_in_biopic.

[43] iThenticate, "About Our Company," www.ithenticate.com/about.html.

[44] Turnitin, "US Court of Appeals Unanimously Affirms Finding of 'Fair Use' for Turnitin," http://turnitin.com/static/media.html.

[45] *Sega Enterprises Ltd. v. Accolade, Inc.*, www.virtualrecordings.com/sega.htm.

[46] Ed Scannell, "1 in 3 IT Shops Uses Combo Proprietary, Open Source Software," *InformationWeek*, March 13, 2009, www.informationweek.com/news/software/open_source/showArticle.jhtml?articleID=215900159&subSection=.

[47] JMRI, "JMRI Defense: Our Story So Far," http://jmri.sourceforge.net/k/updates.html.

[48] Tom Smith, "5 Web Lessons from Google's Analytics Guru," *InformationWeek*, September 17, 2008, www.informationweek.com/blog/main/archives/2008/09/5_web_lessons_f.html;jsessionid=IIA3RJKUPVJEGQSNDLPSKH0CJUNN2JVN.

[49] "P&G, Unilever Reach Spying Settlement," *Business Courier of Cincinnati*, September 6, 2001.

[50] Peter Sayer, "Verizon Wins $31M Judgment in Cybersquatting Case," *Computerworld*, December 26, 2008, www.computerworld.com/action/article.do?command=viewArticleBasic&articleId=9124378.

[51] "Complaint for Damages and Injunctive Relief," United States District Court, Northern District of California, March 22, 2007, www.oracle.com/sapsuit/complaint.pdf.

[52] Linda Tucci, "Oracle/SAP Lawsuit Fuels Third-Party Maintenance Argument," *SearchCIO*, August 9, 2007, http://searchcio.techtarget.com.au/articles/21119-Oracle-SAP-lawsuit-fuels-third-party-maintenance-argument.

[53] Charles Babcock, "Support Comes from the Outside," *InformationWeek*, January 10, 2005, www.informationweek.com/news/global-cio/showArticle.jhtml?articleID=57300371.

[54] Renee Boucher Ferguson, "SAP Buy Targets PeopleSoft Migration," *eWEEK.com*, January 19, 2005, www.eweek.com/c/a/Enterprise-Applications/SAP-Buy-Targets-PeopleSoft-Migration.

233

55 "Complaint for Damages and Injunctive Relief," United States District Court, Northern District of California, March 27, 2007, www.oracle.com/sapsuit/complaint.pdf.

56 "Third Amended Complaint for Damages and Injunctive Relief," United States District Court, Northern District of California, October 8, 2008, www.oracle.com/sapsuit/third-amended-complaint.pdf.

57 Brandon Bailey, "Oracle Expands Theft Allegations Against SAP," *Trade Secrets Vault,* July 30, 2008, www.tradesecretsblog.info/2008/07/oracle_expands_theft_allegatio.html.

58 Rick Whiting, "SAP Shuttering Its TomorrowNow Services Subsidiary," *ChannelWeb*, July 21, 2008, www.crn.com/software/209102245.

59 Linda Tucci, "Interview: Seth Ravin on Third-Party Software Maintenance," *SearchCIO*, August 28, 2007, http://searchcio.techtarget.com/news/interview/0,289202,sid182_gci1269359,00.html.

60 Google Book Search, "Google Books Library Project," http://books.google.com/googlebooks/library.html.

61 Association of Research Libraries, "A Guide for the Perplexed: Libraries and the Google Library Project Settlement," November 13, 2008, http://www.arl.org/bm~doc/google-settlement-13nov08.pdf.

62 Miguel Helft, "Justice Dept. Opens Antitrust Inquiry into Google Books Deal," *New York Times*, April 28, 2009.

63 Association of Research Libraries, "Google Book Search Library Project," November 13, 2008, www.arl.org/pp/ppcopyright/google/index.shtml.

64 Miguel Helft, "Justice Dept. Opens Antitrust Inquiry into Google Books Deal," *New York Times*, April 28, 2009.

65 Miguel Helft, "Justice Dept. Opens Antitrust Inquiry into Google Books Deal," *New York Times*, April 28, 2009.

66 D. J. Power, "A Brief History of Spreadsheets," DSSResources.com, August 30, 2004, http://dssresources.com/history/sshistory.html.

67 William Brandel, "'Look and Feel' Reversal Re-ignites Copyright Fight," *Computerworld*, March 13, 1995.

68 Martin A. Goetz, "Copycats or Criminals?," *Computerworld*, June 12, 1995.

69 The Free Library, "Borland Prevails in Lotus Copyright Suit; Five-Year Old Lawsuit Comes to an End," January 16, 1996, www.thefreelibrary.com/Borland+Prevails+In+Lotus+Copyright+Suit%3B+Five-Year+Old+Lawsuit+Comes...-a017793645.

SOFTWARE DEVELOPMENT

VIGNETTE

Boeing Dreamliner Faces a Few Bumps in the Road

The Boeing Company is the largest manufacturer of commercial jetliners and military aircraft combined.

The company also designs and manufactures defense systems, missiles, satellites, launch vehicles,

and advanced information and communication systems for its customers in more than 90 countries

around the world. Boeing's annual revenue for 2008 was $61 billion.[1]

Boeing's newest commercial airplane is the Boeing 787 Dreamliner—a fast, fuel-efficient midsized

airplane, designed to carry anywhere from 210 to 330 passengers (depending on the model and its

seating configuration). Some models of the Dreamliner will be capable of flying over 9,000 miles without

refueling. As of May 2009, Boeing had orders for 861 Dreamliners valued at $144 billion, making this the

most successful sales launch of a new commercial airplane in Boeing's history. The first flight of seven

test models was planned for June 2009, with delivery of the first Dreamliner scheduled for the first quar-

ter of 2010. Unfortunately, the test flight has been delayed due to a design problem. All has not gone

smoothly with the development of this new aircraft.[2]

The first of the test 787s was originally scheduled to fly in the summer of 2007, but the flight was rescheduled due to major production delays caused by part shortages and by incomplete work on some of the aircraft components supplied by contractors.[3] Then, on September 6, 2008, Boeing assembly-line workers went on strike for 52 days—halting assembly of the test aircraft.

Additional delays resulted from problems with the software that controls and monitors the braking system for the aircraft, which can tip the scales at 540,000 pounds at takeoff. The Boeing 787 will be the first commercial airliner to use electric brakes instead of conventional, hydraulically actuated brakes. The electric brakes are expected to have superior self-checking capabilities and will, therefore, be much more reliable.[4] In the summer of 2008, however, portions of the braking software needed to be rewritten.

"It's not that the brakes don't work, it's to do with the traceability of the software," said Pat Shanahan, Boeing vice president and 787 general manager, back in July 2008. Tracing lines of code back to specific system requirements is a critical part of code verification. Because of design concerns about the brakes that arose during early testing phases, it became necessary for GE Aviation and its subcontractor, Crane Aerospace, to rewrite portions of the software as well as redo verification of the software. There were many delays getting the software verified to meet stringent certification require-ments. Said GE Aviation vice president Peter Woolfrey, "We've had issues to deal with, both thermal issues and brake control, as well as a lot of new technology. It takes a lot to bring [everything] together and to expect it to get it right 100% first time around. But there's nothing fundamentally wrong, and there's nothing that can't be resolved."[5]

In February 2009, Crane Aerospace stated that the original version of the brake control system had been delivered to Boeing and was ready to fly on the upcoming test flights of the Boeing 787.[6] Crane

CEO Eric Fast maintained, however, that Boeing had changed the requirements of the brake control system, resulting in the need to develop a new version of the system. Fast also stated that Crane Company would not undertake additional software development work without new funding from Boeing, and that the cost of the additional work could be substantial.[7]

Boeing's Dreamliner customers are major airline carriers, many of which built pricing and routing strategies based on the use of the new aircraft to replace older aircraft or to increase the size of their current fleets. Boeing's ongoing production problems have delayed the ramp-up of production so that deliveries are stretching years beyond what was anticipated by some carriers.[8] For example, Jetstar and its parent company, Qantas, ordered a total of 65 planes, to be delivered once a month for 65 months starting in August 2008. That schedule has now slipped 21 months, with delivery scheduled to begin in May 2010.[9] These delays have forced Boeing customers to rethink their future pricing and routing strategies. Also, at an average cost of over $150 million each, the slowdown in customer delivery has put pressure on Boeing's cash flow. Furthermore, shrewd customers have a built-in payment-penalty clause for late delivery, which may cost Boeing an additional hundreds of millions of dollars.

Boeing suppliers have also been affected by the missed schedule. Having implemented multiyear financing plans to fund their participation in the Dreamliner program, many are now facing cash flow problems due to the production delays.

Questions to Consider

1. Is this example of software problems holding up the introduction of a major new product and impacting a firm's customers and suppliers unusual, or is it a common occurrence?
2. What can organizations do to reduce the negative consequences of software development problems in the production of their products and the operation of their business processes and facilities?

LEARNING OBJECTIVES

As you read this chapter, consider the following questions:

1. Why do companies require high-quality software in business systems, industrial process control systems, and consumer products?

2. What potential ethical issues do software manufacturers face in making trade-offs between project schedules, project costs, and software quality?

3. What are the four most common types of software product liability claims?

4. What are the essential components of a software development methodology, and what are the benefits of using such a methodology?

5. How can the Capability Maturity Model Integration improve an organization's software development process?

6. What is a safety-critical system, and what special actions are required during its development?

STRATEGIES FOR ENGINEERING QUALITY SOFTWARE

High-quality software systems are easy to learn and use because they perform quickly and efficiently; they meet their users' needs; and they operate safely and reliably so that system downtime is kept to a minimum. Such software has long been required to support the fields of air traffic control, nuclear power, automobile safety, health care, military and defense, and space exploration. Now that computers and software have become integral parts of almost every business, the demand for high-quality software is increasing. End users cannot afford system crashes, lost work, or lower productivity, nor can they tolerate security holes through which intruders can spread viruses, steal data, or shut down Web sites. Software manufacturers face economic, ethical, and organizational challenges associated with improving the quality of their software. This chapter covers many of these issues.

A **software defect** is any error that, if not removed, could cause a software system to fail to meet its users' needs. The impact of these defects can be trivial; for example, a computerized sensor in a refrigerator's ice cube maker might fail to recognize that the tray is full and continue to make ice. Other defects could lead to tragedy—the control system for an automobile's antilock brakes could malfunction and send the car into an uncontrollable spin. The defect might be subtle and undetectable, such as a tax preparation package that makes a minor miscalculation; or the defect might be glaringly obvious, such as a payroll program that generates checks with no deductions for Social Security or other taxes.

Here are some of the most infamous software bugs to appear over the last decade:

- In April 2003, Sallie Mae—the largest U.S. student loan company—notified over 800,000 borrowers that a software installation error had resulted in a miscalculation of the monthly payments on their loans (ultimately, over 1 million loans were affected by the error). Borrowers were notified of an increase in their required payments of anywhere from \$40 to \$100.[10]

- The Los Angeles air traffic control system shut down for part of a day in September 2004, leaving over 800 flights in midair with no contact to air traffic control, because of a software glitch in the control system.[11]
- In December 2005, a critical software bug was detected that affected 39 different Symantec software products—including both home and enterprise versions of its antivirus software. The bug could have been exploited to load unauthorized code onto a computer and to potentially take control of the computer.[12]
- Windows Vista—the operating system from Microsoft—became the subject of much criticism for issues relating to performance, security, and privacy. Indeed, the initial version of the operating system was so unpopular that in mid-2006, Microsoft began allowing PC manufacturers to offer a "downgrade" option to buyers who purchased machines with Vista but who wanted to switch to Windows XP.[13]
- Skype—a company that offers a service to enable its 220 million subscribers to make phone calls over the Internet—suffered a massive outage in August 2007. According to a Skype spokesperson, a "previously unseen software bug within the network resource algorithm" caused the problem.[14]
- Software problems in the automated baggage sorting system at Heathrow airport caused the system to go offline for almost two days in February 2008. As a result, carriers in the terminal were forced to sort baggage manually, and 6,000 passengers experienced delays, flight cancellations, and frustration. The breakdown reportedly occurred during a software upgrade, despite pretesting of the software. The system continued to experience problems in subsequent months.[15]

Software quality is the degree to which a software product meets the needs of its users. **Quality management** focuses on defining, measuring, and refining the quality of the development process and the products developed during its various stages. These products—including statements of requirements, flowcharts, and user documentation—are known as **deliverables**. The objective of quality management is to help developers deliver high-quality systems that meet the needs of their users.

Unfortunately, the first release of any software rarely meets all its users' expectations. A software product does not usually work as well as its users would like it to until it has been used for a while, found lacking in some ways, and then corrected or upgraded.

A primary cause of poor software quality is that many developers do not know how to design quality into software from the very start; some simply do not take the time to do so. In order to develop high-quality software, developers must define and follow a set of rigorous engineering principles and be committed to learning from past mistakes. In addition, they must understand the environment in which their systems will operate and design systems that are as immune to human error as possible.

All software designers and programmers make mistakes in defining user requirements and turning them into lines of code. According to one study, even experienced software developers unknowingly inject an average of one design or implementation defect for every 7 to 10 lines of code.[16] The developers aren't incompetent or lazy—they're just human. Everyone makes mistakes, but in software, these mistakes can result in defects.

The Microsoft Vista operating system took over five years to develop and contains more than 50 million lines of code.[17] Assume that the Microsoft software developers produced code at the accuracy rate mentioned above. Even if 99.9 percent of the defects were identified and fixed before the product was released to the public, there would still be about one bug per 10,000 lines of code, or roughly 5,000 bugs in Windows Vista. Thus, software that is used daily by workers worldwide likely contains thousands of bugs.

Another factor that can contribute to poor-quality software is the extreme pressure that software companies feel to reduce the time to market for their products. They are driven by the need to beat the competition in delivering new functionality to users, to begin generating revenue to recover the cost of development, and to show a profit for shareholders. They are also driven by the need to meet quarterly earnings forecasts used by financial analysts to place a value on the stock. The resources and time needed to ensure quality are often cut under the intense pressure to ship a new product. When forced to choose between adding more user features and doing more testing, most software companies decide in favor of more features. After all, they reason, defects can always be patched in the next release, which will give customers an automatic incentive to upgrade. Additional features make a release more useful and therefore easier to sell to customers.

Customers are stakeholders who are key to a software's success, and they may benefit from new features. However, they also bear the burden of errors that aren't caught or fixed during testing. Thus, many customers challenge whether the decision to cut quality in favor of feature enhancement is ethical.

As a result of the lack of consistent quality in software, many organizations avoid buying the first release of a major software product or prohibit its use in critical systems, their rationale being that the first release often has many defects that cause problems for users. Because of the defects in the first two popular Microsoft operating systems (DOS and Windows), including their tendency to crash unexpectedly, many believe that Microsoft did not have a reasonably reliable operating system until its third major variation— Windows NT.

Even software products that have been reliable over a long period can falter unexpectedly when operating conditions change. For instance, the software in the Cincinnati Bell telephone switch had been thoroughly tested and had operated successfully for months after it was deployed. Later that year, however, when the time changed from daylight saving time to standard time, the switch failed because it was overwhelmed by the number of calls to the local "official time" phone number from people who wanted to set their clocks. The large increase in the number of simultaneous calls to the same number was a change in operating conditions that no one had anticipated.

The Importance of Software Quality

A **business information system** is a set of interrelated components—including hardware, software, databases, networks, people, and procedures—that collects and processes data and disseminates the output. A common type of business system is one that captures and records business transactions. For example, a manufacturer's order-processing system captures order information, processes it to update inventory and accounts receivable, and ensures that the order is filled and shipped on time to the customer. Other examples include an airline's online ticket-reservation system and an electronic-funds transfer system that moves money among banks. The accurate, thorough, and timely processing of business transactions is a key requirement for such systems. A software defect can be devastating, resulting in lost customers and reduced revenue. How many times would bank customers tolerate having their funds transferred to the wrong account before they stopped doing business with that bank?

Another type of business information system is the **decision support system (DSS)**, which is used to improve decision making in a variety of industries. A DSS can help develop accurate forecasts of customer demand, recommend stocks and bonds for an investment portfolio, or schedule shift workers in such a way as to minimize cost while meeting customer service goals. A software defect in a DSS can result in significant negative consequences for an organization and its customers.

Software is also used to control many industrial processes in an effort to reduce costs, eliminate human error, improve quality, and shorten the time it takes to manufacture products. For example, steel manufacturers use process control software to capture data from sensors about the equipment that rolls steel into bars and about the furnace that heats the steel before it is rolled. Without process control computers, workers could react to defects only after the fact and would have to guess at the adjustments needed to correct the process. Process control computers enable the process to be monitored for variations from operating standards (e.g., a low furnace temperature or incorrect levels of iron ore) and to eliminate product defects *before* they affect product quality. Any defect in this software can lead to decreased product quality, increased waste and costs, or even unsafe operating conditions for employees. (Some consequences of these defects are discussed later in this chapter.)

Software is also used to control the operation of many industrial and consumer products, such as automobiles, medical diagnostic and treatment equipment, televisions, radios, stereos, refrigerators, and washers. A software defect could have relatively minor consequences, such as clothes not drying long enough, or it could cause serious damage, such as a patient being overexposed to powerful X-rays.

As a result of the increasing use of computers and software in business, many companies are now in the software business whether they like it or not. The quality of software, its usability, and its timely development are critical to almost everything businesses do. The speed with which an organization develops software can put it ahead of or behind its competitors. Software problems may have caused frustrations in the past, but mismanaged software can now be fatal to a business, causing it to miss product delivery dates, incur increased product development costs, and deliver products that have poor quality.

Business executives frequently face ethical questions of how much effort and money they should invest to ensure high-quality software. A manager who takes a short-term, profit-oriented view may feel that any additional time and money spent on quality assurance will only delay a new product's release, sales revenue, and profits. However, a different manager may consider it unethical not to fix all known problems before putting a product on the market and charging customers for it.

Other key questions for executives are whether their products could cause damage and what their legal exposure would be if they did. Fortunately, software defects are rarely lethal, and few personal injuries are related to software failures. However, the use of software introduces product liability issues that concern many executives.

Software Product Liability

Software product litigation is certainly not new. One lawsuit in the early 1990s involved a financial institution that became insolvent because defects in a purchased software application caused errors in its integrated general ledger system, customers' passbooks, and loan statements. Dissatisfied depositors responded by withdrawing more than $5 million. In another case from 1992, a Ford truck stalled because of a software defect in the truck's fuel injector. In the ensuing accident, a young child was killed.[18] A state supreme court later affirmed an award of $7.5 million in punitive damages against the manufacturer. In October 2008, a faulty onboard computer caused a Qantas passenger flight en route to Perth from Singapore to plunge some 8,000 feet in 10 seconds, injuring 46 passengers. Australian Transport Safety Bureau officials are still investigating the accident, but lawsuits are expected to be filed.[19]

The liability of manufacturers, sellers, lessors, and others for injuries caused by defective products is commonly referred to as **product liability**. There is no federal product liability law; instead, product liability is mainly covered by common law (made by state judges) and Article 2 of the Uniform Commercial Code, which deals with the sale of goods.

If a software defect causes injury or loss to purchasers, lessees, or users of the product, the injured parties may be able to sue as a result. Injury or loss can come in the form of physical mishaps and death, loss of revenue, or an increase in expenses due to a business disruption caused by a software failure. Software product liability claims are typically based on strict liability, negligence, breach of warranty, or misrepresentation—sometimes in combination with one another. Each of these legal concepts is discussed in the following sections.

Strict liability means that the defendant is held responsible for injuring another person, regardless of negligence or intent. The plaintiff must prove only that the software product is defective or unreasonably dangerous and that the defect caused the injury. There is no requirement to prove that the manufacturer was careless or negligent, or to prove who caused the defect. All parties in the chain of distribution—the manufacturer, subcontractors, and distributors—are strictly liable for injuries caused by the product and may be sued.

Defendants in a strict liability action may use several legal defenses, including the doctrine of supervening event, the government contractor defense, and an expired statute of limitations. Under the doctrine of supervening event, the original seller is not liable if the software was materially altered after it left the seller's possession and the alteration caused the injury. To establish the government contractor defense, a contractor must prove that the precise software specifications were provided by the government, that the software conformed to the specifications, and that the contractor warned the government of any known defects in the software. Finally, there are statutes of limitations for claims of liability, which means that an injured party must file suit within a certain amount of time after the injury occurs.

Negligence is the failure to do what a reasonable person would do, or doing something that a reasonable person would not do. When sued for negligence, a software supplier is not held responsible for every product defect that causes customer or third-party loss. Instead, responsibility is limited to harmful defects that could have been detected and corrected through "reasonable" software development practices. Even when a contract is written expressly to protect against supplier negligence, courts may disregard such terms as unreasonable. Software manufacturers or organizations with software-intensive products are frequently sued for negligence and must be prepared to defend themselves.

The defendant in a negligence case may either answer the charge with a legal justification for the alleged misconduct or demonstrate that the plaintiffs' own actions contributed to their injuries (**contributory negligence**). If proved, the defense of contributory negligence can reduce or totally eliminate the amount of damages the plaintiffs receive. For example, if a person uses a pair of pruning shears to trim his fingernails and ends up cutting off a fingertip, the defendant could claim contributory negligence.

A **warranty** assures buyers or lessees that a product meets certain standards of quality. A warranty of quality may be either expressly stated or implied by law. Express warranties can be oral, written, or inferred from the seller's conduct. For example, sales contracts contain an implied warranty of merchantability, which requires that the following standards be met:

- The goods must be fit for the ordinary purpose for which they are used.
- The goods must be adequately contained, packaged, and labeled.
- The goods must be of an even kind, quality, and quantity within each unit.
- The goods must conform to any promise or affirmation of fact made on the container or label.
- The quality of the goods must pass without objection in the trade.
- The goods must meet a fair average or middle range of quality.

If the product fails to meet the terms of its warranty, the buyer or lessee can sue for **breach of warranty**. Of course, most dissatisfied customers will first seek a replacement, a substitute product, or a refund before filing a lawsuit.

Software suppliers frequently write warranties to attempt to limit their liability in the event of nonperformance. Although a certain software may be warranted to run on a given machine configuration, often no assurance is given as to what that software will do. Even if a contract specifically excludes the commitment of merchantability and fitness for a specific use, the court may find such a disclaimer clause unreasonable and refuse to enforce it or refuse to enforce the entire contract. In determining whether warranty disclaimers are unreasonable, the court attempts to evaluate if the contract was made between two "equals" or between an expert and a novice. The relative education, experience, and bargaining power of the parties and whether the sales contract was offered on a take-it-or-leave-it basis are considered in making this determination.

The plaintiff must have a valid contract that the supplier did not fulfill in order to win a breach-of-warranty claim. Because the software supplier writes the warranty, this claim can be extremely difficult to prove. For example, in 1993, M. A. Mortenson Company—one of the largest construction companies in the United States—had a new version of bid-preparation software installed for use by its estimators. During the course of preparing one new bid, the software allegedly malfunctioned several times, each time displaying the same cryptic error message. Nevertheless, the estimator submitted the bid and Mortenson won the contract. Afterward, Mortenson discovered that the bid was $1.95 million lower than intended and filed a breach-of-warranty suit against Timberline Software, makers of the bid software. Timberline acknowledged the existence of the bug. However, the courts ruled in Timberline's favor because the license agreement that came with the software explicitly barred recovery of the losses claimed by Mortenson.[20] Even if breach of warranty can be proven, the damages are generally limited to the amount of money paid for the product.

Intentional misrepresentation occurs when a seller or lessor either misrepresents the quality of a product or conceals a defect in it. For example, if a cleaning product is advertised as safe to use in confined areas and some users subsequently pass out from the product's fumes, they could sue the seller for intentional misrepresentation or fraud.

Advertising, salespersons' comments, invoices, and shipping labels are all forms of representation. Most software manufacturers use limited warranties and disclaimers to avoid any claim of misrepresentation.

Software Development Process

Developing information system software is not a simple process; it requires completing many complex activities, with many dependencies among the various activities. System analysts, programmers, architects, database specialists, project managers, documentation specialists, trainers, and testers are all involved in large software projects. Each of these groups has a role to play, and has specific responsibilities and tasks. In addition, each group makes decisions that can affect the software's quality and the ability of an organization or an individual to use it effectively.

Many software companies have adopted a **software development methodology**—a standard, proven work process that enables systems analysts, programmers, project managers, and others to make controlled and orderly progress in developing high-quality software. A methodology defines activities in the software development process, and the individual and group responsibilities for accomplishing these activities. It also recommends specific techniques for accomplishing the various activities, such as using a flowchart to document the logic of a computer program. A methodology offers guidelines for managing the quality of software during the various stages of development. If an organization has developed such a methodology, it is applied to any software development that the company undertakes.

As with most things, it is usually easier and cheaper to avoid software problems at the beginning than to attempt to fix the damages after the fact. Studies have shown that the cost to identify and remove a defect in an early stage of software development can be up to 100 times less than removing a defect in a piece of software that has been distributed to customers.[21] If a defect is uncovered during a later stage of development, some rework of the deliverables produced in preceding stages will be necessary. The later the error is detected, the greater the number of people who will be affected by the error. Thus, the greater the costs will be to communicate and fix the error. Consider the cost to communicate the details of a defect, distribute and apply software fixes, and possibly retrain end users for a software product that has been sold to hundreds or thousands of customers. Thus, most software developers try to identify and remove errors early in the development process not only as a cost-saving measure but as the most efficient way to improve software quality.

A product containing inherent defects that harm the user may be the subject of a product liability suit. The use of an effective methodology can protect software manufacturers from legal liability in two ways. First, an effective methodology reduces the number of software errors that might occur. Second, if an organization follows widely accepted development methods, negligence on its part is harder to prove. However, even a *successful* defense against a product liability case can cost hundreds of thousands of dollars in legal fees. Thus, failure to develop software carefully and consistently can be serious in terms of liability exposure.

Quality assurance (QA) refers to methods within the development cycle designed to guarantee reliable operation of the product. Ideally, these methods are applied at each stage of the development cycle. However, some software manufacturing organizations without a formal, standard approach to QA consider testing to be their only QA method. Instead of checking for errors throughout the development process, they rely primarily on testing just before the product ships to ensure some degree of quality.

Several types of tests are used in software development, as discussed in the following sections.

Dynamic Software Testing

Software is developed in units called subroutines or programs. These units, in turn, are combined to form large systems. One approach to QA is to test the code for a completed unit of software by actually entering test data and comparing the results to the expected results. This is called **dynamic testing**. There are two forms of dynamic testing:

- **Black-box testing** involves viewing the software unit as a device that has expected input and output behaviors but whose internal workings are unknown (a black box). If the unit demonstrates the expected behaviors for all the input data in the test suite, it passes the test. Black-box testing takes place without the tester having any knowledge of the structure or nature of the actual code. For this reason, it is often done by someone other than the person who wrote the code.

- **White-box testing** treats the software unit as a device that has expected input and output behaviors but whose internal workings, unlike the unit in black-box testing, are known. White-box testing involves testing all possible logic paths through the software unit with thorough knowledge of its logic. The test data must be carefully constructed so that each program statement executes at least once. For example, if a developer creates a program to calculate an employee's gross pay, the tester would develop data to test cases in which the employee worked less than 40 hours, exactly 40 hours, and more than 40 hours (to check the calculation of overtime pay).

Other Types of Software Testing

Other forms of testing include the following:

- **Static testing**—Special software programs called static analyzers are run against the new code. Rather than reviewing input and output, the static analyzer looks for suspicious patterns in programs that might indicate a defect.

- **Integration testing**—After successful unit testing, the software units are combined into an integrated subsystem that undergoes rigorous testing to ensure that the linkages among the various subsystems work successfully.

- **System testing**—After successful integration testing, the various subsystems are combined to test the entire system as a complete entity.

- **User acceptance testing**—Independent testing performed by trained end users to ensure that the system operates as they expect.

Capability Maturity Model Integration

Capability Maturity Model Integration (CMMI)—developed by the Software Engineering Institute at Carnegie Mellon—is a process improvement approach that defines the essential elements of effective processes. The model is general enough to be used to evaluate and improve almost any process, and a specific application of CMMI—**CMMI-Development** (CMMI-DEV)—is frequently used to assess and improve software development practices. CMMI defines five levels of software development maturity (see Table 7-1) and identifies the issues that are most critical to software quality and process improvement. Identifying an organization's current maturity level enables it to specify necessary actions to improve the organization's future performance. The model also enables an organization to track, evaluate, and demonstrate its progress over the years.

TABLE 7-1   CMMI

Maturity level	Definition	Percent of all reporting organizations at this level
Not reported		8.4%
Initial	Process unpredictable, poorly controlled, and reactive	1.9%
Managed	Process characterized for projects and often reactive	33.3%
Defined	Process characterized for the organization and proactive	33.8%
Quantitatively managed	Process measured and controlled	4.4%
Optimizing	Focus on continuous process improvement	18.2%

Source: Carnegie Mellon University, "Capability Maturity Model® Integration (CMMI®) Version 1.2 Overview," Carnegie Mellon University, www.sei.cmu.edu/cmmi/adoption/pdf/cmmi-overview07.pdf.

CMMI-Development is a set of guidelines for 22 process areas related to systems development and maintenance. The premise of the model is that those organizations that do these 22 things well will have an outstanding software development and maintenance process. (A more in-depth discussion of CMMI-Development can be found at *www.sei.cmu.edu/cmmi/models/CMMI-DEV-v1.2.doc*.) The Software Engineering Institute has documented the following results from CMMI-DEV implementations:

- A 33 percent decrease in the cost to fix defects
- A reduction in the number of defects found, from 6.6 to 2.1 per thousand lines of code
- A 30 percent increase in productivity
- An increase in project-schedule milestones met, from 50 percent to 95 percent[22]

After an organization decides to adopt CMMI-DEV, it must conduct an assessment of its software development practices (using outside resources to ensure objectivity) and determine where they fit in the capability model. The assessment identifies areas for improvement and establishes action plans needed to upgrade the development process. Over the course of a few years, the organization can improve its maturity level by executing the action plan.

As the maturity level increases, the organization improves its ability to deliver good software on time and on budget. For example, Lockheed Martin Management and Data Systems converted to CMMI between 1996 and 2002, with outstanding improvements. Lockheed increased software productivity by 30 percent, reduced unit software costs by 20 percent, and cut the costs of finding and fixing software defects by 15 percent.[23] See Table 7-2 for a partial list of organizations that use CMMI.

TABLE 7-2 Partial list of organizations using CMMI

Accenture	Intel
Boeing	Lockheed Martin
DynCorp	Reuters
Federal Aviation Administration	Samsung
General Dynamics	Tata Consultancy
Honeywell	U.S. Army, Air Force, and Navy
IBM Global Services	Wipro

CMMI-DEV can also be used as a benchmark for comparing organizations. In the awarding of software contracts—particularly by the federal government—organizations that bid on a contract may be required to have adopted CMMI and to be performing at a certain level.

KEY ISSUES IN SOFTWARE DEVELOPMENT

Although defects in any system can cause serious problems, the consequences of software defects in certain systems can be deadly. In these kinds of systems, the stakes involved in creating quality software are raised to the highest possible level. The ethical decisions involving a trade-off—if one must be considered—between quality and such factors as cost, ease of use, and time to market require extremely serious examination. The next sections discuss safety-critical systems and the special precautions companies must take in developing them.

Development of Safety-Critical Systems

A **safety-critical system** is one whose failure may cause injury or death. The safe operation of many safety-critical systems relies on the flawless performance of software; such systems control automobiles' antilock brakes, nuclear power plant reactors, airplane navigation, roller coasters, elevators, and numerous medical devices, to name just a few. The process of building software for such systems requires highly trained professionals, formal and rigorous methods, and state-of-the-art tools. Failure to take strong measures to identify and remove software errors from safety-critical systems "is at best unprofessional and [can] at worst lead to disastrous consequences."[24] However, even with these precautions, the software associated with safety-critical systems is still vulnerable to errors that can lead to injury or death. Here are several examples of safety-critical system failures:

- The Mariner I space probe, which was intended to make a close flyby of the planet Venus, was ordered destroyed less than five minutes after launch in July 1962. Faulty software code caused the flight control computer to perform a series of unnecessary course correction signals, which threw the spacecraft dangerously off course.[25]
- Software that operated the Therac-25 radiation device malfunctioned and delivered lethal radiation doses on six occasions and caused serious patient injuries in several other instances in the late 1980s.[26]
- A Royal Air Force helicopter took off from Northern Ireland in June 1994 with 25 British intelligence officials who were heading to a security conference in Inverness. Just 18 minutes into its flight, the helicopter crashed on the peninsula of Kintyre in Argyll, Scotland, killing everyone aboard. The engine management software, which controlled the acceleration and deceleration of the engines, was suspected of causing the crash.[27]
- Between November 2000 and March 2002, therapy planning software at the National Oncology Institute in Panama City, Panama, miscalculated the proper dosage of radiation for patients undergoing therapy; at least eight patients died while another 20 received overdoses that caused significant health problems. Sadly, the developers of this software had not learned from the lessons of the Therac-25 tragedy.
- Fire broke out on a Washington, D.C., six-car Metro train as it pulled out of the L'Enfant Plaza station in April 2007. Fire and smoke were seen underneath the last car, but thankfully, the flames did not penetrate the floor of the car. The train operator stopped and evacuated the passengers. It was determined that the train's brake resistor grid, which checks various subsystems and voltages, overheated and caught fire. Monitoring software failed to perform as expected in detecting and preventing excess power usage in equipment on the passenger rail cars, resulting in overheating and fire.[28]

When developing safety-critical systems, a key assumption must be that safety will *not* automatically result from following an organization's standard development methodology. Safety-critical software must go through a much more rigorous and time-consuming development process than other kinds of software. All tasks—including requirement definition, systems analysis, design, coding, fault analysis, testing, implementation, and change control—require additional steps, more thorough documentation, and vigilant checking and rechecking. As a result, safety-critical software takes much longer to complete and is much more expensive to develop.

The key to ensuring that these additional tasks are completed is to appoint a **project safety engineer**, who has explicit responsibility for the system's safety. The safety engineer uses a logging and monitoring system to track hazards from a project's start to its finish. The hazard log is used at each stage of the software development process to assess how it has accounted for detected hazards. Safety reviews are held throughout the development process, and a robust configuration management system tracks all safety-related documentation to keep it consistent with the associated technical documentation. Informal documentation is not acceptable for safety-critical system development; formal documentation is required, including verification reviews and signatures.

The increased time and expense of completing safety-critical software can draw developers into ethical dilemmas. For example, the use of hardware mechanisms to back up or verify critical software functions can help ensure safe operation and make the consequences of software defects less critical. However, such hardware may make the final product more expensive to manufacture or harder for the user to operate—potentially making the product less attractive than a competitor's. Companies must weigh these issues carefully to develop the safest possible product that also appeals to customers.

Another key issue is deciding when the QA staff has performed sufficient testing. How much testing is enough when you are building a product whose failure could cause loss of human life? At some point, software developers must determine that they have completed sufficient QA activities and then sign off to indicate their approval. Determining how much testing is sufficient demands careful decision making.

When designing, building, and operating a safety-critical system, a great deal of effort must be put into considering what can go wrong, the likelihood and consequences of such occurrences, and how risks can be averted or mitigated. One approach to answering these questions is to conduct a formal risk analysis. **Risk** is the probability of an undesirable event occurring times the magnitude of the event's consequences if it does happen. These consequences include damage to property, loss of money, injury to people, and death. For example, if an undesirable event has a 1 percent probability of occurring and its consequences would cost $1,000,000, then the risk can be calculated as $0.01 \times \$1,000,000$, or $10,000. The risk for this event would be considered greater than that of an event with a 10 percent probability of occurring, at a cost of $100 ($0.10 \times \$100 = \$10$). Risk analysis is important for safety-critical systems but is useful for other kinds of software development as well.

Another key element of safety-critical systems is **redundancy**, the provision of multiple interchangeable components to perform a single function in order to cope with failures and errors. An example of a simple redundant system would be an automobile with a spare tire or a parachute with a backup chute attached. A more complex system used in IT is a redundant array of independent disks (RAID), which is commonly used in high-volume data storage for file servers. RAID systems use many small-capacity disk drives to store large amounts of data to provide increased reliability and redundancy. Should one of the drives fail, it can be removed and a new one inserted in its place. Since the data has also been stored elsewhere, data on the failed disk can be rebuilt automatically without the server ever having to be shut down.

N-version programming is a form of redundancy involving the simultaneous execution of a series of program instructions by two different systems. The two systems use different algorithms to execute instructions that accomplish the same result. The results from the two systems are then compared; if a difference is found, another algorithm is executed to determine which system yielded the correct result. In some cases, instructions for the two systems are written by programmers from two different companies and run on different hardware devices. The rationale behind N-version programming is that both systems are highly unlikely to fail at the same time under the same conditions. Thus, one of the two systems should yield a correct result. IBM employs N-version programming to reduce disk sector failures in data storage devices. Two pieces of code in the same application save a piece of data and then compare the data to ensure that no errors occurred.

During times of widespread disaster, lack of redundant systems can lead to major problems. For example, when Hurricane Katrina knocked out 2.5 million telephone lines, four TV stations, and 36 radio stations, there were inadequate backup communication systems to replace those failed systems.

After an organization determines all pertinent risks to a system, it must decide what level of risk is acceptable. This decision is extremely difficult and controversial because it involves forming personal judgments about the value of human life, assessing potential liability in case of an accident, evaluating the surrounding natural environment, and estimating the system's costs and benefits. System modifications must be made if the level of risk in the design is judged to be too great. Modifications can include adding redundant components or using safety shutdown systems, containment vessels, protective walls, or escape systems. Another approach is to mitigate the consequences of failure by devising emergency procedures and evacuation plans. In all cases, organizations must ask how safe is safe enough if human life is at stake.

Manufacturers of safety-critical systems must sometimes decide whether to recall a product when data indicates a problem. For example, automobile manufacturers have been known to weigh the cost of potential lawsuits against that of a recall. Drivers and passengers in affected automobiles (and, in many cases, the courts) have not found this approach to be ethically sound. Manufacturers of medical equipment and airplanes have had to make similar decisions, which can be complicated if data cannot pinpoint the cause of the problem. For example, there was great controversy in 2000 over the use of Firestone tires on Ford Explorers after numerous tire blowouts and Explorer rollovers caused multiple injuries and deaths. However, it was difficult to determine if the rollovers were caused by poor automobile design, faulty tires, or improperly inflated tires. Consumers' confidence in both manufacturers and their products was nevertheless shaken.

Reliability is the probability of a component or system performing without failure over its product life. For example, if a component has a reliability of 99.9 percent, it has one chance in one thousand of failing over its lifetime. Although this chance of failure may seem low, remember that most systems are made up of many components. As you add more components, the system becomes more complex, and the chance of failure increases. For example, assume that you are building a complex system made up of seven components, each with 99 percent reliability. If none of the components has redundancy built in, the system has a 93.8 percent ($.99^7$) probability of operating successfully with no component malfunctions over its lifetime. If you build the same type of system using 10 components, each with 99 percent reliability, the overall probability of operating without an individual component failure falls to 90 percent. Thus, building redundancy into systems that are both complex and safety critical is imperative.

One of the most important and difficult areas of safety-critical system design is the human interface. Human behavior is not nearly as predictable as the reliability of hardware and software components in a complex system. The system designer must consider what human operators might do to make a system work less safely or effectively. The challenge is to design a system that works as it should and leaves little room for erroneous judgment on the part of the operator. For instance, a self-medicating pain-relief system must allow a patient to press a button to receive more pain reliever, but must also regulate itself to prevent an overdose. Additional risk can be introduced if a designer does not anticipate the information an operator needs and how the operator will react under the daily pressures of actual operation, especially in a crisis. Some people keep their wits about them and perform admirably in an emergency, but others may panic and make a bad situation worse.

Poor design of a system interface can greatly increase risk, sometimes with tragic consequences. For example, in July 1988, the guided missile cruiser USS *Vincennes* mistook an Iranian Air commercial flight for an enemy F-14 jet fighter and shot the airliner down over international waters in the Persian Gulf—killing almost 300 people. Some investigators blamed the tragedy on the confusing interface of the $500 million Aegis radar and weapons control system. The Aegis radar on the *Vincennes* locked onto an Airbus 300, but it was misidentified as a much smaller F-14 by its human operators. The Aegis operators also misinterpreted the system signals and thought that the target was descending, even though the airbus was actually climbing. A third human error was made in determining the target altitude—it was off by 4,000 feet. As a result of this combination of human errors, the *Vincennes* crew thought the ship was under attack and shot down the plane.[29]

Quality Management Standards

The International Organization for Standardization (ISO), founded in 1947, is a worldwide federation of national standards bodies from 161 countries. The ISO issued its 9000 series of business management standards in 1988. These standards require organizations to develop formal quality-management systems that focus on identifying and meeting the needs, desires, and expectations of their customers.

The **ISO 9000 standard** serves as a guide to quality products, services, and management. Approximately 350,000 organizations in more than 150 countries have ISO 9000 certification. Although companies can use the standard as a management guide for their own purposes in achieving effective control, the priority for many companies is having a qualified external agency certify that they have achieved ISO 9000 certification. Many businesses and government agencies specify that a vendor must be ISO 9000 certified to win a contract from them.

To obtain this coveted certificate, an organization must submit to an examination by an external assessor and fulfill the following requirements:

- Have written procedures for all processes
- Follow those procedures
- Prove to an auditor that it has fulfilled the first two requirements; this proof can require observation of actual work practices and interviews with customers, suppliers, and employees.

The various ISO 9000 series of standards address the following software-related activities:

- ISO 9001: Design, development, production, installation, servicing
- ISO 9002: Production, installation, servicing
- ISO 9003: Inspection and testing
- ISO 9000-3: Development, supply, and maintenance of software
- ISO 9004: Quality management and quality systems elements

Failure mode and effects analysis (FMEA) is an important technique used to develop ISO 9000–compliant quality systems by both evaluating reliability and determining the effects of system and equipment failures. Failures are classified according to their impact on a project's success, personnel safety, equipment safety, customer satisfaction, and customer safety. The goal of FMEA is to identify potential design and process failures early in a project, when they are relatively easy and inexpensive to correct. A failure mode describes how a product or process could fail to perform the desired functions described by the customer. An effect is an adverse consequence that the customer might experience. Unfortunately, most systems are so complex that there is seldom a one-to-one relationship between cause and effect. Instead, a single cause may have multiple effects, and a combination of causes may lead to one effect or multiple effects.

LifeScan is part of Johnson & Johnson, and for over 20 years it has developed products for people with diabetes. Every day, more than 3 million people depend on its OneTouch Systems to capture accurate test results of their blood glucose levels.[30] LifeScan uses FMEA methods to test the software for the automated test and assembly stations on the blood glucose meter manufacturing lines.[31]

Table 7-3 provides a useful checklist for an organization interested in upgrading the quality of the software it produces. The preferred response to each question is *yes*.

TABLE 7-3   Manager's checklist for improving software quality

Question	Yes	No
Has senior management made a commitment to develop quality software?		
Have you used CMMI to evaluate your organization's software development process?		
Has your company adopted a standard software development methodology?		
Does the methodology place a heavy emphasis on quality management, and address how to define, measure, and refine the quality of the software development process and its products?		
Are software project managers and team members trained in the use of this methodology?		
Are software project managers and team members held accountable for following this methodology?		
Is a strong effort made to identify and remove errors as early as possible in the software development process?		
Are both static and dynamic software testing methods used?		
Are white-box testing and black-box testing methods used?		
Has an honest assessment been made to determine if the software being developed is safety critical?		
If the software is safety critical, are additional tools and methods employed, and do they include the following: a project safety engineer, hazard logs, safety reviews, formal configuration management systems, rigorous documentation, risk analysis processes, and the FMEA technique?		

Summary

- High-quality software systems are easy to learn and use. They perform quickly and efficiently to meet their users' needs, operate safely and reliably, and have a high degree of availability that keeps unexpected downtime to a minimum.

- High-quality software has long been required to support the fields of air traffic control, nuclear power, automobile safety, health care, military and defense, and space exploration.

- Now that computers and software have become integral parts of almost every business, the demand for high-quality software is increasing. End users cannot afford system crashes, lost work, or lower productivity, nor can they tolerate security holes through which intruders can spread viruses, steal data, or shut down Web sites.

- Software developers are under extreme pressure to reduce the time to market of their products. They are driven by the need to beat the competition in delivering new functionality to users, to begin generating revenue to recover the cost of development, and to show a profit for shareholders.

- The resources and time needed to ensure quality are often cut under the intense pressure to ship the new product. When forced to choose between adding more user features and doing more testing, many software companies decide in favor of more features. After all, they reason, defects can always be patched in the next release, which will give customers an automatic incentive to upgrade. Additional features make the release more useful and therefore easier to sell to customers.

- Software product liability claims are typically based on strict liability, negligence, breach of warranty, or misrepresentation—sometimes in combination.

- A software development methodology defines the activities in the software development process, defines individual and group responsibilities for accomplishing objectives, recommends specific techniques for accomplishing the objectives, and offers guidelines for managing the quality of the products during the various stages of the development cycle.

- Using an effective development methodology enables a manufacturer to produce high-quality software, forecast project-completion milestones, and reduce the overall cost to develop and support software. It also helps protect software manufacturers from legal liability for defective software in two ways: it reduces the number of software errors that could cause damage, and it makes negligence more difficult to prove.

- CMMI defines five levels of software development maturity, and identifies the issues that are most critical to software quality and process improvement. Its use can improve an organization's ability to predict and control quality, schedule, costs, cycle time, and productivity when acquiring, building, or enhancing software systems.

- CMMI also helps software engineers analyze, predict, and control selected properties of software systems.

- A safety-critical system is one whose failure may cause injury or death. In the development of safety-critical systems, a key assumption is that safety will *not* automatically result from following an organization's standard software development methodology.

- Safety-critical software must go through a much more rigorous and time-consuming development and testing process than other kinds of software; the appointment of a project safety engineer and the use of a hazard log and risk analysis are common.

- The International Organization for Standardization (ISO) issued its 9000 series of business management standards in 1988. These standards require organizations to develop formal quality management systems that focus on identifying and meeting the needs, desires, and expectations of their customers.

- The ISO 9000 standard serves as a guide to quality products, services, and management. Approximately 350,000 organizations in more than 150 countries have ISO 9000 certification. Many businesses and government agencies specify that a vendor must be ISO 9000 certified to win a contract from them.

- Failure mode and effects analysis (FMEA) is an important technique used to develop ISO 9000–compliant quality systems. FMEA is used to evaluate reliability and determine the effects of system and equipment failures.

Self-Assessment Questions

The answers to the Self-Assessment Questions can be found in Appendix G.

1. The impact of a software defect can be quite subtle or very serious. True or False?

2. _____ is any error that, if not removed, could cause a software system to fail to meet its users' needs.

3. Which of the following is *not* a major cause of poor software quality?

   a. Many developers do not know how to design quality into software or do not take the time to do it.

   b. Programmers make mistakes in turning design specifications into lines of code.

   c. Software developers are under extreme pressure to reduce the time to market of their products.

   d. Many organizations avoid buying the first release of a major software product.

4. A type of system used to control many industrial processes in an effort to reduce costs, eliminate human error, improve quality, and shorten the time it takes to make products is called _____.

5. There is a federal product liability law governing product liability. True or False?

6. A standard, proven work process for the development of high-quality software is called a(n) _____.

7. The cost to identify and remove a defect in an early stage of software development can be up to 100 times less than the cost of removing a defect in an operating piece of software after it has been distributed to many customers. True or False?

8. Methods within the development cycle designed to guarantee reliable operation of the product are known as _____.

9. A form of software testing that involves viewing a software unit as a device that has expected input and output behaviors but whose internal workings are unknown is known as:

    a. Dynamic testing

    b. White-box testing

    c. Integration testing

    d. Black-box testing

10. An approach that defines the essential elements of an effective process and outlines a system for continuously improving software development is:

    a. ISO 9000

    b. FMEA

    c. CMMI

    d. DO-178B

11. Special measures must be taken in the development of safety-critical systems. True or False?

12. The provision of multiple interchangeable components to perform a single function to cope with failures and errors is called _____.

    a. Risk

    b. Redundancy

    c. Reliability

    d. Availability

13. A reliability evaluation technique that can determine the effect of system and equipment failures is _____.

14. A set of standards requiring organizations to develop formal quality management systems that focus on identifying and meeting the needs, desires, and expectations of their customers are the _____ standards.

15. In a lawsuit alleging _____, responsibility is limited to harmful defects that could have been detected and corrected through "reasonable" software development practices.

Discussion Questions

1. Identify the three criteria you consider to be most important in a quality system. Briefly discuss your rationale for selecting these criteria.

2. Explain why the cost to identify and remove a defect in the early stages of software development might be 100 times less than the cost of removing a defect in software that has been distributed to hundreds of customers.

3. Identify and briefly discuss two ways that the use of an effective software development methodology can protect software manufacturers from legal liability for defective software.

4. Your company is considering using N-version programming with two software development firms and two hardware devices for the navigation system of a guided missile. Briefly describe what this means, and outline several advantages and disadvantages of this approach.

5. Why is the human interface one of the most important but difficult areas of safety-critical systems? What must the system designer consider?

6. Discuss the implications to a project team of classifying a piece of software as safety critical.

7. You have been asked to draft a boilerplate warranty for a software contractor that will absolutely protect the firm from being successfully sued for negligence or breach of contract. Is this possible? Why or why not?

8. Discuss why an organization might elect to use a separate, independent team for quality testing rather than the group of people who originally developed the software.

9. You are considering contracting for the development of new software that is essential to the success of your midsized manufacturing firm. One candidate firm boasts that its software development practices are at level 4 of CMMI. Another firm claims that all its software development practices are ISO 9000 compliant. How much weight should you give to these certifications when deciding which firm to use? Do you think that a firm could lie or exaggerate its level of compliance with these standards?

What Would You Do?

1. Read the fictitious Killer Robot case at the Web site for the Online Ethics Center for Engineering at *www.onlineethics.com/CMS/computers/compcases/killerrobot.aspx*. The case begins with the manslaughter indictment of a programmer for writing faulty code that resulted in the death of a robot operator. Slowly, over the course of many articles, you are introduced to several factors within the corporation that contributed to the accident. After reading the case, answer the following questions:

   a. Responsibility for an accident is rarely defined clearly and is often difficult to trace to one or two people or causes. In this fictitious case, it is clear that a large number of people share responsibility for the accident. Identify all the people you think were at least partially responsible for the death of Bart Matthews, and explain why you think so.

   b. Imagine that you are the leader of a task force assigned to correct the problems uncovered by this accident. Develop a list of the top 10 critical actions to take to avoid future problems. What process would you use to identify the most critical actions?

2. You used a tax preparation software package to prepare your federal tax return this year. Today you received a form letter from the software manufacturer informing all its customers that there was an error in the software that resulted in a substantial underestimation of the amount owed for both those who indicated that they were single and those who were married but filing separate tax returns. In the letter, the software manufacturer suggests that such users of its software promptly file an amended tax return before the IRS sends them a letter informing them that they discovered an error and requesting the payment of fines and interest. What do you do?

3. You are the project manager in charge of developing the latest release of your software firm's flagship product. The product release date is just two weeks away, and enthusiasm for the product is extremely high among your customers. Stock market analysts are forecasting

sales of more than $25 million per month. If so, earnings per share will increase by nearly 50 percent. There is just one problem: two key features promised to customers in this release have several bugs that would severely limit the software's usefulness. You estimate that at least six weeks are needed to find and fix the problems. In addition, even more time is required to find and fix 15 additional, less severe bugs just uncovered by the QA team. What do you recommend to management?

4. You developed a spreadsheet program that helps you perform your role of inventory control manager at a small retail bakery. The software uses historical sales data to calculate expected weekly sales for each of 250 baked goods carried by the store. Based on that forecast, you order the appropriate raw ingredients.

Your store is one of four bakeries in the city all owned by the same person. You sent a copy of the spreadsheet to each of the people responsible for inventory control at the other three retail stores, and they are all now using your software to help them do their jobs. You have started getting complaints that the software is not entirely accurate, and you notice that your own estimates are no longer as accurate as they used to be. What do you do?

5. You have been assigned to manage software that controls the shutdown of the new chemical reactors to be installed at a manufacturing plant. Your manager insists the software is not safety critical. The software senses temperatures and pressures within a 50,000-gallon stainless steel vat and dumps in chemical retardants to slow down the reaction if it gets out of control. In the worst possible scenario, failure to stop a runaway reaction would result in a large explosion that would send fragments of the vat flying and spray caustic liquid in all directions.

Your manager points out that the stainless steel vat is surrounded by two sets of protective concrete walls and that the reactor's human operators can intervene in case of a software failure. He feels that these measures would protect the plant employees and the surrounding neighborhood if the shutdown software failed. Besides, he argues, the plant is already more than a year behind its scheduled start-up date. He cannot afford the additional time required to develop the software if it is classified as safety critical. How would you work with your manager and other appropriate resources to decide whether the software is safety critical?

6. You are a senior software development consultant with a major consulting firm. You have been asked to conduct a follow-up assessment of the software development process for ABCXYZ Corporation, a company for which you had performed an initial assessment using CMMI two years prior. At the initial assessment, you determined the company's level of maturity to be level 1. Since that assessment, the organization has spent a lot of time and effort following your recommendations to raise its level of process maturity. The organization appointed a senior member of its IT staff to be a process management guru and paid him $150,000 per year to lead the improvement effort. This senior member adopted a methodology for standard software development and required all project managers to go through a one-week training course at a total cost of more than $2 million.

Unfortunately, these efforts did not significantly improve process maturity because senior management failed to hold project managers accountable for actually using the standard development methodology in their projects. Too many project managers convinced senior management that the new methodology was not necessary for their particular project and would just slow things down. You are concerned that when senior management learns that no real progress has been made, they will refuse to accept partial blame for the failure and instead drop all attempts at further improvement. You want senior management to ensure that the new methodology is used on all projects—with no exceptions. What would you do?

7. You are the CEO for a small, struggling software firm that produces educational software for high school students. Your latest software is designed to help students improve their SAT and ACT scores. To prove the value of your software, a group of 50 students who had taken the ACT test were retested after using your software for just two weeks. Unfortunately, there was no dramatic increase in their scores. A statistician you hired to ensure objectivity in measuring the results claimed that the variation in test scores was statistically insignificant. You had been counting on touting the results in the promotion of your new software.

A small core group of educators and systems analysts will need at least six months to start again from scratch to design a viable product. Programming and testing could take another six months. Another option would be to go ahead and release the current version of the product and then, when the new product is ready, announce it as a new release. This would generate the cash flow necessary to keep your company afloat and save the jobs of 10 or more of your 15 employees. Given this information about your company's product, what would you do?

Cases

1. Computer Problems at WellPoint Have Serious Impact

WellPoint, Inc., is the second largest health insurance company in the United States, with 2008 annual revenue of $61 billion.[32] The company was formed in December 2004 when Indianapolis-based Anthem, Inc., and California-based WellPoint Health Networks, Inc., merged. In 2005, WellPoint acquired New York–based WellChoice, Inc. WellPoint has more than 42,000 employees, and one out of every nine Americans is a member of a WellPoint-affiliated health plan.[33]

In an attempt to standardize and simplify processes, WellPoint decided to consolidate the claims processing computer systems for the three merged companies into a single system. Unfortunately, the consolidation turned out to be much more difficult than anticipated, and many problems arose—causing significant delays in claims processing.[34]

A claims processing backlog began building in 2007, and by the end of the year, several large-dollar claims had been submitted but not yet paid. Because these large claims were not visible in WellPoint's systems, the company's actuaries underestimated the firm's costs for the year. These artificially low costs were then used to price the firm's health insurance policies for 2008. When the pricing error was discovered, WellPoint was forced to reduce its 2008 profit forecast by 10 percent. First-quarter 2008 earnings came in below even the revised forecasts, and WellPoint was forced to drop its profit forecast by another 6 percent.[35] The impact on the stock price was severe; it plunged from the mid-$70 range in February 2008 to the mid-$40 range by March 2008. Numerous lawsuits from shareholders and participants in the company's 401(k) plan followed. But this was not the end of the problems for WellPoint.[36]

By October 2008, Anthem Blue Cross and Blue Shield (WellPoint's Indiana subsidiary) had a backlog of more than 350,000 claims, resulting in significant delays in payments to hospitals and physicians.[37] The St. Francis hospital system, which operates 10 hospitals in Indiana, sued Anthem because it claimed that the company had taken over 12 months to pay some claims and, in many cases, was not paying fair rates. This was only one of hundreds of complaints to come.

Eventually, Anthem was placed under investigation by the Indiana Department of Insurance after a huge spike in complaints against the company.[38]

More challenges arose for WellPoint in January 2009, when the Centers for Medicare and Medicaid Services (CMS)—the federal agency that administers Medicare, Medicaid, and the Children's Health Insurance Program—banned WellPoint from marketing or selling Medicare health or drug plans. WellPoint's data on who was enrolled and eligible for benefits was not linking correctly to its claims-paying systems. This caused problems for thousands of WellPoint members, who were unable to fill prescriptions for medications for a wide range of illnesses, including such life-threatening ailments as chronic heart failure, asthma, and seizures. WellPoint also erroneously dropped coverage for many beneficiaries while overcharging others. CMS finally tired of WellPoint's failure to follow through on assurances that the problem would be immediately and fully corrected.[39] Federal officials allege that WellPoint has also continued to:

- Charge beneficiaries incorrect amounts for premiums and coinsurance
- Administer the low-income subsidiary improperly
- Fail to coordinate benefits to beneficiaries
- Process appeals and grievances improperly
- Incorrectly reject claims based on medical necessity determinations
- Fail to provide required information to beneficiaries[40]

Discussion Questions

1. "The computer system did it" is an easy excuse for a multitude of problems. Clearly WellPoint's system consolidation effort went poorly, but is it fair to say that this incident raises other more fundamental areas of concern about the management of WellPoint?

2. What can be done to improve the quality of information technology projects at healthcare organizations? Should such system efforts be classified as safety-critical systems and be subjected to more rigorous implementation standards and processes? Should there be oversight of such efforts by either the government or some industry organization?

3. Implementation of electronic healthcare records is a national priority, and much emphasis is being placed on the importance of adopting new information technology. Meanwhile, the fundamental processing systems are falling into disrepair. How should healthcare organizations such as WellPoint prioritize their scarce resources?

2. Prius Plagued by Programming Error

In May 2005, the National Highway Traffic Safety Administration (NHTSA) revealed that it had opened an investigation into problems with the Toyota Prius hybrid, following several complaints of engines stalling. Motorists reported that while they were driving or stuck in traffic, warning lights flashed on and the gas engine suddenly switched off. In all cases, the motorists were able to maneuver to the side of the road safely because the electric motor, steering, and brake system continued to function.[41]

As one of the most fuel-efficient cars in the United States, the Prius delivers better mileage per gallon by shifting between its gas engine and its electric motor in different driving situations. For instance, when the car is stopped in traffic, the gas engine shuts off, and the car is powered by the electric motor. At high speeds, the gas engine takes over to deliver more power to the vehicle. An electronic control unit is designed to ensure a smooth transition between the gas engine and the electric motor. Toyota eventually determined that an error in the software used by the electronic control unit caused it to malfunction and the gas engine to stall.

Today, the average car is equipped with 30 to 40 microprocessors, so software quality is a critical issue for automakers. With 35 million lines of code per car, the potential for error is enormous. IBM automotive software specialist Stavros Stefanis reports that as many as one-third of all warranty claims result from software glitches and electronic defects. "It's a big headache for the automakers," said Stefanis.[42]

The potential for headaches is high because software impacts engine performance and controls many safety-related systems, such as steering, antilock brakes, and air bags. A programming error could have an enormous financial impact on an automaker in terms of major lawsuits and recalls.

When NHTSA opened its investigation, Toyota—a company with an established reputation for reliability—proved extremely cooperative. In fact, the company had already issued a service bulletin to owners of 2004 and 2005 Prius hybrids in October 2004. Toyota conducted an internal investigation, and in October 2005 it recalled 75,000 Prius hybrids sold in the United States; the company recalled an additional 85,000 units sold in Japan, Europe, and other markets.[43]

According to Toyota spokeswoman Allison Takahashi, NHTSA determined that passenger safety was not at issue: "We are voluntarily initiating a customer-service campaign to assure that this unusual occurrence does not cause inconvenience." Following the announcement of the recall campaign, NHTSA closed its investigation.[44]

The investigation, however, did not slow sales of the most popular hybrid on the market. Sales in 2005 were up 200 percent over the previous year. In fact, Toyota had to concentrate a good deal of its software development efforts on projects that would increase production to meet the demand for new hybrids in the United States.

Although the recall's effect on sales was probably negligible, the cost of the recall may be significant. Toyota had to pay for advertising and other costs involved in contacting Prius owners about the recall; the company also had to pay for the repairs. The recall serves as a warning to automakers about the costs of human errors that go undetected during software production.

Discussion Questions

1. Do you agree with NHTSA's assessment that the problem with the Prius was not a safety-critical issue? In such cases, who should decide whether a software bug creates a safety-critical issue—the manufacturer, consumers, government agencies, or some other group?

2. How would the issue have been handled differently if it had been a safety-critical matter? Would it have been handled differently if the costs involved hadn't been so great?

3. As the amount of hardware and software embedded in the average car continues to grow, what steps can automakers take to minimize warranty claims and ensure customer safety?

3. Patriot Missile Failure

The Patriot is an Army surface-to-air mobile missile system that defends against aircraft and cruise missiles and, more recently, short-range ballistic missiles. The system was designed in the 1960s, and in the late 1980s, the short-range antimissile capability was incorporated into the Patriot PAC-2 version of the missile.

Following the Iraqi invasion of Kuwait in August 1990, the United States deployed the Patriot PAC-2 missile to Saudi Arabia during Operation Desert Shield. At the start of Desert Shield, the

U.S. arsenal included only three PAC-2 missiles; however, PAC-2 production was accelerated so that by January 1991, 480 missiles were available. Patriot battalions were deployed to Saudi Arabia and then to Israel to defend key assets, military personnel, and citizens against Iraqi Scud missiles. During the conflict, Iraq launched 81 modified Scud ballistic missiles into Israel and Saudi Arabia.[45]

The Iraqis modified the Scud missile to increase its range and boost its speed by as much as 25 percent. They reduced the weight of the warhead, enlarged the fuel tanks, and modified its flight so that all of the fuel was burned during the early phase of flight rather than throughout the flight's duration. As a result of these modifications, the missiles became structurally unstable and often broke into pieces in the upper atmosphere. This instability made the warhead extremely difficult to intercept; also, radar sometimes mistakenly locked onto pieces of the fragmented missile rather than the actual warhead.

The Patriot missile is launched and guided to the target through three phases. First, the missile guidance system turns the Patriot launcher to face the incoming missile. Second, the computer control system guides the Patriot toward the incoming missile. Third, the Patriot missile's internal radar receiver guides it to intercept the incoming missile.[46]

The Patriot system's radar constantly swept the sky for any object that had the flight characteristics of a Scud missile. Following the detection of an airborne object, the system's range gate—an electronic detection device within the radar system—uses the last observation of the object to forecast an area in the air space where the radar system should next see the object if it truly is a Scud. This forecast is a function of both the object's observed velocity and the time of the last radar detection. If the range gate determines that the detected target was a Scud, and if the Scud was in the Patriot's firing range, the Patriot battery fires its missiles.

On February 11, 1991, the Patriot Project Office received data from Patriots deployed in Israel indicating a 20 percent shift in the system's radar range gate after the system had been running for eight consecutive hours. This shift was significant; it meant that the target was no longer in the center of the range gate, greatly reducing the probability of successfully tracking the target. The Army knew that the Patriot system could not track a Scud with a range gate shift of 50 percent or more.[47]

In the Patriot radar system, time is kept from the moment of system start-up and measured by the system's internal clock in tenths of a second. After looking at the Patriot data from Israel, the Army discovered that the longer the system ran, the less accurate the elapsed time calculation became. Consequently, after the Patriot computer control system ran continuously for extended periods, the range gate made an inaccurate estimate of the area in the air space where the radar system should next see the object. This error could cause the radar to lose the target and fool the system into thinking that there was no incoming Scud.[48]

Army officials assumed that other Patriot users (e.g., Saudi Arabia) were not running their systems for eight hours or more at a time, and that the Israeli experience was an anomaly. However, after analyzing the Israeli data, the Patriot Project Office confirmed some loss in targeting accuracy. As a result, they made a software change to compensate for the inaccurate time calculation. This change was included in a modified software version that was released for distribution to all Patriot missile users on February 16, 1991.[49]

The Patriot Project Office sent a message to all Patriot users on February 21, 1991, informing them that very long run times could cause a shift in the range gate, resulting in difficulty tracking the target. The message also advised users that a software change was on the way that would improve the system's targeting. However, the message did not specify what constituted "very long run times." Patriot Project Office officials assumed that users would not continuously run the batteries for so long that the Patriot would fail to track targets. Therefore, they did not think that more detailed guidance was required.[50]

On February 25, 1991, a Patriot missile defense system operating at Dhahran, Saudi Arabia, failed to track and intercept an incoming Scud. The enemy missile subsequently hit an Army barracks and killed 28 Americans. The ensuing investigation revealed that at the time of the incident, the Patriot battery had been operating continuously for more than 100 hours. The long run time resulted in an inaccurate time calculation, which in turn caused the range gate to shift so much that the system could not track, identify, or engage the incoming Scud.[51]

By cruel fate, modified software to fix the inaccurate time calculation arrived in Dhahran the very next day. Army officials attributed the delay to the difficulties of arranging air and ground transportation during wartime.[52]

The Army did not have the luxury of collecting definitive performance data in Saudi Arabia and Israel. After all, they were operating in a war zone, not a test range. As a result, there was insufficient and sometimes conflicting data on the effectiveness of the Patriot missile. At one extreme was an early report that claimed that the Patriot destroyed 96 percent of the Scuds engaged in Saudi Arabia and Israel. (Presumably, this did not include Scuds the Patriot failed to engage due to the software error.) At the other extreme, observers only reported seeing a Scud destroyed or disabled after a Patriot detonated nearby in about 9 percent of engagements. Of course, some "kills" could have been effected out of the observers' range of vision.[53]

Discussion Questions

1. With the benefit of hindsight, what steps could have been taken during development of the Patriot software to avoid the problems that led to the loss of life? Do you think these steps would have improved the Patriot's effectiveness enough to make it obvious that the missile was a strong deterrent against the Scud? Why or why not?

2. What ethical decisions do you think the U.S. military made in choosing to deploy the Patriot missile to Israel and Saudi Arabia and in reporting the effectiveness of the Patriot system?

3. What key lessons from this example of safety-critical software development could be applied to the development of business information system software?

End Notes

[1] Boeing, "About Us," www.boeing.com/companyoffices/aboutus.

[2] Wendy Kaufman, "Boeing's 787 Prepares for First Test Flight," NPR, May 16, 2009, www.npr.org/templates/story/story.php?storyId=104185867.

[3] "Brake Software Latest Threat to Boeing 787," Baseline, July 15, 2008, www.baselinemag.com/c/a/Application-Development/Brake-Software-Latest-Threat-to-Boeing-787.

[4] Geoffrey Thomas, "Boeing Puts the Brakes on 787 Deliveries," Australian, October 31, 2008, www.theaustralian.news.com.au/story/0,,24577253-23349,00.html?from=public_rss.

[5] Guy Norris, "Brake and Fuselage Completion Watch Items for 787," Aviation Week, July 16, 2008, www.aviationweek.com/aw/generic/story_generic.jsp?channel=businessweekly&id=news/787-0716.xml&headline=Brakes%20and%20fuselage%20completion%20watch%20items%20for%20787.

6 "Crane Says It Must Develop New 787 Brake System as Boeing Changes Requirements," *Seattle Times*, February 19, 2009, http://seattletimes.nwsource.com/html/businesstechnology/2008764380_webcrane19.html.

7 Jon Ostrower, "Crane Co. Reopens 787 Brake Software Questions," Flightglobal, February 18, 2009, www.flightglobal.com/blogs/flightblogger/2009/02/crane-co-reopens-787-brake-sof.html.

8 Geoffrey Thomas, "Boeing Puts the Brakes on 787 Deliveries," *Australian*, October 31, 2008, www.theaustralian.news.com.au/story/0,,24577253-23349,00.html?from=public_rss.

9 Geoffrey Thomas, "The Dream and the Nightmare," *Air Transport World*, December 2008, www.atwonline.com/channels/aircraftEquipment/article.html?articleID=2567.

10 Catherine Tumber, "Sallie Mae Not," *Boston Phoenix*, November 28 to December 4, 2003, www.bostonphoenix.com/boston/news_features/top/features/documents/03351782.asp.

11 Matthew Broersma, "Microsoft Server Crash Nearly Causes 800-Plane Pile-Up," *TechWorld*, September 21, 2004.

12 Will Knight, "Critical Bug Found in Anti-virus Software," *New Scientist*, December 22, 2005, www.newscientist.com/article/dn8505-critical-bug-found-in-antivirus-software.html.

13 Ina Fried, "The XP Alternative for Vista PCs," *CNET*, September 21, 2007, http://news.cnet.com/The-XP-alternative-for-Vista-PCs/2100-1016_3-6209481.html.

14 Wolfgang Gruener, "Software Bug Took Skype Out," *TG Daily*, August 20, 2007, www.tgdaily.com/content/view/33452/103.

15 Michael Krigsman, "System Update Kicks Heathrow Baggage System Offline," *ZDNet Business Network*, Michael Krigsman Blog, February 21, 2008.

16 Noopur Davis and Julia Mullaney, "The Team Software Process in Practice: A Summary of Recent Results," Technical Report CMU/SEI-2003-TR-014, September 2003.

17 Stephen Manes, "Dim Vista," *Forbes*, February 26, 2007, www.forbes.com/forbes/2007/0226/050.html.

18 Cem Kaner, "Quality Cost Analysis: Benefits and Risks," *Software Quality Assurance* 3, no. 1 (1996).

19 eNews 2.0 Staff, "Qantas Dive Blamed on Computer Fault," *eNews* 2.0, October 8, 2008.

20 *M. A. Mortenson Co. v. Timberline Software Co. et al.*

21 Barry W. Boehm, "Improving Software Productivity," *IEEE Computer* 20, no. 8 (1987): 43–58; Capers Jones, "*Software Quality in 2002: A Survey of the State of the Art*," Technical Report, Software Productivity Research, Inc., November 2002.

22 "Demonstrating the Impact and Benefits of CMMI," CMU/SEI-2003-SR-009.

23 Dennis R. Goldenson, Diane L. Gibson, Robert W. Ferguson, "Why Make the Switch? Evidence about the Benefits of CMMI," Carnegie Mellon Software Engineering Institute, SEPG 2004, www.sei.cmu.edu/cmmi/presentations/sepg04.presentations/evidence.pdf.

24 Jonathan P. Bowen, "The Ethics of Safety-Critical Systems," *Communications of the ACM* 43 (2000): 91–97.

25 "NASA, USAF, JPL Announce Mariner I Lost Because Flight Control Computer Generated Incorrect Steering Commands," *New York Times*, July 28, 1962.

26 Nancy Leveson and Clark S. Turned, "An Investigation of the Therac-25 Accidents," *IEEE Computer 6*, no. 7 (1993): 18–41.

27 Peter B. Ladkin and Mike Beims, *The Risks Digest*, January 10, 2002, http://catless.ncl.ac.uk/Risks/21.20.html#subj7.

28 "Surge Caused Fire in Rail Car," *Washington Times*, April 27, 2007, www.washingtontimes.com/news/2007/apr/12/20070412-104206-9871r.

29 George C. Wilson, "Navy Missile Downs Iranian Jetliner," *Washington Post*, July 4, 1988, www.washingtonpost.com/wp-srv/inatl/longterm/flight801/stories/july88crash.htm.

30 LifeScan, "About LifeScan," www.lifescan.com/company/about.

31 Edith Maverick-Folger, "Identifying Critical Requirements Using FMEA" (International Conference on Software Testing, Analysis and Review, San Jose, CA, November 1–5, 1999).

32 Yahoo! Finance, "WellPoint (WLP)," May 21, 2009, http://finance.yahoo.com/q/ct?s=WLP&e=1&i=1&f=2.

33 WellPoint Web site, www.wellpoint.com (accessed May 21, 2009).

34 J. K. Wall, "WellPoint's IT Woes Taking Toll," *Indianapolis Business Journal*, October 18, 2008, www.ibj.com/html/detail_page_Full.asp?content=21925.

35 J. K. Wall, "WellPoint's IT Woes Taking Toll," *Indianapolis Business Journal*, October 18, 2008, www.ibj.com/html/detail_page_Full.asp?content=21925.

36 Yahoo! Finance, "WellPoint (WLP)," June 3, 2009, http://finance.yahoo.com/q/hp?s=WLP&a=00&b=30&c=2008&d=03&e=3&f=2008&g=d.

37 J. K. Wall, "WellPoint's IT Woes Taking Toll," *Indianapolis Business Journal*, October 18, 2008, www.ibj.com/html/detail_page_Full.asp?content=21925.

38 J. K. Wall, "2008 in Review: WellPoint Shares Suffer," *All Business*, December 29, 2008, www.allbusiness.com/insurance/insurance-policies-claims-insurance/11753365-1.html.

39 Vanessa Fuhrmans and Jane Zhang, "WellPoint Penalized for Botching Drug Benefits," *Wall Street Journal*, January 14, 2009, http://online.wsj.com/article/SB123185453162477117.html.

40 Chris Silva, "Medicare Closes Off WellPoint's Drug Plan," *American Medical News*, January 26, 2009, www.ama-assn.org/amednews/2009/01/26/gvsc0126.htm.

41 Matt Nauman, "Toyota to Fix Software Problem on 75,000 Prius Hybrids," *San Jose Mercury News*, October 13, 2005.

42 Mel Duvall, "Software Bugs Threaten Toyota Hybrids," *Baseline*, August 4, 2005.

43 Associated Press, "Toyota Recalls 160,000 Prius Hybrids," *Washington Post*, October 14, 2005.

44 Associated Press, "Toyota Recalls 160,000 Prius Hybrids," *Washington Post*, October 14, 2005.

45 Ralph V. Carlone, "Patriot Missile Defense: Software Problem Led to System Failure at Dhahran, Saudi Arabia," GAO Report B-247094, February 4, 1992, http://archive.gao.gov/t2pbat6/145960.pdf.

46 Ralph V. Carlone, "Patriot Missile Defense: Software Problem Led to System Failure at Dhahran, Saudi Arabia," GAO Report B-247094, February 4, 1992, http://archive.gao.gov/t2pbat6/145960.pdf.

[47] Ralph V. Carlone, "Patriot Missile Defense: Software Problem Led to System Failure at Dhahran, Saudi Arabia," GAO Report B-247094, February 4, 1992, http://archive.gao.gov/t2pbat6/145960.pdf.

[48] Ralph V. Carlone, "Patriot Missile Defense: Software Problem Led to System Failure at Dhahran, Saudi Arabia," GAO Report B-247094, February 4, 1992, http://archive.gao.gov/t2pbat6/145960.pdf.

[49] Ralph V. Carlone, "Patriot Missile Defense: Software Problem Led to System Failure at Dhahran, Saudi Arabia," GAO Report B-247094, February 4, 1992, http://archive.gao.gov/t2pbat6/145960.pdf.

[50] Ralph V. Carlone, "Patriot Missile Defense: Software Problem Led to System Failure at Dhahran, Saudi Arabia," GAO Report B-247094, February 4, 1992, http://archive.gao.gov/t2pbat6/145960.pdf.

[51] "Frontline: The Gulf War: Weapons: MIM-104 Patriot," PBS, www.pbs.org/wgbh/pages/frontline/gulf/weapons/patriot.html, January 20, 1996.

[52] "Frontline: The Gulf War: Weapons: MIM-104 Patriot," PBS, www.pbs.org/wgbh/pages/frontline/gulf/weapons/patriot.html, January 20, 1996.

[53] General Accounting Office, National Security and International Affairs Division, "Operation Desert Storm: Data Does Not Exist to Conclusively Say How Well Patriot Performed," GAO Report B-250335, September 22, 1992, www.fas.org/spp/starwars/gao/b250335.htm.

CHAPTER 8

THE IMPACT OF INFORMATION TECHNOLOGY ON PRODUCTIVITY AND QUALITY OF LIFE

VIGNETTE

Western Cape Striving to Eliminate the Digital Divide

Three languages are mainly spoken by the 5 million people in the Western Cape province of South Africa—55 percent speak Afrikaans, 24 percent isiXhosa, and 19 percent English.[2] Over 1 million children attend the region's 1,570 K–12 schools. The region is poor, with an unemployment rate of 17 percent in 2007.[3] Only about 10 percent of the population have personal computers, and slightly more than that have access to the Internet (often at school, work, or Internet cafés).[4] Most schools in the region cannot afford information and communications technology (ICT). The Western Cape Education Department was concerned that without access to computers and associated curricula, its students would not be ready to take part in the digital economy.[5]

The Department of Education of South Africa is responsible for education across the country as a whole, but each of the nine provinces has its own education department. All schools are expected to abide by certain government principles—education is compulsory for all South Africans from age seven through completion of grade nine, for example. However, the various provinces have flexibility in designing their education programs and initiatives.

In April 2001, the Western Cape province initiated the Khanya Project to address the shortage of teachers in the region, to bridge the digital divide by providing access to computer technology to all schoolchildren and educators, and to prepare the Western Cape for the new "knowledge economy."[6] If the project is successful, "by the start of the 2012 academic year, every educator in every school of the Western Cape will be empowered to use appropriate and available technology to deliver curriculum to each and every learner in the province."[7]

During the initial phase of Khanya, which means "light" in isiXhosa, organizers hope to provide each school with a computer lab consisting of 20 to 40 PCs and peripherals that will be networked and linked to the Internet. At first, donated used PCs were placed in the schools to minimize deployment costs. However, the used equipment proved unreliable and difficult for the schools to maintain.[8]

The Khanya project team eventually decided to partner with NComputing, a for-profit company that specializes in creating "virtual desktops" via software and hardware connected to centralized networks. By using virtual desktop technology, a single computer can be shared among multiple users; each user connects to the computer through a device that is smaller than a paperback book, and each takes turns sharing the computer. This highly efficient solution requires less electricity and less space on each desktop. In addition, "the NComputing solution is simple to install and maintain, with far fewer PCs to manage and update."[9]

In one Khanya pilot project, 10 schools in the area of Cape Town are being supplied with Apple Mac systems in their music and graphic design departments to determine how effective these systems are for learning at the high school level. The goal is to help learners "to be job ready, or at least be at a higher starting point when they enter their tertiary studies in graphic design or music production," says Chas Ahrends, a project manager with Khanya.[10] The three-year pilot project is being done in partnership with Apple, which provided the training sessions for teachers and Khanya staff.[11]

In April 2009, Khanya celebrated having reached 20,000 educators and 770,000 learners at over 1,000 schools—with a total of 40,000 computers installed to date. Computer labs have been sustained at the schools via fund-raising efforts by various local communities.[12]

Kobus van Wyk, program manager of Khanya, cites three critical factors that have contributed to the success of the project: (1) the creation of a separate project team; (2) partnership among the schools, the Centre for E-Innovation (a government agency), and the private sector; and (3) the people on the Khanya team. Van Wyk also commented that "the greatest contribution of Khanya is that it proved that ICT can be implemented successfully on a large scale on this continent."[13]

Questions to Consider

1. How important is access to ICT in children's education?
2. What are the barriers that stand in the way of universal access to ICT for everyone who wants it?

LEARNING OBJECTIVES

As you read this chapter, consider the following questions:

1. What impact has IT had on the standard of living and worker productivity?
2. What is being done to reduce the negative influence of the digital divide?
3. What impact can IT have on improving the quality of health care and reducing its costs?

THE IMPACT OF IT ON THE STANDARD OF LIVING AND WORKER PRODUCTIVITY

The standard of living varies greatly among groups within a country as well as from nation to nation. The most widely used measurement of the material standard of living is gross domestic product (GDP) per capita. National GDP represents the total annual output of a nation's economy. Overall, industrialized nations tend to have a higher standard of living than developing countries.

In the United States and in most developed countries, the standard of living has been improving over time. However, its rate of change varies as a result of business cycles that affect prices, wages, employment levels, and the production of goods and services. Major disasters—such as hurricanes, tsunamis, and war—can negatively impact the standard of living. The worst economic downturn in U.S. history occurred during the Great Depression, when the GDP declined by about 50 percent from 1929 to 1932; by 1932, the unemployment rate had reached 25 percent.[14]

IT Investment and Productivity

Productivity is defined as the amount of output produced per unit of input, and it is measured in many different ways. For example, productivity in a factory might be measured by the number of labor hours it takes to produce one item, while productivity in a service sector company might be measured by the annual revenue an employee generates divided by the employee's annual salary. Most countries have continually been able to produce more goods and services over time—not through a proportional increase in input but by making production more efficient. These gains in productivity have led to increases in the GDP-based standard of living because the average hour of labor produced more goods and services. The Bureau of Labor Statistics tracks U.S. productivity on a quarterly basis. In the United States, labor productivity growth has averaged about 2 percent per year for the past century, meaning that living standards have doubled about every 36 years.[15]

Table 8-1 shows the annual change in U.S. labor productivity since 1839. After averaging 1.9 percent per year from 1889 to 1937, the increase in productivity during the 1950s soared to an average of 3.6 percent per year as modern management techniques and automated technology made workers far more productive. The increase in worker productivity remained at a high level of 2.6 percent during the 1960s. From 2000 to 2008, the increase in productivity was 2.5 percent.[16]

TABLE 8-1   U.S. labor productivity rates (compounded aggregate growth rate)

Years	Rate
1839–1937	1.9%
1950–1960	3.6%
1960–1973	2.6%
1973–1980	1.0%
1980–1990	1.4%
1990–1995	1.1%
1995–2000	2.0%
2000–2008	2.5%

Source: McKinsey Global Institute, "Whatever Happened to the New Economy?," November 2002; United States Department of Labor Bureau of Labor Statistics, "Productivity Change in the Nonfarm Business Sector, 1947–2008," www.bls.gov/lpc/prodybar.htm.

Innovation is a key factor in productivity improvement, and IT has played an important role in enabling innovation. Progressive management teams use IT, as well as other new technology and capital investment, to implement innovations in products, processes, and services.

In the early days of IT in the 1960s, productivity improvements were easy to measure. For example, midsized companies often had a dozen or more accountants focused solely on payroll-related accounting. When businesses learned to apply automated payroll systems, fewer accounting employees were needed. The productivity gains from such IT investments were obvious.

Today, organizations are trying to further improve IT systems and business processes that have already gone through several rounds of improvement. Organizations are also adding new IT capabilities to help workers who already have an assortment of personal productivity applications on their desktop computers, laptops, and personal digital assistants (PDAs). Instead of eliminating workers, companies are saving workers small amounts of time each day. Whether these saved minutes actually result in improved worker productivity is a matter for debate. Many analysts argue that workers merely use the extra time to do some small task they didn't have time to do before, such as respond to e-mail they would have ignored. These minor gains make it harder to quantify the benefits of today's IT investments on worker productivity.

The relationship between investment in information technology and U.S. productivity growth is more complex than you might think. Consider the following facts:

- The rate of productivity from 1995 to 2005 is only slightly higher than the long-term U.S. rate and not nearly as high as it was during the two decades following World War II. So, although the recent increase in productivity is welcome, it is not statistically significant.

- Labor productivity in the United States remained relatively high despite a reduced level of investment in IT from 1999 to 2004. If there were a simple, direct relationship, the productivity rate should have decreased.

One possible explanation for the previous points is that there is a lag time between the application of innovative IT solutions and the capture of significant productivity gains. IT can enhance productivity in fundamental ways by allowing firms to make radical changes in work processes, but such major changes can take years to complete because firms must make substantial complementary investments in retraining, reorganizing, changing reward systems, and the like. Furthermore, the effort to make such a conversion can divert resources from normal activities, which can actually reduce productivity—at least temporarily. For example, researchers examined data from 527 large U.S. firms from 1987 to 1994 and found that it can take five to seven years for IT investment to result in a substantial increase in productivity.[17]

Another explanation for the complex relationship between IT investment and U.S. productivity growth lies in the fact that many other factors influence worker productivity rates besides IT. Table 8-2 summarizes fundamental ways in which companies can increase productivity.

TABLE 8-2 Fundamental drivers for productivity performance

Reduce the amount of input required to produce a given output by:	Increase the value of the output produced by a given amount of input by:
Consolidating operations to better leverage economies of scale	Selling higher-value goods
Improving performance by becoming more efficient	Selling more goods to increase capacity and use of existing resources

The following list summarizes additional factors that can affect national productivity rates:

- Labor productivity growth rates differ according to where a country is in the business cycle—expansion or contraction. Times of expansion enable firms to gain full advantage of economies of scale and full production. Times of contraction present fewer investment opportunities.
- Outsourcing can skew productivity if a contracting firm has different productivity rates than the outsourcing firm.
- U.S. regulations make it easier for companies to hire and fire workers and to start and end business activities compared to many other industrialized nations. This flexibility makes it easier for markets to relocate workers to more productive firms and sectors.
- More competitive markets for goods and services can provide greater incentives for technological innovation and adoption as firms strive to keep ahead of competitors.

- In today's service-based economy, it is difficult to measure the real output of such services as accounting, customer service, and consulting.
- IT investments don't always yield tangible results, such as cost savings and reduced head count; instead, many produce intangible benefits, such as improved quality, reliability, and service.

As you can see, it is difficult to quantify how much the use of IT has contributed to worker productivity. Ultimately, however, the issue is academic. There is no way to compare organizations that don't use IT with those that do, because there is no such thing as a noncomputerized airline, financial institution, manufacturer, or retailer.

Businesspeople analyze the expected return on investment to choose which IT option to implement, but at this point, trying to measure its precise impact on worker productivity is like trying to measure the impact of telephones or electricity.

Telework

Telework (also known as telecommuting) is a work arrangement in which an employee works away from the office—at home, at a client's office, in a hotel—literally, anywhere. In telework, an employee uses various forms of electronic communication, including e-mail, audio conferencing, video conferencing, and instant messaging. Teleworkers access the Internet via cell phone, PDA, laptop, or similar device to retrieve computer files; log on to software applications; and communicate with fellow employees, managers, customers, and suppliers. The goal of telework is to allow employees to be effective and productive from wherever they are. It is estimated that there are roughly 14 million employees who telework more than eight hours per week.[18]

Factors that have increased the prevalence of telework include advances in technology that enable people to communicate and access the Internet from almost anywhere, the increasing number of broadband connections in homes and retail locations, high levels of traffic congestion, rising gasoline prices, and growing concern over the effects of automobile CO_2 emissions. Another key factor is that "scarce, highly skilled workers have begun to demand more flexible work arrangements, especially as they choose to live farther and farther from their employers."[19]

A number of states have passed laws to encourage telework. For example, in Virginia, 20 percent of eligible state workers must be teleworking by 2010. The state defines someone who works out of the office at least once per week or 32 hours per month as a teleworker. Several states—California and Virginia, among others—and the federal government have based their pandemic-preparedness plans on the widespread use of telework.[20]

Organizations should prepare guidelines and policies to define the types of positions and workers who represent ideal telework opportunities. Clear guidelines must be set for how and when work will be given to and collected from teleworkers. If there are certain hours during which the teleworker must be available, these, too, must be defined. Employee work expectations and performance criteria must also be delineated.

Some positions—such as management positions or those in which face-to-face communications with other employees or customers is required—may not be well suited for telework. In addition, some individuals are not well suited to be teleworkers. Telework opportunities need to be weighed based on the characteristics of the individual as well as the requirements of the position. Tables 8-3 and 8-4 list some of the advantages and disadvantages of telework from the perspectives of employees and organizations, respectively.

TABLE 8-3   Advantages/disadvantages of teleworking for employees

Advantages	Disadvantages
People with disabilities who otherwise find public transportation and office accommodations a barrier to work may now be able to join the workforce.	Some employees are unable to be productive workers away from the office.
Teleworkers avoid long, stressful commutes and gain time for additional work or personal activities.	Teleworkers may suffer from isolation and may not really feel "part of the team."
Telework minimizes the need for employees to take time off to stay home to care for a sick family member.	Workers who are out of sight tend to also be out of mind. The contributions of teleworkers may not be fully recognized and credited.
Teleworkers have an opportunity to experience an improved work/family balance.	Teleworkers must guard from working too many hours per day since work is always there.
Telework reduces ad hoc work requests and disruptions from fellow workers.	The cost of the necessary equipment and communication services can be considerable if the organization does not cover these.

TABLE 8-4   Advantages/disadvantages of teleworking for organizations

Advantages	Disadvantages
As more employees telework, there is less need for office and parking space; this can lead to lower costs.	Allowing teleworkers to access organizational data and systems from remote sites creates potential security issues.
Allowing employees to telework can improve morale and reduce turnover.	Informal, spontaneous meetings become more difficult if not impossible.
Telework allows for the continuity of business operations in the event of a local or national disaster, and supports national pandemic-preparedness planning.	Managers may have a harder time monitoring the quality and quantity of the work performed by teleworkers, wondering, for instance, if they really "put in a full day."
The opportunity to telework can be seen as an additional perk that can help in recruiting.	Increased planning is required by managers to accommodate and include teleworkers.
There may be an actual gain in worker productivity.	There are additional costs associated with providing equipment, services, and support for people who work away from the office.
Telework can decrease an organization's carbon footprint by reducing daily commuting.	Telework increases the potential for lost or stolen equipment.

The Digital Divide

When people talk about standard of living, they are often referring to a level of material comfort measured by the goods, services, and luxuries available to a person, group, or nation—factors beyond the GDP-based measurement of standard of living. Some of these indicators are:

- Average number of calories consumed per person per day
- Availability of clean drinking water
- Average life expectancy
- Literacy rate
- Availability of basic freedoms
- Number of people per doctor
- Infant mortality rate
- Crime rate
- Rate of home ownership
- Availability of educational opportunities

Another indicator of the standard of living is the availability of technology. The **digital divide** is a term used to describe the gulf between those who do and those who don't have access to modern information and communications technology such as cell phones, personal computers, and the Internet. The digital divide shows up clearly when one examines the rates of PC ownership and Internet use in various countries around the world. For example, of the roughly 1 billion Internet users worldwide, only 20 million (2 percent) are estimated to live in less developed nations. Table 8-5 highlights some startling statistics relating to the estimated rate of PC ownership in the United States and other areas of the world. While 812 out of every 1,000 people in the United States own a computer, that figure drops to 21 out of 1,000 in the Middle East and Africa. Table 8-6 shows similar gaps in the rate of Internet access.

TABLE 8-5 Personal computer ownership rates

Country/Region	Estimated number of personal computers per 1,000 people
United States	812
China	111
Middle East and Africa	21

Source: "Bridging the Global Digital Divide, One Laptop at a Time," Knowledge@Wharton, June 11, 2008, http://knowledge.wharton.upenn.edu/article.cfm?articleid=1978.

TABLE 8-6   Rates of Internet access

Geographic area	Percent of population using the Internet (2008)
North America	74.4%
Oceania/Australia	60.4%
Europe	48.9%
Latin America/Caribbean	29.9%
Middle East	23.3%
Asia	17.4%
Africa	5.6%
World average	23.8%

Source: Miniwatts Marketing Group, "Internet Usage Statistics: The Internet Big Picture," www.internetworldstats.com/stats.htm.

The digital divide exists not only between more and less developed countries but also within countries—among age groups, economic classes, and people who live in cities versus those in rural areas.

- About 20 million homes in the United States were without Internet access in 2007 and nearly 20 percent of all U.S. heads-of-household have never used e-mail. Even in the United States, the digital divide remains.[21]
- There is a digital divide among the nearly 500 million people of the European Union, where 8 percent of the population is not connected to the Internet. The two extremes are the Netherlands, where over 80 percent of the population regularly uses the Internet, and Poland, where less than 40 percent of the population are Internet users.[22]
- There is no economical way to provide telecommunications service to the 700 million rural people in India who have an annual GDP of less than $500 per capita. In many of India's rural communities, one must travel more than 5 miles to reach the nearest telephone.[23]
- Many parts of Africa lack basic facilities, such as drinking water, roads, and electricity. Easy access to modern computers and information technology is simply not realistic.

Many people believe that the digital divide must be bridged for a number of reasons. Clearly, health, crime, and other emergencies could be resolved more quickly if a person in trouble had easy access to a communications network. Access to IT and communications technology can also greatly enhance learning and provide a wealth of educational and economic opportunities. Much of the vital information people need to manage their career, retirement, health, and safety is increasingly provided by the Internet. High-tech access can give a country's industries a competitive advantage over its less fortunate neighbors.

The E-Rate and Ed-Tech programs were designed to help eliminate the digital divide within the United States. These programs, as well as the availability of low-cost computers and cell phones, are discussed in the following sections.

E-Rate Program

The **Education Rate (E-Rate) program** was created through the Telecommunications Act of 1996. One of E-Rate's goals is to help schools and libraries obtain access to state-of-the-art services and technologies at discounted rates. The program's discounts range from 20 percent to 90 percent for eligible telecommunications services, depending on location (urban or rural) and economic need. The level of discount is based on the percentage of students eligible for participation in the National School Lunch Program. The Federal Communications Commission (FCC) ruled that the program would be supported with up to $2.25 billion per year from a fee charged to telephone customers. The FCC also established the private, nonprofit Universal Service Administrative Company (USAC) to administer the program.[24]

Unfortunately, the program has not gone well. Following a yearlong investigation, a House subcommittee in 2005 approved a bipartisan staff report detailing abuse, fraud, and waste in E-Rate. In one infamous example, USAC disbursed $101 million between 1998 and 2001 to equip Puerto Rico's 1,540 schools with high-speed Internet access, but a review found that very few computers were actually connected to the Internet. In fact, $23 million worth of equipment was found in unopened boxes in a warehouse.[25]

A University of Chicago study examined the impact of the E-Rate program in California and found that the number of students in poor schools going online had indeed increased dramatically. However, the study found no evidence that the program had any effect on students' performance on any of the six subjects (math, reading, science, language, spelling, and social studies) covered in the Stanford Achievement Test. Researchers concluded that either the schools did not know how to make effective use of the Internet or that use of the Internet was simply not a productive way to boost test scores.[26] The E-Rate program continues today; funding for the 2007–2008 school year was $1.94 billion.[27]

Ed-Tech Program

The goal of the No Child Left Behind Act (NCLB), which was signed into law in 2002, is to end the achievement gap between rich and poor students and between white and minority students, while improving the academic performance of all students. Recognizing that the ability to use computers and access the Internet is a requirement for succeeding in the U.S. educational system and the global workforce, NCLB requires each state to have an **Enhancing Education Through Technology (Ed-Tech)** program. The Ed-Tech program has the following goals:

- Improve student academic achievement through the use of technology in schools
- Assist children in crossing the digital divide by ensuring that every student is technologically literate by the end of eighth grade
- Encourage the effective integration of technology with teacher training and curriculum development to establish successful, research-based instructional methods

To help achieve these goals, the U.S. Department of Education awards up to $270 million in grants to state education agencies each year.[28]

States have begun innovative and successful programs to help achieve the Ed-Tech program's goals, but there have also been some unsuccessful initiatives. Overall, more than 25 percent of all K–12 public schools now provide some form of individual online instruction to supplement regular classes or provide for special needs.[29]

Low-Cost Computers

It is estimated that more than 1 billion personal computers will be connected to the Internet by 2010. While that number is impressive, it still leaves more than 5.5 billion people unconnected. What most of those 5.5 billion people have in common is low income (see Table 8-7). Increasing the availability of low-cost computers could help reduce this huge digital divide.

TABLE 8-7   Percent of people worldwide with access to a computer

Annual income (USD)	Number of people earning this much	Percent with access to a computer
Under $1,000	1.0 billion	0%
$1,000 to $25,000	4.7 billion	10%
Over $25,000	0.8 billion	70%

Source: Michael Kanellos, "Intel's Bridge for the Digital Divide," *CNET*, June 15, 2006, http://news.cnet.com/Intels-bridge-for-the-digital-divide/2100-1005_3-6084250.html.

One Laptop per Child (OLPC) is a nonprofit organization whose goal is to provide children around the world with low-cost (less than $100) laptop computers to aid in their education. The first version of its laptop, OLPC XO, was made available to third-world countries in 2007 and came with a hand crank for generating power in places where electricity is not readily available. The cost was around $200. The next version (OLPC XO-2)—expected to arrive in 2010—will come with two touch screens, one of which can be used as a touch-sensitive keyboard. This version will be half the size of the original XO and will require just one watt of electricity to run. OLPC will market it as a feature-packed e-book reader with a 500-book capacity at a hoped-for cost of around $75 (see Figure 8-1).[30] In India, two government entities and one private-sector group placed combined orders for 250,000 OLPC computers for 1,500 schools. A human-rights organization plans to provide 5,000 of the machines to Sierra Leone.[31]

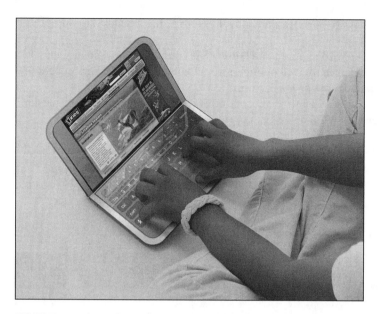

FIGURE 8-1   The OLPC XO-2 laptop computer

Intel created a competitor to the OLPC called the classmate PC. The first generation of this notebook computer cost under $400 and was designed for use in kindergarten through high school classrooms in developing countries. The computer began shipping in early 2007 to 25 countries, including Brazil, Chile, Nigeria, China, India, and Vietnam. A second-generation machine is now available. The top of the line of this model comes with a 9-inch LCD screen, 3–5 hours of battery life, 512 MB of RAM, a 30 GB hard drive, and an integrated Webcam.[32] Governments in 60 countries have created programs for people to buy the personal computers on credit. Intel does not manufacture the computers but provides free blueprints to manufacturers and independent dealers.[33]

ASUS is a Taiwanese multinational manufacturer of computers and computer components. In late 2007, the company launched its Eee notebook computer, which ranges in cost from $200 to $400.[34] At this price, the computer is a competitor to the OLPC and Intel's classmate PC for widespread deployment in developing countries.

Mobile Phone the Tool to Bridge the Digital Divide?

Even though the mobile phone cannot do all that a personal computer can, many industry observers think that it will be the cell phone that will ultimately bridge the digital divide. Cell phones have the following advantages over PCs in developing countries:

- Cell phones come in a wide range of capabilities and costs, but are cheaper than personal computers. For example, one can purchase a cell phone in India for the equivalent of $40 or less, and 500 minutes of use cost about $10 per month. The OLPC XO-2 laptop computer will cost around $75.
- In developing countries, many more people have access to cell phones than they do PCs (see Table 8-8).

- Cell phones are more portable and convenient than the smallest laptop computer.
- Cell phones come with an extended battery life (much longer than any PC battery), which makes the cell phone more reliable in regions where access to electricity is inadequate or nonexistent.
- The infrastructure needed to connect wireless devices to the Internet is easier and less expensive to build.
- There is almost no learning curve required to master the use of a cell phone.
- Cell phones require no costly or burdensome applications that must be loaded and updated.
- There are essentially no technical-support challenges to overcome when using a cell phone.

TABLE 8-8   Comparison of PCs and cell phones per 1,000 people

Country	Number of PCs per 1,000 people	Number of cell phone subscribers per 1,000 people
Afghanistan	0.4	143
Cambodia	2	181
China	122	411
India	36	203
Kenya	15	301
Pakistan	5	524
South Africa	N/A	966
Swaziland	27	337

Source: Mobile Phone Subscribers per 1,000 People, www.sitesatlas.com/Thematic-Maps/Mobile-phone-subscribers-per-capita.html.

When IT is available to everyone—regardless of economic status, geographic location, language, or social status—it can enhance the sharing of ideas, culture, and knowledge. How much will the benefits of IT raise the standard of living in underdeveloped countries? Could the end of the digital divide change the way people think about themselves in relation to the rest of the world? Could such enlightenment, coupled with a better standard of living, contribute to a reduction in violence, poverty, poor health, and even terrorism?

THE IMPACT OF IT ON HEALTHCARE COSTS

The rapidly rising cost of health care is one of the major challenges of the 21st century. The United States spent an estimated $2.4 trillion on health care in 2008 versus $1.7 trillion in 2003. The share of gross domestic product spent on national health care has grown from

7.2 percent in 1970 to an estimated 16.6 percent in 2008.[35] Current estimates are that national healthcare spending will more than double to over $4 trillion per year by 2016, with one out of every five dollars spent in the United States going toward health care.[36]

The development and use of new medical technology, such as new diagnostic procedures and treatments (see Figure 8-2), has increased spending and "accounts for one-half to two-thirds of the increase in healthcare spending in excess of general inflation."[37] Although many new diagnostic procedures and treatments are at least moderately more effective than their older counterparts, they are also more costly. In addition, even if new procedures and treatments cost less (for example, magnetic resonance imaging), they may stimulate much higher rates of use because they are more effective or cause less discomfort to patients.

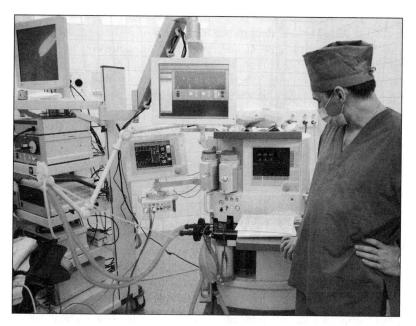

FIGURE 8-2   The development and use of new technology has increased healthcare spending

Patients sometimes overuse medical resources that appear to be free or almost free thanks to the share of medical bills that is paid by third parties, such as insurance companies and government programs. A patient who doesn't have to pay for a medical test or procedure is probably less likely to consider its cost-to-benefit ratio. Attempts by insurance companies to rein in those costs have led to a blizzard of paperwork but have proven ineffective.

To really gain control over soaring healthcare costs, patient awareness must be raised and technology costs must be managed more carefully. In the meantime, however, the improved use of IT in the healthcare industry can lead to significant cost reductions in a number of ways.

Electronic Health Records

Although the healthcare industry depends on highly sophisticated technology for diagnostics and treatment, it has been slow to implement IT solutions to improve productivity and efficiency. The healthcare industry invests about $3,000 in IT for each worker, compared with about $7,000 per worker in private industry generally and nearly $15,000 per worker in the banking industry.[38]

One tremendous opportunity for improving health care through the use of IT is in the process of capturing and recording patient data. Before seeing a physician, many patients are given a clipboard and pen with a standard form to complete. Some people must wonder: "This is the same form I filled out last time; what did they do with the data from my last visit?"

It is nearly impossible to pull together the paper trail created by a patient's interactions with various healthcare entities to create a clear, meaningful, consolidated view of that person's health history. This lack of patient data transparency can result in diagnostic and medication errors as well as the ordering of duplicate tests, which dramatically increase healthcare costs. It can even compromise patient safety. For example, physicians in the emergency room must often treat a patient who is unconscious and incapable of providing essential medical information, such as the name of his or her primary care physician, information about recent illnesses or surgeries, medications taken, allergies, and other useful data. Without such data, the ER physician is essentially taking a gamble in treating the patient. If the United States had a comprehensive healthcare information network, such medical data could be readily available for all patients at any medical facility.

A 1999 report by the Institute of Medicine (IOM) found that as many as 98,000 Americans die annually due to preventable medical errors. In addition, a 2006 IOM report concluded that more than 1.5 million preventable medication errors per year cost the United States about $3.5 billion annually.[39] A 2009 Consumers Union report claims that we have made no real progress toward reducing the number of deaths. However, as far back as 2004, healthcare experts agreed that "going digital" could eliminate many of these needless deaths.[40]

An **electronic health record (EHR)** is a summary of health information generated by each patient encounter in any healthcare delivery setting. An EHR includes patient demographics, medical history, immunization records, laboratory data, problems, progress notes, medications, vital signs, and radiology reports. EHRs could incorporate data from any healthcare entity a patient uses and make the data easily accessible to other healthcare professionals. Healthcare professionals can use an EHR to generate a complete electronic record of a clinical patient encounter (see Figure 8-3).

FIGURE 8-3   An EHR is a summary of health information generated by each patient encounter

A study based on data collected in 2008 revealed that of the 3,000 U.S. hospitals surveyed, less than 2 percent use comprehensive EHRs and only about 8 percent use basic EHRs.[41] A separate study reported slightly higher rates, with 13 percent of U.S. physicians using basic EHRs and only 4 percent using "fully functional" EHRs.[42]

Effective use of EHR systems has been shown to improve patient care and reduce costs. In a Commonwealth Fund study of 41 Texas hospitals that treat a diverse group of patients suffering from a variety of medical conditions, researchers found that a 10 percent increase in the use of electronic notes and medical records was associated with a 15 percent reduction in the likelihood of patient death. When physicians electronically entered patient care instructions, there was a 55 percent reduction in the likelihood of death related to some procedures.[43]

A study by the RAND Corporation concluded that if 90 percent of doctors and hospitals adopted EHR systems, the United States would save at least $81 billion per year in health-care costs—with $77 billion coming from a reduction in hospital stays as well as a drop in duplicate and unnecessary testing. The other $4 billion in savings would come from decreased medication errors and treatment of adverse side effects.[44] The RAND Corporation also estimated that it would take 15 years and cost hospitals roughly $98 billion and physicians about $17 billion to implement EHR systems. The hospitals and physicians who must make the investment will not, however, reap the full amount of savings from this new technology. The RAND Corporation estimated that Medicare would receive about $23 billion of the savings each year, and private insurers about $31 billion a year.[45]

A much more conservative estimate by the federal government sees a savings of $15.5 billion in healthcare spending from 2016 to 2019 ($5.1 billion a year) due to improved efficiencies and reduced reimbursements to healthcare providers who are "not meaningful" users of EHR.[46]

As part of the $787 billion 2009 economic stimulus plan, the federal government earmarked $33 billion in incentives for healthcare providers to implement government-certified, interoperable EHR systems by 2015. The PricewaterhouseCoopers LLP Health Research Institute estimates that a 500-bed hospital could receive $6.1 million in incentives to purchase, deploy, and maintain such an EHR system. On the other hand, failure to implement such a system by 2015 could cause the hospital to lose $3.2 million in funding annually, depending on the hospital's volume of Medicare, Medicaid, and charity-care patients. A typical three-physician practice will need to spend between $173,000 and $296,000 to purchase and maintain an EHR system. Individual physicians, not practices, will receive up to $44,000 for adopting certified EHR systems.[47]

One cannot help but wonder where we will be in 10 years in terms of healthcare spending. Will we have made a meaningful reduction in the number of avoidable deaths? Will we have earned a worthwhile return on the investment in EHR?

Use of Mobile and Wireless Technology in the Healthcare Industry

Although slow to invest in IT, the healthcare industry was actually a leader in adopting mobile and wireless technology, perhaps because of the frequent urgency of communications with doctors and nurses who are almost always on the move. For example, doctors were among the first large groups to start using PDAs on the job. Other common uses of wireless technology in the healthcare field include:

- Providing a means to access and update EHRs at patients' bedsides to ensure accurate and current patient data
- Enabling nurses to scan bar codes on patient wristbands and on medications to help them administer the right drug in the proper dosage at the correct time of day (an attached computer on a nearby cart is linked via a wireless network to a database containing physician medication orders)
- Using wireless devices to communicate with healthcare employees wherever they may be

Telemedicine

Telemedicine employs modern telecommunications and information technologies to provide medical care to people who live far away from healthcare providers. This technology reduces the need for patients to travel for treatment and allows healthcare professionals to serve more patients in a broader geographic area. There are two basic forms of telemedicine: store-and-forward and live.

Store-and-forward telemedicine involves acquiring data, sound, images, and video from a patient and then transmitting everything to a medical specialist for later evaluation. This type of monitoring does not require the presence of the patient and care provider at the same time, and having access to such information can enable healthcare professionals to recognize problems and intervene before high-risk situations become life threatening. For example, patients who have chronic diseases often don't recognize early warning signs that indicate an impending health crisis. A sudden weight gain by a patient who has suffered congestive heart failure could indicate retention of fluids, which could lead to a traumatic trip to the emergency room or even loss of life. A physician who uses telemedicine to keep tabs on such patients could make a vital difference.

Each year some 24,000 people in the United States become legally blind from diabetic retinopathy—damage to the blood vessels of the retina. Early detection and treatment for this disease can reduce vision loss by 90 percent. The University of California at Berkeley has developed a store-and-forward telemedicine system that can be used to screen diabetes patients who might not otherwise get a recommended annual exam. It has proven to be an effective and low-cost method that can be used in a community clinic without the need for an eye specialist to be present at the time of the screening.[48]

Live telemedicine requires the presence of patients and healthcare providers at the same time and often involves a video conference link between the two sites. The University of Rochester's Golisano Children's Hospital in New York uses the Health-e-Access telemedicine program to screen children in inner-city schools for asthma. This program uses video conferencing to allow a primary-care telemedicine clinician to look at sick children and communicate with the school nurse. Results to date show that 90 percent of the parents of the participants in this program avoided a primary care or emergency room visit. Expansion of the program to include access for developmentally challenged children and adults, tele-dentistry, and elder care is under consideration.[49]

The use of telemedicine does raise some new issues. Must the physicians providing advice to patients at a remote location be licensed to perform medicine at that location—perhaps a different state or country? Must a healthcare system be required to possess a license from a state in which it has a "virtual" facility, such as a video conferencing room? Will the various states require some form of assurance that minimum technological standards (such as the minimum resolution of network-transmitted images) are being met?

Medical Information Web Sites for Laypeople

Healthy people as well as those who suffer from illness need reliable information on a wide range of medical topics to learn more about healthcare services and to take more responsibility for their health. Clearly, laypeople cannot become as informed as trained medical practitioners, but a tremendous amount of healthcare information is available via the Web. These sites have a critical responsibility to publish current, reliable, and objective information. Table 8-9 provides just a small sample of Web sites that offer information on a variety of medical-related topics.

TABLE 8-9   Health information Web sites

URL	Site
www.americanheart.org	American Heart Association
www.cancer.org	American Cancer Society
www.cdc.gov	Centers for Disease Control and Prevention
www.diabetes.org	American Diabetes Association
www.heartburn.about.com	Information on what causes heartburn and how to prevent it
www.heartdisease.about.com	Basic information about heart disease and cardiology
www.medicinenet.com	Source for medical information on a variety of topics, including symptoms, procedures, tests, and medications, as well as a medical dictionary
www.nia.nih.gov/Alzheimers/	Alzheimer's Disease Education and Referral Center
www.niddk.nih.gov	National Institute of Diabetes and Digestive and Kidney Diseases
www.oncolink.upenn.edu	Abramson Cancer Center of the University of Pennsylvania
www.osteo.org	The NIH Osteoporosis and Related Bone Diseases—National Resource Center
www.urologychannel.com	Information about urologic conditions, including erectile dysfunction, HIV, AIDS, kidney stones, and STDs; site contains overviews, symptoms, causes, diagnostic procedures, and treatment options
www.webmd.com	Access to medical reference material and online professional publications

The contents of a medical information Web site, such as text, graphics, and images, are for informational purposes only. These Web sites are not intended to be substitutes for professional medical advice, diagnosis, or treatment. Individuals should always seek the advice of a physician or other qualified healthcare provider with any questions regarding a medical condition. A patient should never disregard professional medical advice or delay seeking it because of something he or she reads on a medical information Web site.

In addition to publicly available information on the Web, some healthcare providers and employers offer useful online tools to members and employees that go beyond basic health information. For example, WellPoint—a leading health benefits company—formed a partnership with technology provider Subimo LLC to offer online healthcare support tools. (In December 2006, WebMD acquired Subimo, which is now called WebMD Health Services.) With this technology, WellPoint members can "go online to compare, among other things, quality, safety, and cost information on hospitals nationwide; quality and risk indicators for specific health treatment options; nationwide average prices of drugs and treatment options; and coverage and costs for treatments by in-network and out-of-network healthcare providers."[50]

Subimo is responsible for capturing data and updating its databases as needed to provide accurate and current medical information. For example, a WellPoint member who needs a hip replacement can go online and find information about the surgery; other available treatment options; a list of questions to ask the physician; potential risks; nearby hospitals that perform the surgery; and quality-of-service information about the hospitals, such as the number of reported postoperative infections and other complications.[51]

Summary

- The most widely used measurement of the material standard of living is gross domestic product (GDP) per capita.

- In the United States, as in most developed nations, the standard of living has been improving over time. However, its rate of change varies as a result of business cycles that affect prices, wages, employment levels, and the production of goods and services.

- Productivity is defined as the amount of output produced per unit of input.

- Most countries have continually been able to produce more goods and services over time—not through a proportional increase in input but by making production more efficient. These gains in productivity have led to increases in the GDP-based standard of living because the average hour of labor produced more goods and services.

- Progressive management teams use IT, other new technology, and capital investment to implement innovations in products, processes, and services.

- It can be difficult to quantify the benefits of IT investments on worker productivity because there can be a considerable lag between the application of innovative IT solutions and the capture of significant productivity gains. In addition, many other factors influence worker productivity rates besides IT.

- Many organizations offer telework opportunities to their employees as a means of reducing costs; increasing productivity; reducing the organization's carbon footprint; and preparing for potential local or widespread disasters, including pandemics.

- Telework opportunities provide many advantages for employees, such as avoiding long, stressful commutes and providing more flexibility to balance the needs of work and family life.

- The *digital divide* is a term used to describe the gulf between those who do and those who don't have access to modern information and communications technology, such as computers and the Internet.

- The digital divide exists not only between more and less developed countries but also within countries—among age groups, economic classes, and people who live in cities versus those in rural areas.

- New information technologies can be used with little capital cost to reduce the digital divide. These technologies can make computing and communication better, cheaper, faster, and more available to larger segments of the world's population.

- Healthcare costs are soaring out of control, with the share of GDP spent on national health care growing from 7.2 percent in 1970 to 16.6 percent in 2008.

- Many experts point to two primary causes for cost increases: the use of more expensive technology and the shielding of patients from the true costs of medical care.

- Improved use of IT in the healthcare industry can lead to significantly reduced costs in a number of ways: electronic health records (EHRs) of patient information can be generated from each patient visit in every healthcare setting, and wireless technology can be used to access and update EHRs at patients' bedsides, match bar-coded patient wristbands and medication packages to physician orders, and communicate with healthcare employees wherever they may be.

- Telemedicine employs modern telecommunications and information technologies to provide medical care to people who live far away from healthcare providers. This technology reduces the need for patients to travel for treatment and allows healthcare professionals to serve more patients in a broader geographic area.

- Web-based health information can help people inform themselves about medical topics.

Self-Assessment Questions

The answers to the Self-Assessment Questions can be found in Appendix G.

1. Which of the following statements about the standard of living is *not* true?
   a. It varies greatly from nation to nation.
   b. It varies little among groups within the same country.
   c. Industrialized nations generally have a higher standard of living than developing countries.
   d. It is frequently measured using the GDP per capita.

2. The amount of output produced per unit of input is called _____.

3. The rate of productivity growth from 1995 to 2005 is only slightly higher than the long-term U.S. rate and not nearly as high as it was for the two decades following World War II. True or False?

4. A study of 527 large U.S. firms from 1987 to 1994 found that the benefits of applying IT grow over time and can take _____ to _____ years to fully realize.

5. _____ is a term used to describe the gulf between those who do and those who don't have access to modern information and communications technology such as cell phones, personal computers, and the Internet.

6. It is difficult to quantify how much the use of IT has contributed to worker productivity. True or False?

7. Which of the following statements about the digital divide is *not* true?
   a. It exists not only between more and less developed countries but also within countries—among age groups, economic classes, and people who live in cities versus those in rural areas.
   b. Only about 20 million of the world's 1 billion Internet users live in less developed nations.
   c. The number of PCs per 1,000 people in the Middle East and Africa is more than in China.
   d. In many of India's rural communities, one must travel more than 5 miles to the nearest telephone.

The Impact of Information Technology on Productivity and Quality of Life

8. Which of the following is a valid reason for trying to reduce the digital divide?

   a. Health, crime, and other emergencies could be resolved more quickly if people in trouble had access to a communications network.

   b. Much of the vital information that people need to manage their retirement, health, and safety is increasingly provided by the Internet.

   c. Ready access to information and communications technology can provide a country with a wealth of economic opportunities and give its industries a competitive advantage.

   d. All of the above.

9. The No Child Left Behind Act requires that each state have an _____ program to improve academic achievement through the use of technology in schools.

10. It is estimated that more than 1 billion personal computers will be connected to the Internet by 2010, leaving more than 5.5 billion people unconnected. _____ is the standard of living characteristic that most of the 5.5 billion people have in common.

11. The new version of the OLPC XO is expected to arrive in 2010 with an expected cost of around _____ .

12. Which of the following statements about healthcare spending is *not* true?

    a. The United States spent $2.4 trillion on health care in 2008, and that number is expected to rise to over $4.0 trillion by 2016.

    b. U.S. spending on health care increased from $1.7 billion in 2003 to $2.4 trillion in 2008.

    c. U.S. spending on health care in 2008 represented 16.6 percent of its GDP.

    d. The development and use of new medical technology in the United States has clearly led to a reduction in healthcare costs.

13. Two main reasons have been advanced as the cause of rising healthcare costs: the use of more expensive technology and the _____ of patients from the true costs of medical care.

14. The healthcare industry spends less per employee on IT than private industry. True or False?

15. An Institute of Medicine study found that about _____ patients die annually from preventable medical errors.

16. A study based on data collected in 2008 revealed that just _____ percent of the 3,000 hospitals surveyed use comprehensive EHRs.

17. Unfortunately, there is no real data to support the premise that EHRs will improve patient care. True or False?

Discussion Questions

1. Why is it so difficult to quantify the benefits of IT investments on worker productivity? Discuss the question fully.

2. Explain how increases in worker productivity can lead to an increase in the standard of living.

3. Identify three factors that you think are significant indicators of the standard of living. Find current measures of these three factors for five different countries. Based on your indicators, which of the five countries would you say has the highest standard of living?

4. What do you see as the most significant advantages and disadvantages of telework?

5. Would you accept a telework position in which you would work from home four days per week? Why or why not?

6. As mentioned in the text, a University of Chicago study examined the impact of the E-Rate program and found that the number of poor schools going online had increased dramatically. However, the study found no evidence that the program had any effect on students' performance on the Stanford Achievement Test. Should this be sufficient evidence to discontinue this program, which costs U.S. taxpayers roughly $2 billion per year?

7. Explain how the development and use of new medical technology, such as new diagnostic procedures and treatments, has increased spending and accounts for one-half to two-thirds of the increase in healthcare spending in excess of general inflation.

8. Should the medical industry place more emphasis on using older medical technologies and containing medical costs?

9. Medical information that you obtain from Web sites must be accurate and reliable. How can you ascertain the credibility of a Web site's information?

What Would You Do?

1. Many people believe that the "haves" in society should help the "have-nots." One idea is to help others gain high-speed, reliable access to the Internet. If you were a U.S. senator, would you support the following suggestion to help close the digital divide? Why or why not? Discuss some of the problems—technical and nontechnical—in providing this type of access:

   Every time anyone buys a personal computer, signs up for a cell phone, or contracts for cable or satellite TV service, he or she pays a $20 "Internet enablement" fee. The hardware vendors and service providers match the fee. The resulting money is paid into a fund managed by a nonprofit organization and is used to provide computers and Internet access to U.S. residents who cannot afford them.

2. You have been invited to accept a six-month internship at the Media Lab in Kharagpur, which is in the Midnapore West district of the state of West Bengal in India. Your modest living expenses will be paid, and you will receive US$6,000 at the end of your internship. The lab's goal is to bring the benefits of new information and communications technologies to millions of people in Asia, Africa, and Latin America. Kharagpur was chosen to have the first campus of the prestigious Indian Institutes of Technology (IITs). The IITs are the premier technical education institutes in India and are internationally recognized for their academic and technical excellence. Assume that you decide to accept this challenging opportunity. In what research and other activities would you try to participate to help bring the benefits of information and communications technology to India? Do you think that the Media Lab can make a positive impact? Why or why not?

3. You are a midlevel manager at a major metropolitan hospital and are responsible for capturing and reporting statistics regarding the cost and quality of patient care. You believe in a strict interpretation when defining various reportable incidents; as a result, your hospital's rating on a number of quality issues has declined in the six months you have held the position. Your predecessor was more lenient and was inclined to let minor incidents go unreported or to classify some serious incidents as less serious. The quarterly quality meeting is next week, and you know that your reporting will be challenged by the chief of staff and other members of the quality review board. How should you prepare for this meeting? Should you defend your strict reporting procedures or revert to the former reporting process for the "sake of consistency in the numbers," as several people have urged?

4. As a second-year teacher at a lowly rated inner-city elementary school, you have been asked to form and lead a three-person committee to define and obtain funding for an Ed-Tech program for your school. How would you go about fulfilling this responsibility?

5. You are on the administrative staff of a large, midwestern hospital that is taking its first steps to implement a comprehensive EHR system. You have been asked to survey the medical personnel to determine their level of acceptance regarding the implementation of the system and to help reduce any resistance. How would you propose surveying the staff? What sort of resistance would you expect to encounter?

6. You have volunteered to lead a group of citizens in approaching the board of directors of the nearest hospital (55 miles away) about establishing some sort of telemedicine-based monitoring of 50 or so chronically ill people in your small community. What sort of facts do you need to gather to make a sound recommendation to the board?

7. You have been offered a position as a software support analyst. If you accept, you will have three weeks of on-site training, after which you will work from your home full-time, answering customer service calls. What questions would you want answered before you decide whether or not to take this position?

Cases

1. IT Enables a Smarter Power Grid

The U.S. electrical power infrastructure, also called the grid, delivers electricity from points of generation to consumers around the country. The electric delivery network functions via two primary systems: the transmission system and the distribution system. The transmission system delivers electricity from power plants to distribution substations, while the distribution system delivers electricity from distribution substations to consumers.[52]

The Department of Energy has been charged with leading the wholesale modernization of the U.S. electrical power grid. This transition is likely to take decades and cost hundreds of billions of dollars. The first phase of the transition involves the implementation of a smart grid, which will convert the power grid from a centralized generation and distribution model to one that is more distributed and diverse. Peaks in demand on the smart grid will be partially met by distributed renewable power sources—such as solar panels and windmills—feeding power into the grid. For example, Pacific Gas & Electric, which serves Southern California, would be able to draw energy from some 15,000 solar installations in its service area rather than fire up an auxiliary power plant in the middle of a hot afternoon.

The smart grid will use networks and switches for power management, data collection, communications, and real-time monitoring, as well as building sensors and applications for meter reading and optimizing energy use and distribution. The smart grid will create a *two-way* flow of information from power plant to power plug. Billions of power meter devices, each with its own address—much like every computer device has its own Internet address—will be connected to the smart grid. This two-way communications link between power consumer and power provider will give both parties much greater control over power consumption.[53] The smart grid could become enormous in terms of the number of devices connected to it—as big as 100 Internets—and require spending as much as $100 billion on routers, switches, and secure Internet-based communications.[54]

With the smart grid, "a power company can optimize grid performance, prevent outages, restore outages faster, and allow consumers to manage energy use right down to the individual networked appliance. Smart grids can also incorporate new sustainable energies such as wind and solar generation, interact locally with distributed power sources, and even accept power stored in the batteries of consumers' plug-in electric vehicles."[55]

Of course, there are a number of unresolved issues with the smart grid, including costs, benefits, value proposition to consumers, implementation, and deployment. To gain answers to these questions, many organizations are forming partnerships and conducting pilot projects:

- Google formed partnerships with eight national and international energy companies to allow consumers to access data about their energy use through Google's PowerMeter software application. The software, which is currently only available to a limited number of customers, displays data about home energy use provided by the new generation of network-ready smart power meters that are being installed by various utilities around the world.[56]
- Cisco, General Electric, and Florida Power & Light plan to deploy 1 million advanced wireless smart meters to every home and most businesses in Miami-Dade County.[57]
- The United Kingdom has plans to install devices in every household in Britain that will enable consumers to monitor their own energy use and make reductions in energy consumption and carbon emissions as a result.[58]
- Duke Energy plans to build a smart grid in Cincinnati, where it serves 700,000 customers—spending about $1 billion on sensors, intelligent meters, and other system upgrades.[59]

In addition to these pilot projects, part of a 2009 U.S. federal stimulus bill was aimed at giving other smart-grid initiatives a jump start by allocating $11 billion to power utilities to shift their energy supply networks to digital technology.[60]

Smart systems have the potential to save billions of dollars a year. Based on initial results from various pilot projects, consumers' power cost reductions are expected to be in the range of 5 percent to 15 percent.[61] Such savings estimates are based on the assumption that consumers will use information about their energy consumption and costs to modify their power-usage habits to save energy and money. If consumers do this, it will enable power utilities to operate more efficiently.

San Diego Gas & Electric will install more than 1.4 million smart meters in the San Diego region by the end of 2011. Hal D. Snyder, vice president of customer solutions, claims: "Just getting this information in the hands of our customers, these customers could reduce energy usage by 5 percent to 10 percent.... Once we get this information into the home, the next step is the automation of home energy usage."[62]

Indeed it is the prospect of "automation of home energy usage" that has many consumers worried. To what extent will the opportunity to conserve energy power become pressure to do so?

Will utilities charge a premium for specific power-usage patterns or specific appliances? Will onerous utility or government power-usage guidelines be mandated (e.g., thermostats cannot be set below 76 degrees in the summer or above 70 degrees in the winter)? Will utilities automatically and with no advance warning dial down consumers' power consumption to allow confined brownouts but prevent widespread outages?

Discussion Questions

1. What do you see as the major advantages of implementing a smart power grid?

2. What are the key issues and barriers associated with implementation of a smart power grid?

3. If the government provides much of the money required to implement a smart power grid, does that give it the right to control how the power grid is operated?

2. Does IT Investment Pay Off?

Tween Brands, Inc., sells fashion merchandise and accessories for girls aged 7–14 through its 900 Limited Too and Justice stores located in the United States, Puerto Rico, Europe, and the Middle East. (Tween Brands is scheduled to convert all Limited Too stores to Justice stores in 2009.) It also offers its fashions through each brand's Web site. Recent annual revenue was just under $1 billion.[63] The company recently implemented a "put-to-light" system that improved the productivity of warehouse workers by 25 percent. Workers scan identifiers on products received at the warehouse and then follow flashing light signals at stocking locations that indicate where to put the scanned product. In many instances, the incoming product can be unloaded from one truck and loaded directly onto another outbound truck headed for a store. Such cross-docking opportunities eliminate the need to store product in inventory and greatly streamline operations. Just prior to installing the "put-to-light" system, Tween Brands upgraded its warehouse management system, which tracks inventory levels for each item.[64]

MIT economist Erik Brynjolfsson has reported a strong correlation between IT capital per worker and a company's productivity. He says there is a growing consensus that IT is the most important factor in increasing productivity.[65] Considering the impact that technology has had on the workplace, this correlation seems obvious.

Or is it? What about the cost of training Tween Brands workers to use the new inventory and warehouse management systems? What about the costs incurred when the hardware breaks down? What if there is a bug in the software? What if a new system makes the current one obsolete in two or three years?

Paul Strassmann, a former CIO of both Xerox and NASA, disagrees with Brynjolfsson and argues that IT has not improved labor productivity at all. Back in the 1990s, Strassmann developed a method to measure increased productivity based on microeconomics instead of GDP and other large-scale measurements. Strassmann argues that productivity should be taken as the ratio of the cost of goods to transaction costs, which includes administrative and other costs not directly related to the production or delivery of a good or service. Strassmann and others at Alinean LLC have found that this ratio has remained constant for the past 10 years.[66]

In response, macroeconomists argue that Strassmann's calculations do not account for increases in customer value, such as timely delivery or the creation of innovative products and services. In addition, they say technology has allowed companies to meet regulatory standards and reporting requirements established by government agencies.

Strassmann and other proponents of the microeconomic approach don't deny this oversight and don't argue against IT investment. No company can remain competitive and fail to innovate. Strassmann and his colleagues are simply hoping to do away with short "build and junk" cycles, in which new information systems are scrapped before ever breaking even, and avoid ill-conceived IT investments. They want to develop microeconomic parameters that CIOs can take to their boards of directors as proof that they are cutting costs.[67] Thus, while economists still debate the impact of IT investment, the discussions are at least producing tools that help businesses improve their IT decision making and perhaps even increase their productivity.

Discussion Questions

1. Apart from the annual rate of output per worker, what are other ways of measuring labor productivity?

2. What factors determine whether a new information system will increase or decrease labor productivity?

3. Why is it so difficult to determine whether IT has increased labor productivity?

3. Technological Advances Create Digital Divide in Health Care

Healthcare costs are skyrocketing, with premiums rising annually at double-digit rates. A host of factors are to blame, including rising prescription and hospital costs, expanding coverage mandated by state and federal governments, excessive administrative expenses, and—last but not least—expensive technological advances. Yet, some of these advances are designed to decrease healthcare costs—at least in theory.

Diabetic patients can purchase home monitoring systems such as glucose tests and a device called Health Buddy, which collects the readings and transmits them via phone line to the Health Hero database; the data is then available for review by the patient's healthcare provider.[68] The goal of the program is to avoid health crises and reduce total expenses, but an important question is: How will it affect the medical expenses of the individual participants?[69] Will the patients have to pay for all or part of the Health Buddy? Will the system increase or decrease annual patient costs?

The irony of technological advances in medicine is that while they can prolong and improve the quality of life, they also create new standards and expectations, which could ultimately either push the healthcare system toward a financial crisis or increase the disparity between rich and poor. Patients who cannot pay for these new technologies fall through the gap of the healthcare system's very own "digital divide." The success of programs such as Health Buddy depends not only on whether they avoid health crises and decrease Medicare expenses but also on whether they lower patient costs.

Discussion Questions

1. Can you provide examples that either refute or confirm the idea that a gap exists between the kinds of healthcare services available to the wealthy and the poor in the United States?

2. Should healthcare organizations make major investments in telemedicine to provide improved services that only the wealthy can afford?

3. What are the drawbacks of telemedicine? What situations might not lend themselves to telemedicine solutions?

End Notes

1. Philip Maddocks, "Struggling to Find a New Meaning to Its Existence, the Dow Hires a Self-Help Guru," *Wicked Local Natick*, March 6, 2009, www.wickedlocal.com/natick/news/lifestyle/columnists/x949368656/MADDOCKS-Struggling-to-find-a-new-meaning-to-its-existence-the-Dow-hires-a-self-help-guru.

2. MediaClubSouthAfrica.com, "South Africa's Languages," www.mediaclubsouthafrica.com/index.php?option=com_content&view=article&id=80:languages&catid=33:land_bg&Itemid=70.

3. South African Institute of Race Relations, "Unemployment Still High Across All Provinces—24th April 2008," www.sairr.org.za/press-office/archive/unemployment-still-high-across-all-provinces-24th-april-2008.html.

4. Cheryl Brown, Herbert Thomas, Antoinette van der Merwe, and Liezl van Dyk, "The Impact of South Africa's ICT Infrastructure on Higher Education," Proceedings of the 3rd International Conference on e-Learning, June 26, 2008, http://sun025.sun.ac.za/portal/page/portal/Administrative_Divisions/SOL/All%20shared%20documents/Dokumente/Brown.pdf.

5. NComputing, "South Africa's Western Cape Schools Bridge the Digital Divide," www.ncomputing.com/Portals/0/New_CS/newcasestudies/CS_EDU_WESTERNCAPE_US_EN_REV4_060408_WEB.pdf.

6. Western Cape Education Department Technology in Education Project, "Summary of the Project," Khanya, www.khanya.co.za/projectinfo/?catid=32.

7. Western Cape Education Department Technology in Education Project, Home Page, Khanya, www.khanya.co.za.

8. NComputing, "South Africa's Western Cape Schools Bridge the Digital Divide," www.ncomputing.com/Portals/0/New_CS/newcasestudies/CS_EDU_WESTERNCAPE_US_EN_REV4_060408_WEB.pdf.

9. NComputing, "South Africa's Western Cape Schools Bridge the Digital Divide," www.ncomputing.com/Portals/0/New_CS/newcasestudies/CS_EDU_WESTERNCAPE_US_EN_REV4_060408_WEB.pdf.

10. Ryan Hoffmann, "E-Learning Boost in Western Cape," *The Teacher*, April 9, 2009, www.theteacher.co.za/article/elearning-boost-in-western-cape.

11. Monique Duval, Western Cape Education Department Technology in Education Project, "Learners Make Music Thanks to New Laboratory," Khanya, April 16, 2009, www.khanya.co.za/news/media/?pageid=491.

12. Moira De Roche, Western Cape Education Department Technology in Education Project, "Khanya Computerises over 1000 Schools," Khanya, April 17, 2009, www.khanya.co.za/news/media/?pageid=486.

13. Moira De Roche, Western Cape Education Department Technology in Education Project, "Khanya Computerises over 1000 Schools," Khanya, April 17, 2009, www.khanya.co.za/news/media/?pageid=486.

14. Kimberly Amadeo, "The Great Depression of 1929—Could It Happen Again?," About.com, http://useconomy.about.com/od/grossdomesticproduct/p/1929_Depression.htm.

15. Stephen D. Oliner and William L. Wascher, "Is a Productivity Revolution Underway in the United States?," *Challenge*, November–December 1995, www.questia.com/googleScholar.qst;

jsessionid=K2sGFsv9Dtjpghh25XQYNGcZcWNVv4hGWLJJ1sLX0wJtGGX2RZZq!
1481560549!1622387428?docId=5000361656.

16 McKinsey Global Institute, "Whatever Happened to the New Economy?," November 2002.

17 Erik Brynjolfsson and Lorin M. Hitt, "Computing Productivity: Firm-Level Evidence," November 2002, http://opim.wharton.upenn.edu/~lhitt/cpg.pdf.

18 Eve Tahmincioglu, "The Quiet Revolution: Telecommuting," MSNBC.com, October 5, 2007, www.msnbc.msn.com/id/20281475.

19 Lisa Shaw, *Telecommute! Go to Work Without Leaving Home* (New Jersey: Wiley, 1996).

20 Matt Williams, "Swine Flu: Agencies Scramble to Update Telecommuting Policies," *Government Technology*, April 30, 2009, www.govtech.com/gt/652450.

21 "Digital Divide in U.S. Closing Slowly but Still 'Significant,'" *Baltimore Business Journal*, May 14, 2008, www.bizjournals.com/baltimore/stories/2008/05/12/daily26.html.

22 Mark Bradshaw, "Europe's Digital Divide," *Krakow Post*, April 28, 2008, www.krakowpost.com/article/1069.

23 "Gunjan Bagla, "Bringing IT to Rural India One Village at a Time," *CIO*, March 2005.

24 U.S. Department of Education, "E-Rate Program—Discounted Telecommunications Services," www.ed.gov/about/offices/list/oii/nonpublic/erate.html.

25 Andrew T. LeFevre, "Report Finds Fraud, Waste, and Abuse in Federal E-Rate Program," Heartland Institute, January 1, 2006, www.heartland.org/publications/school%20reform/article/18217/Report_Finds_Fraud_Waste_and_Abuse_in_Federal_ERate_Program.html.

26 Antone Gonsalves, "Study: Internet Has No Impact on Student Performance," *Information-Week*, November 21, 2005, www.informationweek.com/news/global-cio/showArticle.jhtml?articleID=174400767.

27 Bridge Multimedia, "EdTech Online: U.S. Department of Education Technology Grant Programs," www.edtechonline.org/grantsataglance.php.

28 Bridge Multimedia, "EdTech Online: U.S. Department of Education Technology Grant Programs," www.edtechonline.org/grantsataglance.php.

29 U.S. Department of Education, "Tear Down Those Walls: The Revolution Is Underway," www.ed.gov/about/offices/list/os/technology/plan/2004/site/theplan/edlite-TearDownThoseWalls.html.

30 Joanna Stern, "First Look: OLPC XO-2," *Laptop*, May 20, 2008, http://blog.laptopmag.com/first-look-olpc-xo-generation-20.

31 John Ribeiro and Olusegun Ogundeji, "India and Sierra Leone Place OLPC Orders," *Industry Standard*, April 24, 2009, www.thestandard.com/news/2009/04/24/india-sierra-leone-place-olpc-orders.

32 Intel Corporation, "Product Brief: The Classmate PC Powered by Intel," www.intel.com/intel/worldahead/pdf/CMPCbrochure.pdf.

33 Michael Kanellos, "Intel's Bridge for the Digital Divide, *CNET*, June 15, 2006, http://news.cnet.com/Intels-bridge-for-the-digital-divide/2100-1005_3-6084250.html.

34 ASUS, "ASUS Introduces the Highly-Anticipated Eee PC Mobile Internet Gadget," October 16, 2007, http://usa.asus.com/news_show.aspx?id=8684.

35 Kaiser Family Foundation, "Health Care Costs and Election 2008," October 17, 2008, www.kff.org/insurance/h08_7828.cfm.

36 Amanda Gardner, "U.S. Health-Care Costs to Top $4 Trillion by 2016," *Washington Post*, February 21, 2007, www.washingtonpost.com/wp-dyn/content/article/2007/02/21/AR2007022100524.html.

37 Len M. Nichols, "Can Defined Contribution Health Insurance Reduce Cost Growth?," Employee Benefit Research Institute Issue Brief, no. 246, June 2002, http://papers.ssrn.com/sol3/papers.cfm?abstract_id=318824.

38 Steve Lohr, "Government Wants to Bring Health Records into Computer Age," *New York Times*, July 21, 2004, www.nytimes.com/2004/07/21/business/government-wants-to-bring-health-records-into-computer-age.html.

39 Consumers Union, "To Err Is Human—to Delay Is Deadly," SafePatientProject.org, May 2009, www.safepatientproject.org/safepatientproject.org/pdf/safepatientproject.org-ToDelayIsDeadly.pdf.

40 Reuters, "US Pushes Digital Medical Records," *Boston Globe*, July 22, 2004, www.boston.com/business/technology/articles/2004/07/22/us_pushes_digital_medical_records.

41 "US Hospital Use of Electronic Health Records Abysmally Low, Says New Study," *e! Science News*, March 25, 2009, http://esciencenews.com/articles/2009/03/25/us.hospital.use.electronic.health.records.abysmally.low.says.new.study.

42 Todd Park and Peter Basch, "A Historic Opportunity: Wedding Health Information Technology to Care Delivery Innovation and Provider Payment Reform," *Center for American Progress*, May 2009, www.americanprogress.org/issues/2009/05/pdf/health_it.pdf.

43 "Physician Use of HIT in Hospitals Linked to Fewer Deaths and Complications, Lower Costs," *e! Science News*, January 27, 2009, http://esciencenews.com/articles/2009/01/27/physician.use.hit.hospitals.linked.fewer.deaths.and.complications.lower.costs.

44 Lauran Neergaard, "Study: Doctors Resist Use of Electronic Medical Records," *SignOnSanDiego.com*, September 13, 2005, www.signonsandiego.com/news/health/20050913-2252-computerizedmedicine.html.

45 Lauran Neergaard, "Study: Doctors Resist Use of Electronic Medical Records," *SignOnSanDiego.com*, September 13, 2005, www.signonsandiego.com/news/health/20050913-2252-computerizedmedicine.html.

46 John Commins, "Hospital CIOs: EHR Carrot Too Small, Stick Too Big," *HealthLeaders Media*, April16, 2009, www.healthleadersmedia.com/content/231613/topic/WS_HLM2_TEC/Hospital-CIOs-EHR-Carrot-Too-Small-Stick-Too-Big.html.

47 "Rock and a Hard Place: An Analysis of the $36 Billion Impact from Health IT Spending," PricewaterhouseCoopers' Health Research Institute, accessed at http://pwchealth.com/cgi-local/hregister.cgi?link=reg/HIT_incentives_rock_and_a_hard_place.pdf.

48 Jorge Cuadros, "Preventing Diabetic Blindness with EyePACS, a Low-Cost Store-and-Forward Telemedicine System," February 6, 2008, www.citris-uc.org/RE-Feb06.

49 Carolyn Bloch, "Telemedicine Session Held," *Federal Telemedicine News*, September 14, 2008, http://telemedicinenews.blogspot.com/2008/09/telemedicine-session-held.html.

298

50 Marianne Kolbasuk McGee, "WellPoint Offers New Online Tools to Help Consumers Make Smarter Health-Care Decisions," *InformationWeek*, September 21, 2005, www.information week.com/news/internet/ebusiness/showArticle.jhtml?articleID=171000842&cid=RSSfeed_ IWK_Authors.

51 Marianne Kolbasuk McGee, "WellPoint Offers New Online Tools to Help Consumers Make Smarter Health-Care Decisions," *InformationWeek*, September 21, 2005, www.information week.com/news/internet/ebusiness/showArticle.jhtml?articleID=171000842&cid=RSSfeed_ IWK_Authors.

52 U.S. Department of Energy, "Smart Grid," www.oe.energy.gov/smartgrid.htm.

53 Martin LaMonica, "Will Anyone Pay for the 'Smart' Power Grid?," *CNET*, May 16, 2007, http://news.cnet.com/Will-anyone-pay-for-the-smart-power-grid/2100-11392_3-6184046.html.

54 Cora Nucci, "Cisco: Smart Grid's a $100 Billion Baby," *InformationWeek*, May 18, 2009, www.informationweek.com/blog/main/archives/2009/05/cisco_smart_gri.html; jsessionid=31E1ST5JYPSMUQSNDLPSKHSCJUNN2JVN.

55 IBM, "Smart Grid," www.ibm.com/ibm/ideasfromibm/ca/en/smartplanet/topics/utilities/ 20090219/index.shtml.

56 Thomas Claburn, "Google, Utilities Bringing Energy Usage Data to Consumers," *Information-Week*, May 20, 2009, www.informationweek.com/news/internet/google/showArticle.jhtml? articleID=217600324.

57 Cora Nucci, "Smart Grid Opposed by AARP," *InformationWeek*, May 20, 2009, www.informationweek.com/blog/main/archives/2009/05/smartgrid_oppos.html;jsessionid=31 E1ST5JYPSMUQSNDLPSKHSCJUNN2JVN.

58 Cora Nucci, "Smart Grid Opposed by AARP," *InformationWeek*, May 20, 2009, www.informationweek.com/blog/main/archives/2009/05/smartgrid_oppos.html;jsessionid=31 E1ST5JYPSMUQSNDLPSKHSCJUNN2JVN.

59 K. C. Jones, "IT Gives Smart Grid Initiatives a Jolt," *InformationWeek*, March 21, 2009, www.informationweek.com/news/government/stimulus/showArticle.jhtml? articleID=215901346.

60 K. C. Jones, "IT Gives Smart Grid Initiatives a Jolt," *InformationWeek*, March 21, 2009, www.informationweek.com/news/government/stimulus/showArticle.jhtml? articleID=215901346.

61 K. C. Jones, "IT Gives Smart Grid Initiatives a Jolt," *InformationWeek*, March 21, 2009, www.informationweek.com/news/government/stimulus/showArticle.jhtml? articleID=215901346.

62 Thomas Claburn, "Google, Utilities Bringing Energy Usage Data to Consumers," *Information-Week*, May 20, 2009, www.informationweek.com/news/internet/google/showArticle.jhtml? articleID=217600324.

63 Tween Brands, Inc., "Tween Brands 2008 Annual Report," http://phx.corporate-ir.net/External. File?item=UGFyZW50SUQ9MTE3NHxDaGlsZEIEPS0xfFR5cGU9Mw==&t=1.

64 "Web-Enabled Fulfillment System Increases Productivity 25% at Tween Brands," *Internet Retailer*, January 8, 2007, www.internetretailer.com/internet/marketing-conference/436365475- web-enabled-fulfillment-system-increases-productivity-25-at-tween-brands.html.

[65] Erik Brynjolfsson, "The IT Productivity Gap," *InformationWeek's Optimize* 21 (July 2003).

[66] Paul A. Strassmann, *Information Productivity: Assessing the Information Management Costs of U.S. Industrial Corporations* (New Canaan, CT: Information Economics Press, 1999).

[67] Paul A. Strassmann, *Information Productivity: Assessing the Information Management Costs of U.S. Industrial Corporations* (New Canaan, CT: Information Economics Press, 1999).

[68] Seemeen Mirza, "Overview of the U.S. Diabetes Remote Patient Monitoring Devices Market," *Frost & Sullivan*, April 29, 2004, www.frost.com/prod/servlet/market-insight-top.pag?docid=18086184.

[69] McKesson, "Health Buddy Appliance," www.mckesson.com/en_us/McKesson.com/For%2BHealthcare%2BProviders/Home%2BCare/Telehealth%2BSolutions/McKesson%2BTelehealth%2BAdvisor%2BComponents/Health%2BBuddy%2BAppliance.html.

300

SOCIAL NETWORKING

VIGNETTE

Twitter Emerges as News Source for Iran Protesters

Twitter is a social networking service that enables its users to send and read each other's text-based

posts, known as *tweets*. A tweet is supposed to answer the question "What are you doing?" in 140

characters or less. Many people use Twitter as a means of staying connected to friends, relatives, and

coworkers; others use it for professional networking. Twitter users can restrict delivery of their messages

to those in their circle of friends or allow access to all Twitter users. Users can also control whose tweets

they receive, when they receive them, and on what devices.

The service works over multiple networks with various devices in countries around the world.

Messages can be sent using the Twitter Web site (*www.twitter.com*) or via cell phones employing Short

Message Service (SMS).

While Twitter does not release information about the number of active accounts, Nielsen Online

estimated that there were 7 million unique U.S. visitors to the company's Web site in February 2009;

each visitor spent an average of about six minutes per day on the service. The year over year growth in number of users was estimated to be an amazing 1,382 percent.[1]

Twitter was created by Jack Dorsey in 2006. The company is backed by a number of venture capitalists and is headquartered in San Francisco. Twitter has had some problems related to its rapid growth. The service experienced almost three and a half days of downtime during 2008, resulting from traffic overloads.[2] In January 2009, 33 celebrity Twitter accounts were compromised, and bogus tweets—with drug-related and sexually explicit contents—were sent from the accounts.[3] A 2009 study by Nielsen Online estimated that more than 60 percent of Twitter users stop using the service within one month of joining.[4]

In June 2009, Iran held its presidential election. Within hours of the polls closing, the interior ministry proclaimed that the incumbent, Mahmoud Ahmadinejad, had defeated the reformist candidate, Mir Hossein Mousavi, by a margin of over 11 million votes. Many Iranians, reasoning that it was not possible for over 39 million paper ballots to have been counted so quickly, were extremely upset; they felt that the election had been rigged in favor of Ahmadinejad. Thousands of protesters soon took to the streets of Iran to demonstrate their frustration.

Twitter users in Iran soon began using the service to share commentary about the ongoing protests and to plan future rallies and protests. The tweets chronicling the events in Iran numbered in the tens of thousands and were highly subjective, unverifiable, and impossible to tie to sources. However, they provided an interesting insight into events taking place in Iran, especially after the Iranian authorities ordered foreign journalists to halt coverage of demonstrations, and reporters with temporary visas were ordered to leave the country.[5] Shortly after the election, the U.S. State Department contacted Twitter and asked the company to delay a scheduled software upgrade that would have cut off daytime service to Iran. "We highlighted to them that this was an important form of communication," according to an unidentified State Department official.[6]

Throughout the period of intense protests immediately following the election, there were numerous rumors that Iranian authorities were trying to disrupt Twitter traffic. Because Internet communications in and out of Iran flow through a very small number of servers, the government could fairly easily monitor these servers and block IP addresses delivering tweets. Messages sent out via the Twitter network can be blocked in a similar fashion. In response to the rumors, supporters outside Iran set up proxy servers to relay Twitter Internet content through unblocked network addresses. It was reported that in addition to attempting to block Twitter traffic, the Iranian government began to make use of Twitter to communicate its own message.[7]

While Twitter was not responsible for the protest movement in Iran, some observers felt that Twitter "emboldened the protesters, reinforced their conviction that they [were] not alone, and engaged populations outside Iran in an emotional, immediate way that was never possible before."[8]

Questions to Consider

1. Many people question the value of Twitter's goal of helping people stay connected in real time. Do the events in Iran illustrate the potential value to society of social networking tools such as Twitter? Are there other ways that Twitter and other social networks could be used by people and organizations to add real value?
2. How trustworthy is the information one gleans from social networks such as Twitter?

LEARNING OBJECTIVES

As you read this chapter, consider the following questions:

1. What are social networks, how do people use them, and what are some of their practical business uses?
2. What are some of the key ethical issues associated with the use of social networking Web sites?
3. What is a virtual life community, and what are some of the ethical issues associated with such a community?

WHAT IS A SOCIAL NETWORKING WEB SITE?

A **social networking Web site** is a site whose purpose is to create an online community of Internet users that enables members to break down barriers created by time, distance, and cultural differences. Social networking Web sites allow people to interact with others online by sharing opinions, insights, information, interests, and experiences. Members of an online social network may use the site to interact with friends, family members, and colleagues—people they already know—but they may also wish to develop new personal and professional relationships.

With over 1.6 billion Internet users, there is an endless range of interests represented online, and a correspondingly wide range of social networking Web sites catering to those interests.[9] In the United States, total minutes per month spent on social networking Web sites increased 83 percent from April 2008 to April 2009. Total minutes on Facebook grew from 1.7 billion in April 2008 to 13.9 billion in April 2009, making Facebook the number one social networking Web site when ranked by total minutes per month.[10] There are thousands of social networking Web sites worldwide; Table 9-1 lists some of the most popular ones, based on the number of unique visitors per month.

TABLE 9-1 Popular social networking Web sites

Social networking Web site	Description	Number of unique visitors in January 2009
Facebook.com	Largest social networking Web site based on the number of unique visitors per month; used by members to keep up with friends, upload photos, share links and videos, and learn more about the people they meet	69 million
MySpace.com	General social networking Web site used by teenagers and adults worldwide; designed to allow members to communicate with friends via personal profiles, blogs, and groups, as well as post photos, music, and videos to their personal pages	59 million
Classmates.com	Networking site designed to help members find and keep in touch with people they knew in grade school, high school, college, and the military	17 million
Reunion.com	Site that helps members find and keep in touch with old friends, relatives, and loved ones	14 million
LinkedIn.com	Business-oriented Web site used for professional networking; users create a network made up of people they know and trust in business	11 million
imeem.com	Music sharing site that enables members to watch video clips, stream music, view photos, post to blogs and forums, join groups, and browse profiles	9 million

TABLE 9-1 Popular social networking Web sites (*continued*)

Social networking Web site	Description	Number of unique visitors in January 2009
Flixster	Networking site geared toward people interested in discussing movies and actors with other members; the site has an extensive database of information about movies and actors, and recommends new friends based on similar tastes in movies	8 million
Twitter.com	Service for friends, family members, and coworkers looking to stay connected through the frequent, quick exchange of messages that are a maximum of 140 characters	6 million

Source: Andy Kazeniac, "Social Networks: Facebook Takes Over Top Spot, Twitter Climbs," *Compete,* February 9, 2009, http://blog.compete.com/2009/02/09/facebook-myspace-twitter-social-network.

According to the Pew Internet & American Life Project, about 35 percent of U.S. Internet users age 18 and older have a profile on an online social networking Web site. Meanwhile, 65 percent of teenagers use social networking sites.[11] Table 9-2 shows some of the top social networks for adults in the United States as of May 2008.

TABLE 9-2 Most popular U.S. social networks for Internet users age 18 and older

Social network	Percent of all U.S. adult social network users with a profile on this site as of May 2008
MySpace	50%
Facebook	22%
LinkedIn	6%
Yahoo	2%
YouTube	1%
Classmates.com	1%
Others (BlackPlanet, Orkut, hi5, and Match.com)	10%

Source: Amanda Lenhart, "Adults and Social Network Websites," Pew Internet & American Life Project, January 14, 2009, www.pewinternet.org/Reports/2009/Adults-and-Social-Network-Websites.aspx.

BUSINESS APPLICATIONS OF ONLINE SOCIAL NETWORKING

While social networking Web sites are primarily used for nonbusiness purposes, a number of forward-thinking organizations are employing this technology to advertise, assess job candidates, and sell products.

Social Network Advertising

Social network advertising involves the use of social networks to inform, promote, and communicate the benefits of products and services. Social network advertising has become big business in the United States, with 2009 ad spending on Facebook estimated to be $230 million, and ad spending on MySpace estimated to be $495 million.[12]

Organizations are increasingly looking to new forms of advertising to reach their target markets. As Michael Wiley, director of new media at General Motors Communications, puts it, "The existing advertising paradigm sucks. It's woefully inefficient."[13] Two significant advantages of social networking advertising over more traditional advertising media (e.g., radio, TV, and newspapers) are: (1) advertisers can create an opportunity to generate a conversation with viewers of the ad, and (2) ads can be targeted to reach people with the desired demographic characteristics.

There are several social network advertising strategies, and organizations may employ one or more of the following:

Direct Advertising

Direct advertising involves placing banner ads on a social networking Web site. An ad can either be displayed to each visitor to the Web site or, by using the information in user profiles, be directed toward those members who would likely find the product most appealing. Thus, an ad for a new magazine on mountain biking could be directed to individuals on a social networking Web site who are male, who are 18 to 35 years old, and who express an interest in mountain biking. Others on the social networking Web site would not see it.

Advertising Using an Individual's Network of Friends

Companies can use social networking Web sites to advertise to an individual's network of contacts. When you sign on to your favorite social networking Web site, you might see a message saying, "Jared [your friend] just went to see Transformers II—awesome, he says!" This can be an extremely persuasive message, as people frequently make decisions to do something or purchase something based on input from their close group of friends. This might be a spontaneous message sent by Jared, or Jared might be getting paid by an online promotion firm to send messages about certain products. There are certainly ethical issues with this approach, as some people consider this to be exploiting an individual's personal relationships for the financial benefit of a company.

Indirect Advertising Through Groups

Innovative companies are also making use of a new marketing technique by creating a group on a social networking Web site that interested users can join by becoming "fans." These groups can quickly grow in terms of numbers of fans to become a very effective marketing tool for a company looking to market contests, promote new products, or simply increase brand awareness.

In its ongoing fight for market share in the beverage industry, Coca-Cola has implemented a number of social networking initiatives to promote its brands. Coke has its own corporate blog called Coca-Cola Conversations that covers its brand history and provides information about Coca-Cola collectibles. In April 2007, Coke started a competition for the residents of the Second Life virtual world, challenging them to design a vending machine that dispenses the essence of Coke. The company also placed a video on YouTube called "Mean Joe Greene—The Making of the Commercial," documenting the making of one of Coke's most famous TV commercials.

In August 2008, two fans of Coke, neither of whom had any official connection to the company, launched a Coca-Cola Facebook page. Within a few weeks, the page had attracted over 750,000 fans. As the number of fans grew into the millions, the page's creators agreed to turn over administration of the page to Coca-Cola.[14] The site is monitored by software filters for offensive words and phrases, and live moderators check its pages for anything truly offensive. Other than that, Coca-Cola managers pretty much let Facebook fans say what they want on the site.[15] The result has been nothing short of amazing. The Facebook fan page quickly grew to over 3 million members worldwide.[16] It is now the second most popular page on Facebook, with 3.3 million members as of March 2009.[17]

Company-Owned Social Networking Web Site

A variation on the above approach is for a company to form its own social networking Web site. Dell created its own social networking Web site, IdeaStorm, as a means for its millions of customers in more than 100 countries to talk about what new products, services, or improvements they would like to see Dell develop. Since its launch in February 2007, the Dell community has suggested 11,996 ideas and posted 84,851 comments; Dell has implemented 350 customer-submitted ideas.[18]

Viral Marketing

Viral marketing encourages individuals to pass along a marketing message to others, thus creating the potential for exponential growth in the message's exposure and influence as one person tells two people, each of those two people tell two or three more people, and so on. The goal of a viral marketing campaign is to create a buzz about a product or idea that spreads wide and fast. A successful viral marketing campaign requires little effort on the part of the advertiser; however, the success of such campaigns can be very difficult to predict.

Hotmail created what is recognized by many as the most successful viral marketing campaign ever when it first launched its service in 1996. Every e-mail sent by a Hotmail user contained a short message at the end of the e-mail that promoted Hotmail's free e-mail service. As a result, almost 12 million new users signed up for Hotmail over a period of 18 months.[19]

The Use of Social Networks in the Hiring Process

Employers can and do look at the social networking profiles of job candidates when making hiring decisions (see Figure 9-1). According to a recent survey by CareerBuilder.com, 22 percent of hiring managers use social networking Web sites as a source of information about candidates, and an additional 9 percent are planning to do so. Of those managers who use social networking Web sites to screen candidates, 34 percent have found information that made them drop a candidate from consideration. Companies may reject candidates who post information about their drinking or drug use habits or those who post provocative or inappropriate photos. Candidates are also sometimes rejected due to postings containing discriminatory remarks relating to race, gender, and religion or because of postings that reveal confidential information from previous employers.[20]

FIGURE 9-1   Employers often use social networking Web sites as a source of information about job candidates

Employers can legally reject a job applicant based on the contents of the individual's social networking profile as long as the company is not violating federal or state discrimination laws. For example, an employer cannot legally screen applicants based on race or ethnicity. Or suppose that by checking a social networking Web site, a hiring manager finds out that a job candidate is pregnant and makes a decision not to hire that person based on that information. Refusing to hire on the basis of pregnancy is prohibited by the Pregnancy Discrimination Act, which amended Title VII of the Civil Rights Act of 1964. The employer would be at risk of a job employment discrimination lawsuit.

Members of social networking Web sites frequently provide sex, age, marital status, sexual orientation, religion, and political affiliation data in their profile. Users who upload personal photos may reveal a disability or their race or ethnicity; therefore, without even thinking about it, an individual may have revealed data about personal characteristics that are protected by civil rights legislation. Some human resource executives feel that they can use social networking Web sites to "learn a little about the candidate's cultural fit and professionalism."[21]

Another survey done by CollegeGrad.com revealed that 47 percent of college graduates who use social networking Web sites change the contents of their pages as a result of their job search.[22] Many jobseekers delete their Facebook or MySpace account altogether because they know employers check such sites. More graduates are beginning to realize that pictures and words posted online, once intended for friends only, are reaching a much larger audience and could have an impact on their job search.[23]

Social Shopping Web Sites

A **social shopping Web site** brings shoppers and sellers together in a social networking environment in which members can share information and make recommendations while shopping online. Thus, these sites combine two highly popular online activities—shopping and social networking. Social shopping Web site members can typically build their own pages to collect information and photos about items in which they are interested. On many social shopping Web sites, users can offer opinions on other members' purchases or potential purchases. The social shopping Web site Stuffpit has implemented a reward system for members, in which they are paid a commission each time another shopper acts on their recommendation to purchase a specific item.[24]

There are numerous social shopping Web sites, a few of which are summarized in Table 9-3.

TABLE 9-3 Sample of social shopping Web sites

Social shopping site	Brief description
Buzzillions	Product review Web site that collects thousands of product reviews from the Web sites of various retailers
Crowdstorm	Shopping resource that aggregates product information from various online buyers guides, reviews, and blog postings
Kaboodle	Site where members can discover and recommend new products; get discounts; and locate bargains
OSOYOU	UK-based social shopping site for women with an interest in fashion and beauty products
ZEBO	Site that allows members to create a personal profile about what they own, want, and love to shop for; members can check out one another's profiles, provide shopping tips, and chat online to ask questions and get advice

Social shopping Web sites generate revenue through retailer advertising. Some also earn money by sharing with retailers data about their members' likes and dislikes.

Social shopping Web sites can be a great way for small businesses to boost their sales. Amenity Home—a tiny start-up with just three products, four employees, and no advertising budget—became a retailer on ThisNext.com, a social shopping Web site whose goal is to link shoppers with hard-to-find products. Shoppers at ThisNext.com found the Amenity Home products, copied photos of the products to their own blog pages, and brought the tiny firm some much-needed recognition—Amenity Home products started getting more and more hits on ThisNext.com.[25]

Retailers can purchase member data and comments from some social shopping Web sites to find out what consumers like and don't like, and what they are looking for in items sold by the retailer. This can help the retailer design product improvements and come up with ideas for new product lines.

When you have a community of tens of millions of users, not everyone is going to be a good "neighbor" and abide by the rules of the community. Many will stretch or exceed the bounds of generally accepted behavior. Some of the common ethical issues that arise for members of social networking Web sites are cyberbullying, cyberstalking, encounters with sexual predators, and the uploading of inappropriate material.

Cyberbullying

Cyberbullying is the harassment, torment, humiliation, or threatening of one minor by another minor or group of minors via the Internet or cell phone. According to a recent survey of over 800 students ages 13–17, about 43 percent had experienced cyberbullying in the past year. Cyberbullying is more common among females and among 15- and 16-year-olds (see Figure 9-2).[26]

FIGURE 9-2   Cyberbullying is more common among teenage females

Cyberbullying has sometimes become so intense that some children have committed suicide as a result. Ryan Halligan, a 13-year-old boy in Vermont, committed suicide after bullying from his schoolmates and cyberbullying online. Megan Meier, a 13-year-old girl from Missouri, also committed suicide when she was harassed by a fictitious boy created by neighbors with whom she had had a falling out (see Case #1 at the end of this chapter).[27] As a result of these incidents, several states—including Arkansas, Delaware, Idaho, Iowa, Michigan, Minnesota, Nebraska, New Jersey, Oklahoma, Oregon, South Carolina, and Washington—have enacted laws to curb cyberbullying by calling on school districts to develop policies regarding cyberbullying detection and punishment.[28]

There are numerous forms of cyberbullying, such as the following:

- Sending mean-spirited or threatening messages to the victim
- Sending thousands of text messages to the victim's cell phone and running up a huge cell phone bill
- Impersonating the victim and sending inappropriate messages to others
- Stealing the victim's password and modifying his or her profile to include racist, homophobic, sexual, or other inappropriate data that offends others or attracts the attention of undesirable people
- Posting mean, personal, or false information about the victim in the cyberbully's blog
- Creating a Web site whose purpose is to humiliate or threaten the victim
- Taking inappropriate photos of the victim and either posting them online or sending them to others via cell phone
- Setting up an Internet poll to elicit responses to embarrassing questions, such as "Who's the biggest geek in Miss Adams's homeroom?" and "Who is the biggest loser in the senior class?"
- Sending inappropriate messages while playing interactive games that enable participants to communicate with one another

Because cyberbullying can take many forms, it can be difficult to identify and stop. Ideally, minors would inform their parents if they became a victim of cyberbullying. Unfortunately, this often does not happen. When school authorities do get involved in an effort to discipline students for cyberbullying, they are sometimes sued for violating the student's right to free speech, especially if the activity occurred off school premises. As a result, some schools have modified their discipline policy to reserve the right to punish a student for actions taken off school premises if they adversely affect the safety and well-being of a student while in school.

All children should be educated about the potential serious impacts of cyberbullying, how to identify cyberbullying, and why it is important for them to refrain from cyberbullying. Children should be encouraged not to retaliate to mean-spirited messages, as doing so may cause the harassment to increase. Children need to understand that they can become inadvertent cyberbullies if they fail to think through the consequences of their actions. They should also be counseled against posting any data that is too personal, such as phone numbers, their home address, their school, or any other information that could allow a stranger to locate the child.

Cyberstalking

Cyberstalking is threatening behavior or unwanted advances directed at an adult using the Internet or other forms of online and electronic communications; it is the adult version of cyberbullying. Online stalking can be a serious problem for victims, terrifying them and causing mental anguish. It is not unusual for cyberstalking to escalate into abusive or excessive phone calls, threatening or obscene mail, trespassing, vandalism, physical stalking, and even physical assault. Over a dozen states have passed laws prohibiting cyberstalking. "Estimates from Internet safety groups such as Working to Halt

Online Abuse, SafetyEd, and Cyber Angels reveal an increasing number of cyberstalking reports, with 50 to 500 requests per day for help from victims of cyberstalking."[29] Many researchers feel that it is likely that the true extent of cyberstalking has been underestimated, since the number of people online is increasing each year and many cases still go unreported.[30]

In 2007, a New Mexico woman, Devon Lynn Townsend, was convicted of hacking into a Yahoo! server to gain access to the personal e-mail account of Chester Bennington, lead singer of the band Linkin Park. Townsend was able to secretly change the password to Bennington's account so that she could view all the messages sent and received from the account. This enabled her to view photos of Bennington's children; read business correspondence; see information about his family and the band's travel plans; learn the date and location of the Benningtons' childbirth classes; and view information about a new home the Benningtons had purchased, including the home inspection report and photos of the house. Townsend was also able to gain access to Bennington's voice mail and listen to all the messages left on his phone answering service. She used this information to download copies of the Benningtons' wedding pictures, contact the Benningtons and their friends anonymously, and travel to locations where Chester Bennington might be. Townsend was sentenced to two years in a federal prison camp in Arizona.[31]

The National Center for Victims of Crime offers a detailed set of recommended actions for combating cyberstalking, including the following:

- When the offender is known, victims should send the stalker a written notice that their contact is unwanted and that all further contact must cease.
- Evidence of all contacts should be saved.
- Victims of cyberstalking should inform their ISP provider as well as the stalker's ISP, if possible.
- Victims should consider speaking to law enforcement officers.
- Above all else, victims of cyberstalking should never agree to meet with the stalker to "talk things out."[32]

Encounters with Sexual Predators

Some social networking Web sites have been criticized for not doing enough to protect minors from encounters with sexual predators. MySpace spent two years purging potential problem members from its site, including 90,000 registered sex offenders, who were banned from its site in early 2009. (It is estimated that there are over 700,000 sex offenders in the United States.)[33] "Almost 100,000 convicted sex offenders mixing with children on MySpace ... is absolutely appalling and totally unacceptable," stated Attorney General Richard Blumenthal of Connecticut, who is pushing social networking Web sites to adopt stronger safety measures.[34]

Uploading of Inappropriate Material

Most social networking Web sites have policies against uploading videos depicting violence or obscenity. Facebook, MySpace, and most other social networking Web sites have terms of use agreements, a privacy policy, or a content code of conduct that summarizes key legal aspects regarding use of the Web site. Typically, the terms state that the Web site has the right to delete material and terminate user accounts that violate the site's policies. The policies set specific limits on content that is sexually explicit, defamatory, hateful, or violent, or that promotes illegal activity.[35]

Policies do not stop all members of the community from attempting to post inappropriate material, and most Web sites do not have sufficient resources to review all material submitted for posting. For example, about 10 hours of media is being uploaded to YouTube every minute.[36] Quite often, it is only after other members of a social networking Web site complain about objectionable material that such material is taken down. This can be days or even weeks. Ideally, reviewers would also look at the text content submitted to a networking site—not just photos and videos. A posting to a teenage-oriented Web site may advocate underage drinking, sex, and drug use without the use of photos or videos.

Individuals who appear in photos or videos doing inappropriate things may find themselves in trouble with authorities if those photos and videos end up on the Internet. Several students at East Grand Rapids High School in Michigan were given two-week suspensions from sports, a school-sponsored dance, and other extracurricular activities when some parents reported seeing online photos of students drinking alcohol at parties. While none of the parties happened on school grounds, the school has a student conduct code that states that students involved in extracurricular activities can be disciplined for code violations.[37]

In April 2008, six teenagers recorded their beating of 16-year-old Victoria Lindsay and planned to post the video on MySpace and YouTube. The beating was severe enough that Victoria suffered temporary damage to her sight and hearing. According to YouTube, the video was never uploaded, and a YouTube spokesperson stated that "if a video shows someone getting hurt, attacked, or humiliated, it will be removed." A MySpace employee confirmed that the video was never uploaded to that site. MySpace has a dedicated content review team that views every video before it is posted to ensure that the poster is not violating the MySpace terms of use.[38]

On May 20, 2009, an organized group of users uploaded video clips of explicit adult content to YouTube. The video clips were uploaded with no warning that they were for adults only. Even worse, they were tagged with child-friendly identifiers, such as "Jonas Brothers." YouTube worked quickly to remove the video clips; however, the video search results and their explicit thumbnails showed up in searches for days.[39]

ONLINE VIRTUAL WORLDS

An **online virtual world** is a computer-simulated world in which a visitor can move in three-dimensional space, communicate and interact with other visitors, and manipulate elements of the simulated world. Virtual worlds are usually thought of as alternative worlds where visitors go to entertain themselves and interact with others. A visitor to a virtual world represents him- or herself through an **avatar** (see Figure 9-3), a character usually in the form of a human but sometimes in some other form. Avatars can typically communicate with each other via text chat or via voice using Voice over IP.

FIGURE 9-3   An avatar is a representation of a virtual world visitor

Avatars in many virtual worlds can shop, hold jobs, run for political office, develop relationships with other avatars, take a test drive in a virtual world car, and even engage in criminal activities. Avatars may promote events and hold them in the virtual world (e.g., garage sales or concerts). Avatars can even start up new businesses and create or purchase new entities, such as houses, furnishings for their houses, clothing, jewelry, and other products. Avatars use the virtual world's currency to purchase goods and services in the virtual world. The ownership of such items is recognized by other avatars in the virtual world—for example, this is John's house; others may not occupy it without his permission.

Avatars can earn virtual world money by performing tasks in the virtual world, or their owners can purchase virtual world money for them using real world cash. In some virtual worlds, avatars can convert their virtual world money back into real dollars at whatever the going exchange rate is by using their credit card at online currency exchanges. Virtual world items may also be sold to other virtual world players for real world money. A virtual world may also support e-commerce and allow users to sell their own real products (e.g., autos and time-share vacations) within the real world.

Most virtual worlds have rules against offensive behavior in public, such as using racial slurs or performing overtly sexual actions. However, consenting adults can travel to private areas and engage in all sorts of socially unacceptable behavior.

Table 9-4 lists some of the most popular online virtual worlds.

TABLE 9-4 Popular virtual worlds

Virtual world	Description
Coke Studios	Virtual world in which teens and young adults can create customized music mixes in a virtual music studio, play them for other members, and receive ratings for each mix; favorable ratings allow members to earn points that can be used to purchase virtual furnishings for their personal studio areas
Disney's Toontown Online	Disney's virtual world, designed for children as young as seven years old; visitors cans play games, dress their cartoon avatars, and communicate with other visitors through a drop-down menu of phrases
Habbo Hotel	Virtual world for teens, built around a hotel theme; visitors can purchase furniture to decorate their guest rooms or mingle in the lobby
Second Life	Highly imaginative three-dimensional world that is appealing to young adults; avatars are very customizable, and users can change every micro-pixel of their avatar's shape, size, and color
The Sims Online	A networked version of the Sims computer game, targeted at teens and young adults; the object of the game is to earn as much currency as possible to furnish one's virtual home with beautifully rendered household objects

Source: Virtual Worlds Review, www.virtualworldsreview.com/index.shtml.

Crime in Virtual Worlds

Virtual worlds raise many interesting questions regarding what is a criminal act and whether law enforcement—real or virtual—should get involved in acts that occur in virtual worlds. Some virtual activities are clear violations of the law—for example, trafficking in actual drugs or stolen credit cards. Other virtual activities, such as online muggings and sex crimes, can cause real life anguish for the human owners of the avatars involved but may or may not rise to the level of a real life crime.

The following list includes examples of unethical and criminal activities in a variety of virtual worlds:

- Authorities in Germany investigated an incident of virtual abuse in Second Life after they were sent photos of an animated child character engaging in simulated sex with an animated adult figure. Such activity could violate German laws against child pornography even though both animated characters were created by adults.[40]
- In The Sims Online, a number of players formed a "Sim Mafia" and attacked the avatars of unpopular players. The group would send a flood of insulting messages to the targeted character, then trash the target's virtual property, and finally demand that the target delete his or her avatar from the game.[41]
- Some regions in the World of Warcraft, an extremely popular online game, have become so lawless that gangs of animated characters frequently attack lone avatars, stealing their virtual belongings and sometimes murdering them.[42]
- A man playing the Lineage II online game was arrested by Japanese officials for using software to mug avatars and then sell the stolen virtual items for real money.[43]

- A Chinese player of the online game Mir 3 murdered another player because that player had borrowed a high-value saber and then sold it at an online auction site.[44]
- A player of the online game MapleStory became so upset about her sudden divorce from her online husband that she logged onto the game with her virtual world husband's ID and password and killed his avatar. The woman was arrested by Japanese police and jailed on suspicion of illegally accessing a computer and manipulating electronic data. Under Japanese law, she could be imprisoned up to five years or fined up to $5,000.[45]

Bad deeds done online can often be mediated by the game administrators, who can take action according to the rules of the game and with consequences internal to the game. It is only when the harm reaches the real world that victims should look to criminal law to protect them.

Educational and Business Uses of Virtual Worlds

Virtual online worlds are also being used for education and business purposes. The New Media Consortium (NMC) is an international consortium of almost 300 organizations focused on exploring the use of new media and technologies to improve teaching, learning, and creative expression. The NMC attracts nearly 1,500 unique visitors to its Second Life campus each week. Princeton, Yale, the University of Southern California, Ball State, and New York University are just a few of the schools with a campus in Second Life.[46]

Media Grid's Immersive Education Initiative is an international collaboration of universities, research institutes, and companies to define and develop open standards, best practices, platforms, and game-based learning and training systems. Amherst College, Boston College, the City of Boston, Loyola Marymount University, Massachusetts Institute of Technology, and Seton Hall University are some of the members of the Media Grid.[47]

Members of the New Media Consortium and the Media Grid can conduct classes and meetings from within a growing number of virtual learning worlds. They can also build custom virtual learning worlds, simulations, and learning games. The virtual reality experience provides participants with a real sense of being there when attending a virtual class or conference. Experienced designers can develop virtual classes that immerse and engage students in the same way that today's video games grab and keep the attention of players.

Second Life is used by the marketing firm Rivers Run Red as a virtual meeting place to display ads, posters, and other designs in three-dimensional settings for clients and partners around the world. Using virtual meeting space can cut weeks off the time that would be required to transport the actual materials back and forth for a real world meeting.[48]

Northrop Grumman Corporation (NGC) is a global security company with 120,000 employees.[49] As part of its efforts to explore emerging technologies to improve collaboration, NGC built a simulation of its Combat Information Center (CIC) in Second Life to provide a real-to-life training experience for its future operators. The CIC is located inside a scale virtual model of the USS *Blue Ridge*—the command and control ship for the Seventh Fleet. In the real world, the CIC consists of numerous powerful computers distributed throughout the ship.

Data from worldwide sources are entered into a single integrated database that provides a complete tactical picture of air, surface, and subsurface contacts—enabling the fleet commander to quickly assess and concentrate on any situation that might arise.[50] Up to 10 trainees sit side-by-side, practicing their functions simultaneously using the Second Life simulation. The trainees can run through a number of different scenarios, ranging from an attack on a member of the fleet by an enemy submarine, to a mission to provide support for a Navy Seal team conducting a search and rescue mission. Trainees do not need to be transported to the real USS *Blue Ridge* to train, and the impact of a training mistake is minimized with no risk to life or equipment.[51]

Summary

- A social networking Web site creates an online community of Internet users that enables members to break down barriers created by time, distance, and cultural differences; such a site allows people to interact with others online by sharing opinions, insights, information, interests, and experiences.

- Some 35 percent of U.S. Internet users age 18 and older have a profile on an online social networking Web site, while 65 percent of U.S. teenagers use social networking Web sites.

- Social network advertising uses social networks to inform, promote, and communicate the benefits of products and services.

- Social network advertising enables advertisers to generate a conversation with viewers of their ads, and to target ads to reach people with the desired demographic characteristics.

- There are several social network advertising strategies, including direct advertising, advertising using an individual's network of friends, indirect advertising through social networking groups, advertising via company-owned social networking Web sites, and viral marketing.

- Employers often look at the social networking Web site profiles of job candidates when making hiring decisions.

- Students who use social networking Web sites should review and make appropriate changes to their profiles before starting a job search.

- A social shopping Web site brings shoppers and sellers together in a social networking environment in which members share information and make recommendations while shopping online.

- Cyberbullying is the harassment, torment, humiliation, or threatening of one minor by another minor or group of minors via the Internet or cell phone. About 43 percent of 13- to 17-year-olds have experienced cyberbullying in the past year.

- There are about 700,000 registered sex offenders in the United States; 90,000 of them were onetime members of MySpace.

- Cyberstalking is threatening behavior or unwanted advances directed at an adult using the Internet or other forms of online and electronic communications; it is the adult version of cyberbullying.

- Although many social networking Web sites have policies against uploading violent or obscene material, these policies are difficult to enforce.

- An online virtual world is a computer-simulated world in which a visitor can move in three-dimensional space, communicate and interact with other visitors, and manipulate elements of the simulated world.

- Virtual worlds raise many interesting questions regarding what is a criminal act and whether law enforcement, real or virtual, should get involved in acts that occur in virtual worlds.

Self-Assessment Questions

The answers to the Self-Assessment Questions can be found in Appendix G.

1. According to the Pew Internet & American Life Project, approximately what percentage of U.S. adult Internet users age 18 and older have a profile on an online social network site?
   a. less than 25 percent
   b. around 35 percent
   c. more than 50 percent
   d. none of the above

2. Twitter is the social networking Web site with the largest number of adult U.S. Internet users. True or False?

3. _____ is a popular business-oriented Web site that is used by professionals for networking.

4. Social network advertising has become big business, with some social networking Web sites earning more than $200 million in ad revenue. True or False?

5. Which of the following approaches to social network advertising involves placing a banner ad on a Web site?
   a. direct advertising
   b. viral marketing
   c. indirect advertising through groups
   d. company-owned social networking Web site

6. Hotmail employed a _____ campaign when it first launched its service, with each e-mail sent by a Hotmail user containing a short message that promoted Hotmail's free e-mail service.

7. Employers can legally reject a job applicant based on the content of the individual's social networking Web site as long as the company is not violating discrimination laws. True or False?

8. What percentage of those managers who use information from social networking Web sites to screen candidates have found information that made them drop a candidate from consideration?
   a. less than 22 percent
   b. around 34 percent
   c. about 55 percent
   d. nearly 74 percent

9. There are around 700,000 registered sex offenders in the United States, and 90,000 of them were found on and subsequently banned from the social networking Web site _____.

10. Cyberbullying is more common among 15- and 16-year-old males than any other group of social networking users. True or False?

11. It is not unusual for cyberstalking to escalate to vandalism, physical stalking, and even
_____.

12. Which of the following measures is employed by social networking Web sites to avoid the posting of objectionable material?
    a. The terms of use agreement for most social networking Web sites states that the Web site reserves the right to delete material or terminate user accounts that violate the site's policies.
    b. The Web sites employ people to review material submitted.
    c. Other users sometimes report objectionable material.
    d. All of the above.

13. To date, no practical business applications of online virtual worlds have been implemented. True or False?

14. Social shopping Web sites generate money primarily through advertising and by selling
_____.

Discussion Questions

1. Do you think that college instructor–student friendships on social networking Web sites are appropriate? Why or why not?

2. Discuss the following idea: The information posted on social networking Web sites about news events occurring in foreign countries is an excellent source of up-to-the-minute news.

3. Keep track of the time that you spend on social networking Web sites for one week. Do you think that this is time well spent? Why or why not?

4. Identify two significant advantages that social network advertising has over other forms of advertising.

5. What advice would you give a friend who is the victim of cyberstalking?

6. Under what conditions can school students be punished for actions shown on videos posted to a social networking Web site?

7. Do you believe that viewing videos depicting violence can incite others to behave violently?

8. Do a search of the Web and develop a list of 10 companies that have created their own social networking Web sites.

9. What type of online information about a candidate should employment managers consider when screening candidates for a job interview? Give three examples of information that should automatically disqualify a candidate from a job offer. Give three examples of online information that should increase a candidate's chances of a job offer.

10. Review your user profile on your most frequently used social networking Web site. Do you think you need to make any changes to this profile? If so, what changes?

11. Some people immerse themselves in online virtual worlds for dozens of hours per week. What are the social implications of spending so many hours inside a virtual world?

12. Can role-playing illegal and violent fantasies in a virtual world affect individuals and society in the real world? Should limits be placed on what players can do in virtual worlds?

13. Check out the privacy policy of three social shopping Web sites to see if they say anything about selling user data to retailers. Write a couple of sentences summarizing your findings.

What Would You Do?

1. A coworker who is a recruiter told you that she is going to drop a job candidate because she feels that he is totally irresponsible. She found out through research on Facebook that the candidate married and divorced his high school sweetheart before graduating from college and once had his car repossessed. What would you do?

2. You are a new hire at a firm that manufactures and markets athletic equipment. You and several other new hires have been asked by your employer to begin posting positive messages about your firm's latest product—a $200 running shoe with state-of-the-art features—on Facebook, MySpace, and Twitter. While you are interested in the product, you cannot afford to buy it and have never tried it. Your manager says not to worry, the marketing department will provide you with prewritten statements for you to post. What would you do?

3. You are extremely disappointed to receive a rejection letter from the ABCXYZ Company. The job interviewer for the firm had told you that you "aced" the job interview and that you should expect a job offer after the formality of a background check was completed. What would you do?

4. You have just received a second invitation to join John's friend list. You met John three weeks ago at a group study session prior to last semester's Calculus II finals. He came across as very quiet and sort of strange. You did nothing to encourage his attention, but now you keep running into him at the oddest places and oddest times—at the self-service car wash, the 24-hour gym at 1:00 a.m., and the bakery at 7:00 a.m. He always flashes you a smile but has nothing to say. You think you've caught him taking snapshots of you a couple of times with his cell phone. He is starting to creep you out. What would you do?

5. You are shocked to read several anonymous messages posted on MySpace accusing one of your employees of inflating his expense reports to steal money from the company. You are responsible for approving his expense reports. What would you do?

6. A friend of yours has asked you to help him and a group of three or four others shoot a video and upload it to MySpace. The subject of the video is "Happenings at Work," and it will include several vignettes about funny things that happen at work. What would you do?

7. You are a new player in Second Life and are surprised when another avatar asks if you want to buy some drugs. You are not sure if the person is merely role-playing or is serious. What would you do?

Cases

1. Cyberbullying Results in Death

Megan Meier was a seventh grader who had been diagnosed with depression and attention deficit disorder; she had been under the care of a psychiatrist since the third grade. As a result of her depression, she had suffered from some suicidal thoughts. Megan was delighted when 16-year-old Josh Evans began exchanging messages with her through MySpace. They became online friends even though they never met face-to-face or spoke to one another. For about six weeks things went well, and Megan's family noticed that the online friendship seemed to have lifted her spirits.

But one day the tone of Josh's messages changed. He sent a message to Megan saying, "I don't know if I want to be friends with you anymore because I've heard that you are not very nice to your friends." Josh followed up with similar messages, and he shared some of Megan's messages to him with others. Quickly, other people began posting abusive messages and online bulletins about Megan.

Megan told her mother that she had suddenly become the target of an increasing number of mean online messages. When Megan's mother saw some of the replies that Megan had posted, they got into an argument over the vulgar language that Megan had used in some of her responses. Her mother was also upset to find that Megan had not logged off the computer when she had been told to do so earlier that afternoon. Megan ran upstairs and shut herself in her room. Twenty minutes later her mother found her dead; she had committed suicide by hanging.

The last message Megan received from Josh read, "You are a bad person and everybody hates you. Have a shitty rest of your life. The world would be a better place without you." Megan had responded, "You're the kind of boy a girl would kill herself over."

Six weeks after Megan's death, her parents were informed by a neighbor that Lori Drew, a 49-year-old mother who lived just four doors down from the Meier family, had created a fake "Josh Evans" MySpace account. The neighbor said that "Lori laughed about it" and said she intended to use it to "mess with Megan." Megan and Lori's 13-year-old daughter, Sarah, had been friends but had had a falling out because Megan allegedly had spread gossip about Sarah.[52]

The local police and the FBI asked the Meier family not to say anything publicly to keep the Drew family from knowing that they were being investigated. In December 2007, some 13 months after the suicide, Jack Banas, the county's prosecuting attorney, held a press conference. He stated that Ashley Grills, an 18-year-old employee in Lori Drew's advertising business, admitted that she wrote most of the messages to Megan, including the final message. After Megan's death, Grills was hospitalized and underwent psychiatric treatment for her involvement in the cyberbullying incident. Banas went on to say that Sarah Drew had moved and was now attending a new school in a different city. Lori Drew would not reveal where her daughter was living out of fear of what might happen to her. Neighbors avoided the Drews, and business advertisers in Lori Drew's coupon book business were shunned. After reviewing the case, the county prosecutor decided not to file any criminal charges related to the hoax.[53]

In January 2008, a federal grand jury began issuing subpoenas to MySpace and other witnesses as federal prosecutors considered whether to bring federal charges against Lori Drew. The grand jury in Los Angeles was given jurisdiction because MySpace has its headquarters in Beverly Hills.[54] The U.S. attorney granted immunity to Ashley Grills in exchange for her testimony against Lori Drew.[55] The grand jury indicted Lori Drew on one count of conspiracy and three counts of unauthorized access for accessing protected computers—a violation of the Computer Fraud and Abuse Act. The case was to go forward not as a homicide but as a computer fraud prosecution.

The Computer Fraud and Abuse Act is intended to prosecute those who hack into what are known as protected computers. It applies to cases with a compelling federal interest, in which computers of the federal government or certain financial institutions are involved, the crime itself is interstate in nature, or the computers are used in interstate and foreign commerce. Section (b) of the act punishes not just anyone who commits or attempts to commit an offense under the Computer Fraud and Abuse Act but also those who conspire to do so.

The basis for the charges was Lori Drew's alleged violation of the MySpace terms of service agreement by creating a fake profile for the nonexistent Josh Evans. Prosecutors argued that this was the legal equivalent of hacking into a protected computer. Drew's attorneys filed motions to dismiss the case on the basis that violating the MySpace terms of service did not constitute illegal access of a protected computer. Judge George Wu ruled against the motion and ordered the case to go to court.[56]

The trial began in November 2008. The prosecution's case against Drew was dealt a setback when Ashley Grills admitted that it was her idea to create the MySpace account and that it was she who clicked to agree to the MySpace terms of service. According to Grills, neither she nor Lori—or even Sarah Drew—looked at the terms of service. In addition, Grills had previously admitted that she sent most of the messages from Josh Evans. However, several witnesses said that Drew admitted that she had created the MySpace account with Grills and Sarah, and that she had sent some of the messages herself. Grills did state that Drew had encouraged her to create the fake profile and told her not to worry because "people do it all the time."[57]

Drew's attorney acknowledged that the case was indeed sad but that "it doesn't amount to a violation of a computer statute. The reality is that there's lots of blame to go around in this case."[58]

After just one day of deliberation, the jury was deadlocked on the conspiracy charge. They acquitted Drew on the three felony counts of accessing a computer without authorization to inflict emotional harm. The jury did convict Drew of three lesser charges of accessing computers without authorization. Probation authorities recommended probation and a $5,000 fine, while the prosecution asked for a sentence of three years and a $300,000 fine.[59]

Drew's attorney asked Judge George Wu to grant a motion for a directed acquittal based on the defense's view that the prosecution failed to prove that Drew knew that the MySpace terms of service existed, and that she knew what they said and intentionally violated them.[60] Wu postponed his ruling until July 2, 2009, at which time he dismissed the case. Wu expressed concern that if Drew's conviction was allowed to stand, it would create a precedent that violating a Web site's terms of service agreement is the legal equivalent of computer hacking and could result in criminal prosecution for what would have been in the past a misdemeanor for civil breach of contract.[61]

Discussion Questions

1. Do you believe that knowingly violating the terms of a Web site's service agreement should be punishable as a serious crime, with potential penalties of substantial fines and jail time? Why or why not?

2. Imagine that you are the defense counsel for Lori Drew. Present your strongest argument for why your defendant should not be convicted for violation of the Computer Fraud and Abuse Act. Now imagine that you are the prosecutor. Present your strongest argument for why Drew should be convicted. Which argument do you believe is stronger?

3. Do you think that Lori Drew was responsible for Megan's death? Do you think that justice was served in this case? Should new laws be created to address similar future cases?

2. Kaboodle—A Successful Social Shopping Web Site

Kaboodle is a social shopping Web site created in 2005. The idea for the Web site grew out of the frustration that Manish Chandra and his wife, Asha, experienced when remodeling their home. The process of finding products that matched their taste, identifying the retailers of those products, and searching for bargain prices was tedious and time consuming. As a result, Chandra created Kaboodle, an online social shopping community where members can discover and recommend new products. Kaboodle members can create shopping lists, read product reviews, and get help making purchasing decisions.[62]

Kaboodle members can install a Kaboodle browser toolbar, which includes an "Add to Kaboodle" button. As they are browsing the Web, members can click the "Add to Kaboodle" button on their toolbar when they see a product that interests them. (Retailers can also easily add an "Add to Kaboodle" button to their individual product Web pages.) With a click of the button, a summary of the product is added to the member's Kaboodle page—including a description, a picture, and pricing information. The member can add comments to the product snapshot, save it in a customized list, and share it with others to invite their input. And because all members' Kaboodle pages are public, other members can copy information about items they've already found onto other members' pages to help them in their shopping research.

In addition to the advantages that Kaboodle offers to shoppers, the site offers two compelling benefits for retailers. First, customers simply need to review their personal Kaboodle pages to remember what they wanted to buy from the retailer. Second, when Kaboodle members add the retailer's product to their list, others can discover that item and add it to their lists.[63]

In just two short years, Chandra and his team created a highly successful Web site—so successful, in fact, that Hearst Corporation purchased it in August 2007 for about $40 million. It is now part of Hearst Magazines Digital Media, which manages 24 Web sites and nine mobile sites.[64] At the time of its purchase, Kaboodle had about 2 million unique monthly visitors.[65] By the end of 2008, the number of unique monthly visitors had exploded to nearly 10 million.[66]

Discussion Questions

1. Register as a new user at Kaboodle (it's free) and experiment with the social shopping Web site. Briefly summarize how easy or difficult it was to join, discover products you were interested in, create a list, use the "Add to Kaboodle" button, and look at similar lists created by other members.

2. Based on your experience, what recommendations for improvements or new features do you have?

3. Would you expect the number of members in social shopping Web sites to continue to grow over the next few years? Why or why not?

3. Social Networking Disaster for Domino's

The video posted to YouTube showing a Domino's employee making sandwiches was disgusting. First, the employee sprays snot on the bread, then he sticks cheese up his nose before putting it on a piece of bread, and finally he passes gas on a slice of salami destined for the revolting sandwiches. A woman in the background exclaims: "In about five minutes, they'll be sent out to delivery, where somebody will be eating these, yes, eating them. And little did they know that cheese was in his nose and that there was some lethal gas that ended up on their salami. That's how we roll at Domino's."[67]

Within a few days, the video had been viewed over a million times on YouTube. People were also commenting on it on Twitter and other social networking Web sites.[68] The video inflicted tremendous brand damage on Domino's and the fast-food industry in general. Indeed, a national online survey by HCD Research found that 65 percent of respondents who would previously order from or visit Domino's Pizza were less likely to do so after seeing the video.[69]

Although the video was posted on a Monday night, the company became aware of it only after a blogger alerted them on Tuesday. Domino's executives initially decided to take no action, hoping that the hoopla over the video would soon subside. They did not want to attract even more people to the video by issuing a formal press release to the news media. By Wednesday it became apparent that the matter would not simply go away and that just the opposite had occurred—there were many posts on Twitter demanding Domino's reaction. The company was finally forced to take action.[70]

The employees responsible for the video were fired from Domino's. Even though they claimed that the video was just a joke and that the food was never delivered, they were charged with food tampering and face future legal proceedings. The restaurant in Conover, North Carolina, where the video was made, was closed for a day so that it could be completely sanitized—all food and paper products that were not sealed were thrown out.

Domino's ran into an unexpected delay in getting the video pulled from the YouTube site—it was told by YouTube that the signature of one of the employees who created the video was needed, since the account holder was the copyright owner of the video.[71] Domino's posted an apology to its customers on its Web site and asked each employee with a Twitter account to tweet a link to it. Domino's also created its own Twitter account so it could reassure customers that this was a unique incident. Finally, Domino's U.S. president, Patrick Doyle, uploaded his own video to YouTube. In the video, he apologizes for the initial repulsive video and clarifies his company's commitment to food safety.[72]

Discussion Questions

1. Some observers believe that if an organization does not respond to an attack on its brand within the first 24 hours, then the damage has been done—lack of management response is judged as an admission of guilt. Others feel that some time is required to gather facts and figure out what happened before responding. With the advantage of 20/20 hindsight, how might Domino's have reacted more effectively?

2. Do you find it unusual that Domino's response was primarily through the online media rather than the usual printed press releases? Does this seem an effective and appropriate way to respond under these circumstances? Why or why not? Does Domino's use of the online media set a precedent for others to follow in the future?

3. Identify three lessons that other companies could learn from Domino's experience.

End Notes

[1] Adam Ostrow, "Twitter Now Growing at a Staggering 1,382 Percent," *Mashable*, March 16, 2009, http://mashable.com/2009/03/16/twitter-growth-rate-versus-facebook/.

[2] "Overview: Twitter.com," Pingdom, www.pingdom.com/reports/wx4vra365911/check_overview/?name=Twitter.com.

[3] Michael Arrington, "Twitter Gets Hacked, Badly," *TechCrunch*, January 5, 2009, www.techcrunch.com/2009/01/05/twitter-gets-hacked-badly.

[4] Belinda Goldsmith, "Many Twitters Are Quick Quitters: Study," Reuters, April 29, 2009, www.reuters.com/article/deborahCohen/idUSTRE53S1A720090429.

5   Warren P. Strobel, "Iran Expels Many Foreign Reporters as Violence Looms," *Sacramento Bee*, June 16, 2009.

6   Sue Pleming, "U.S. State Department Speaks to Twitter over Iran," Reuters, June 16, 2009, www.reuters.com/article/rbssTechMediaTelecomNews/idUSWBT01137420090616.

7   Lev Grossman, "Iran Protests: Twitter, the Medium of the Movement," *Time*, June 17, 2009, www.time.com/time/world/article/0,8599,1905125,00.html.

8   Lev Grossman, "Iran Protests: Twitter, the Medium of the Movement," *Time*, June 17, 2009, www.time.com/time/world/article/0,8599,1905125,00.html.

9   "Internet World Stats," www.internetworldstats.com/stats.htm.

10 "Time Spent on Facebook up 700%, but MySpace Still Tops for Video," Nielsen Wire, June 2, 2009, http://blog.nielsen.com/nielsenwire/online_mobile/time-spent-on-facebook-up-700-but-myspace-still-tops-for-video.

11 Amanda Lenhart, "Adults and Social Network Websites," Pew Internet & American Life Project, January 14, 2009, www.pewinternet.org/Reports/2009/Adults-and-Social-Network-Websites.aspx.

12 Stan Schroeder, "Social Network Ad Spending to Fall, but It's Not All About the Money," *Mashable*, May 13, 2009, http://mashable.com/2009/05/13/social-network-ad-spending.

13 Bruce Nussbaum, "Social Networking, Advertising and Innovation," *BusinessWeek*, June 22, 2009, www.businessweek.com/innovate/NussbaumOnDesign/archives/2006/06/social_networki_1.html.

14 Joe Guy Collier, "Coke Fans' Facebook Page Draws Millions of Users," *Atlanta Journal-Constitution*, March 30, 2009, www.ajc.com/business/content/business/coke/stories/2009/03/30/coke_facebook_page.html.

15 Theresa Howard, "Seeking Teens, Marketers Take Risks by Emulating MySpace," *USA Today*, May 22, 2006, www.usatoday.com/tech/news/2006-05-01-myspace-marketers_x.htm.

16 dna13, "How Fortune 1000 Companies are Harnessing the Power of Social Media," White Paper, 2009, www.dna13.com/company/downloads/white-papers/how-f1000-leverage-social-media.

17 Rosalie Marshall, "Social Networking Adds Extra Fizz to Coca-Cola," *V3.co.uk*, March 10, 2009, www.v3.co.uk/vnunet/analysis/2238220/analysis-coca-cola-adds-fizz.

18 Dell, Inc., "About IdeaStorm," IdeaStorm, www.ideastorm.com/ideaAbout?pt=About+IdeaStorm.

19 Dr. Ralph F. Wilson, "The Six Simple Principles of Viral Marketing," *Web Marketing Today*, February 1, 2005, www.wilsonweb.com/wmt5/viral-principles.htm; "Viral Marketing Is Nothing New," *AllBusiness*, October 1, 2000, www.allbusiness.com/marketing-advertising/4448168-1.html.

20 CareerBuilder.com, "One-in-Five Employers Use Social Network Sites to Research Job Candidates, CareerBuilder.com Survey Finds," www.careerbuilder.com/share/aboutus/press releasesdetail.aspx?id=pr459&sd=9%2F10%2F2008&ed=12%2F31%2F2008&cbRecursionCnt=1&cbsid=31c1a270495b477ca4a0a09de676e5e3-299863612-x1-6&ns_siteid=ns_us_g_Only_One_in_Five_Empl_.

21 Mike Hargis, "Social Networking Sites Dos and Don'ts," *CNN*, November 5, 2008, www.cnn.com/2008/LIVING/worklife/11/05/cb.social.networking/index.html.

22 Mallory Terrence, "Employers View Social Networking Sites Before Hiring Employees," *Loquitur*, April 17, 2008, www.theloquitur.com/media/storage/paper226/news/2008/04/17/News/Employers.View.Social.Networking.Sites.Before.Hiring.Employees-3341565.shtml.

23 Mallory Terrence, "Employers View Social Networking Sites Before Hiring Employees," *Loquitur*, April 17, 2008, www.theloquitur.com/media/storage/paper226/news/2008/04/17/News/Employers.View.Social.Networking.Sites.Before.Hiring.Employees-3341565.shtml.

24 Stuffpit, "Earn Money Recommending Products," www.stuffpit.com/stuff/earn.

25 Bob Tedeschi, "Like Shopping? Social Networking? Try Social Shopping," *New York Times*, September 11, 2006, www.nytimes.com/2006/09/11/technology/11ecom.html.

26 January W. Payne, "What to Do If Your Child Is Bullied Online," *U.S. News and World Report*, December 4, 2007, http://health.usnews.com/articles/health/2007/12/04/what-to-do-if-your-child-is-bullied-online.html.

27 "How Lori Drew Became America's Most Reviled Mother," *The Age*, December 1, 2007, www.theage.com.au/articles/2007/11/30/1196394672124.html.

28 Jessica Calefati, "California Law Targets Cyberbullying," *U.S. News and World Report*, January 7, 2009, www.usnews.com/blogs/on-education/2009/01/07/california-law-targets-cyberbullying.html.

29 Claire M. Renzetti and Jeffrey L. Edleson, eds., *Encyclopedia of Interpersonal Violence* (Thousand Oaks, CA: Sage Publications, 2008), 164.

30 Claire M. Renzetti and Jeffrey L. Edleson, eds., *Encyclopedia of Interpersonal Violence* (Thousand Oaks, CA: Sage Publications, 2008), 164.

31 "Cyberstalker Gets Two Years in Prison; Ex-Sandia Labs Worker Sentenced," *Albuquerque Journal*, February 21, 2008; Sharon Gaudin, "Sandia National Labs Worker Pleads to Cyberstalking Linkin Park Singer," *InformationWeek*, July 5, 2007, www.informationweek.com/news/internet/showArticle.jhtml?articleID=200900509.

32 The National Center for Victims of Crime, "Cyberstalking," www.ncvc.org/ncvc/main.aspx?dbName=DocumentViewer&DocumentID=32458.

33 Jenna Wortham, "MySpace Turns Over 90,000 Names of Registered Sex Offenders," *New York Times*, February 3, 2009, www.nytimes.com/2009/02/04/technology/internet/04myspace.html?fta=y.

34 Jenna Wortham, "MySpace Turns Over 90,000 Names of Registered Sex Offenders," *New York Times*, February 3, 2009, www.nytimes.com/2009/02/04/technology/internet/04myspace.html?fta=y.

35 Citizen Media Law Project, "Evaluating Terms of Service," www.citmedialaw.org/legal-guide/evaluating-terms-service.

36 Larry Fiorino, "Commentary: Tech Talk: Backlash Against Violent Content Poses," *Baltimore Daily Record*, April 18, 2008.

37 "Teens Nabbed for Drinking After Report in Blogs," MSNBC.com, February 8, 2006, www.msnbc.msn.com/id/11234450.

327

38 K. C. Jones, "Videotaped Florida Teen Beating Prompts Calls to Block Violent Content," *InformationWeek*, April 8, 2008, www.informationweek.com/news/internet/security/showArticle. jhtml?articleID=207100417.

39 Jacqui Cheng and Ken Fisher, "4chan, eBaum's World Carpet Bombing YouTube with Porn Videos," *Ars Technica*, May 20, 2009, http://arstechnica.com/web/news/2009/05/4chan-ebaumsworld-carpet-bombing-youtube-with-porn-videos.ars.

40 Alan Sipress, "Does Virtual Reality Need a Sheriff?," *Washington Post*, June 2, 2007, www.washingtonpost.com/wp-dyn/content/article/2007/06/01/AR2007060102671.html.

41 Elliot Feldman, "Real Crime in Virtual Worlds," *Associated Content*, August 1, 2007, www.associatedcontent.com/article/329951/real_crime_in_virtual_worlds.html?cat=19.

42 Alan Sipress, "Does Virtual Reality Need a Sheriff?," *Washington Post*, June 2, 2007, www.washingtonpost.com/wp-dyn/content/article/2007/06/01/AR2007060102671.html.

43 Alan Sipress, "Does Virtual Reality Need a Sheriff?," *Washington Post*, June 2, 2007, www.washingtonpost.com/wp-dyn/content/article/2007/06/01/AR2007060102671.html.

44 Elliot Feldman, "Real Crime in Virtual Worlds," *Associated Content*, August 1, 2007, www.associatedcontent.com/article/329951/real_crime_in_virtual_worlds.html?cat=19.

45 "Jilted Woman Kills Man's Virtual Avatar in Japan—Goes to Jail," *Prairie Chicken*, October 24, 2008, http://prairiechicken.blogspot.com/2008/10/jilted-woman-kills-mans-virtual-avatar.html.

46 Linden Lab, "Case Study: Developing New Learning and Collaboration Environments for Educators: The New Media Consortium (NMC) in Second Life," http://secondlifegrid.net.s3. amazonaws.com/docs/Second_Life_Case_NMC_EN.pdf.

47 Media Grid: Immersive Education, Home Page, www.immersiveeducation.org.

48 "It's Not All Fun and Games," *BusinessWeek*, May 1, 2006, www.businessweek.com/magazine/content/06_18/b3982007.htm.

49 Northrop Grumman Corporation, "About Us," www.northropgrumman.com/about_us/index. html.

50 Unofficial US Navy Site, "USS Blue Ridge," http://navysite.de/ships/lcc19.htm.

51 Linden Lab, "Case Study: Simulation Training and Prototyping in Virtual Worlds: Northrop Grumman in Second Life," http://secondlifegrid.net.s3.amazonaws.com/docs/Second_Life_Case_NGC_EN.pdf.

52 "How Lori Drew Became America's Most Reviled Mother," *The Age*, December 1, 2007, www.theage.com.au/articles/2007/11/30/1196394672124.html.

53 Stephen Pokin, "UPDATE: No Charges to Be Filed over Meier Suicide," *Suburban Journals*, December 3, 2007.

54 Scott Glover and P. J. Huffstutter, "L.A. Grand Jury Issues Subpoenas in Web Suicide Case," *Los Angeles Times*, January 9, 2008, www.latimes.com/news/local/la-me-myspace9-jan09,0,5809715.story.

55 Jonann Brady, "Exclusive: Teen Talks About Her Role in Web Hoax That Led to Suicide," Good Morning America, April 1, 2008, http://abcnews.go.com/GMA/story?id=4560582&page=1.

56 Kim Zetter, "Judge Postpones Lori Drew Sentencing, Weighs Dismissal," *Wired*, May 18, 2009, www.wired.com/threatlevel/2009/05/drew_sentenced.

57 Kim Zetter, "Judge Postpones Lori Drew Sentencing, Weighs Dismissal," *Wired*, May 18, 2009, www.wired.com/threatlevel/2009/05/drew_sentenced.

58 Kim Zetter, "Prosecution: Lori Drew Schemed to Humiliate Teen Girl," *Wired*, November 25, 2008, www.wired.com/threatlevel/2008/11/defense-lori-dr.

59 Kim Zetter, "Judge Postpones Lori Drew Sentencing, Weighs Dismissal," *Wired*, May 18, 2009, www.wired.com/threatlevel/2008/11/defense-lori-dr.

60 Kim Zetter, "Can Lori Drew Verdict Survive the 9th Circuit Court?," *Wired*, December 1, 2008, www.wired.com/threatlevel/2008/12/can-lori-drew-v.

61 Kim Zetter, "Can Lori Drew Verdict Survive the 9th Circuit Court?," *Wired*, December 1, 2008, www.wired.com/threatlevel/2008/12/can-lori-drew-v; Alexandra Zavis, "Judge Tentatively Dismisses Case in MySpace Hoax That Led to Teenage Girl's Suicide," *Los Angeles Times*, July 2, 2009, http://latimesblogs.latimes.com/lanow/2009/07/myspace-sentencing.html.

62 Kaboodle, "About Kaboodle," www.kaboodle.com/zm/about.

63 Kaboodle, "Retailers: Get the 'Add to Kaboodle' Button for Your Online Store!," www.kaboodle.com/zm/retailers.

64 Kaboodle, "Kaboodle Appoints Shari Gunn Vice President of Advertising and Business Development," Press Release, October 28, 2008, www.kaboodle.com/zm/pr10.

65 Loren Baker, "Kaboodle Social Shopping Bought by Hearst," *Search Engine Journal*, August 8, 2007, www.searchenginejournal.com/kaboodle-social-shopping-bought-by-hearst/5451.

66 Kaboodle, "Kaboodle Appoints Shari Gunn Vice President of Advertising and Business Development," Press Release, October 28, 2008, www.kaboodle.com/zm/pr10.

67 Sean Gregory, "Domino's YouTube Crisis: 5 Ways to Fight Back," *Time*, April 18, 2009, www.time.com/time/nation/article/0,8599,1892389,00.html.

68 Vquence, "The Domino's Incident," April 19, 2009, www.vquence.com.au/blog/2009/04/19/the-dominos-incident.

69 Raymund Flandez, "Domino's Response Offers Lessons in Crisis Management," *Wall Street Journal*, April 20, 2009, http://blogs.wsj.com/independentstreet/2009/04/20/dominos-response-offers-lessons-in-crisis-management.

70 Raymund Flandez, "Domino's Response Offers Lessons in Crisis Management," *Wall Street Journal*, April 20, 2009, http://blogs.wsj.com/independentstreet/2009/04/20/dominos-response-offers-lessons-in-crisis-management.

71 Raymund Flandez, "Domino's Response Offers Lessons in Crisis Management," *Wall Street Journal*, April 20, 2009, http://blogs.wsj.com/independentstreet/2009/04/20/dominos-response-offers-lessons-in-crisis-management.

72 Sean Gregory, "Domino's YouTube Crisis: 5 Ways to Fight Back," *Time*, April 18, 2009, www.time.com/time/nation/article/0,8599,1892389,00.html.

CHAPTER 10

ETHICS OF IT ORGANIZATIONS

VIGNETTE

Problems with Suppliers

Many computer hardware manufacturers rely on foreign companies to provide raw materials; build computer parts; and assemble hard drives, monitors, keyboards, and other components. While there are many advantages to dealing with foreign suppliers, hardware manufacturers may find certain aspects of their business (such as quality and cost control, shipping, and communication) much more complicated when dealing with a supplier in another country.

In addition to these fairly common business problems, hardware manufacturers are sometimes faced with serious ethical issues relating to their foreign suppliers. Two such issues that have recently surfaced involve (1) suppliers who run their factory in a manner that is unsafe or unfair to their workers and (2) raw materials suppliers who funnel money to groups engaged in armed conflict, including some that commit crimes and human rights abuses.

In February 2009, alarming information came to light about the Meitai Plastics and Electronics factory in Dongguan City, in China's Guangdong province. This factory, in fact, represents an extreme example of a supplier who runs its factory in an unsafe and unfair manner. Meitai Plastics employs 2,000 workers, mostly young women, who make computer equipment and peripherals—such as printer cases and keyboards—for Dell, IBM, Lenovo, Microsoft, and Hewlett-Packard products.[1] Based on research conducted between June 2008 and January 2009, the National Labor Committee—a human rights organization based in the United States—published a report in February 2009 highly critical of the work environment at the factory.[2] According to the report, young workers sit on hard wooden stools for 12 hours a day, working on an assembly line that never stops. Workers are prohibited from talking, listening to music, raising their heads from their work, or putting their hands in their pockets. Employees are fined for being even one minute late, not trimming their fingernails, or stepping on the grass of the factory grounds. A worker who needs to use the restroom must wait until there is a group break. The average workweek consists of 74 hours, with a take-home pay of $57.19—well below the amount necessary to meet subsistence-level needs. If a worker takes a Sunday off, she is docked one-and-a-half-day's wages. Workers are housed 10 to 12 per dorm room. There is no air conditioning, and temperatures in the dorm rooms can reach the high 90s in the summer. Workers must walk down several floors to get hot water in a small bucket to use for personal hygiene.[3]

Manufacturers who use rare raw materials face another ethical issue related to the use of foreign suppliers: how to ensure that their suppliers do not funnel money to groups that engage in armed conflict or commit crimes and human rights abuses. Manufacturers of computers, digital cameras, cell phones, and other electronics frequently purchase rare minerals such as gold, tin, tantalum, and tungsten for use in their products. Unfortunately, some of these purchases are helping to finance the deadliest war in the world today—the war in the Democratic Republic of Congo. The war began in 1998 and has dragged on long after a peace agreement was signed in 2003. During the war and its aftermath, over 5 million people have died—mostly from disease and starvation—making it the deadliest conflict since World War II.[4] In the Congo, many mines are controlled by groups that engage in armed conflict and inflict human rights abuses on local populations. The Enough Project's "Raise Hope for Congo" campaign is trying to get large electronics firms to trace and audit their supply chains to ensure that their suppliers do not source minerals from mines in the Congo that are controlled by armed groups. This is often easier said than done because of the long, complex supply chain and often disreputable middlemen involved in the minerals trade. As manufacturers struggle with these issues, some are trying to use their influence to demand that their suppliers stop sourcing from mines that continue to fund violence in the Congo and elsewhere.[5]

Questions to Consider

1. How can an organization ensure that all the members of its supply chain will behave ethically?
2. What responsibility does an organization have to ensure that its suppliers and business partners behave ethically?

LEARNING OBJECTIVES

As you read this chapter, consider the following questions:

1. What are contingent workers, and how are they employed in the information technology industry?
2. What key ethical issues are associated with the use of contingent workers, including H-1B visa holders and offshore outsourcing companies?
3. What is whistle-blowing, and what ethical issues are associated with it?
4. What is an effective whistle-blowing process?
5. What measures are members of the electronics manufacturing industry taking to ensure the ethical behavior of the many participants in their long and complex supply chains?

KEY ETHICAL ISSUES FOR ORGANIZATIONS

This chapter will touch on the following ethical topics that are pertinent to organizations in the IT industry, as well as to organizations that make use of IT:

- The use of nontraditional workers, including temporary workers, contractors, consulting firms, H-1B visa workers, and outsourced offshore workers, gives an organization more flexibility in meeting its staffing needs, often at a lower cost. The use of nontraditional workers also raises ethical issues for organizations. When should such nontraditional workers be employed, and how does such employment affect an organization's ability to grow and develop its own employees? How does the use of such resources impact the wages of the organization's employees?
- **Whistle-blowing** is an effort to attract public attention to a negligent, illegal, unethical, abusive, or dangerous act by a company or some other organization. It is an important ethical issue for individuals and organizations. How does one safely and effectively report misconduct, and how should managers handle a whistle-blowing incident?
- **Green computing** is a term applied to a variety of efforts directed toward the efficient design, manufacture, operation, and disposal of IT-related products, including personal computers, laptops, servers, printers, and printer supplies. Computer manufacturers and end users are faced with many questions about when and how to transition to green computing, and at what cost.
- The electronics and information and communications technology (ICT) industry recognizes the need for a code to address ethical issues in the areas of worker safety and fairness, environmental responsibility, and business efficiency. What has been done so far, and what still needs to be done?

Let's begin with a discussion of the use of nontraditional workers and the ethical issues raised by this practice.

THE NEED FOR NONTRADITIONAL WORKERS

According to a March 2008 survey by the Computing Research Association, the number of declared undergraduate computer science majors at doctoral-granting computer science departments continued its seven-year decline, with the number of students enrolled in the fall of 2007 half of what it was in the fall of 2000.[6] In 2008, however, enrollment numbers increased slightly, which may indicate that a turnaround has begun (see Figure 10-1).[7] The decline in enrollment took place in spite of the forecast for an increased need for workers in this field. The Bureau of Labor Statistics has forecasted that "employment in computer system design and related services will grow by 38.3 percent" between 2006 and 2014 and "that this employment growth will be driven by the increasing reliance of businesses on information technology and the continuing importance of maintaining system and network security."[8] Jobs related to networking and data communications analysis had the highest forecasted growth rate at 50 percent.[9] As a result, IT firms and organizations that use IT products and services are concerned about a shortfall in the number of U.S. workers to fill these positions.

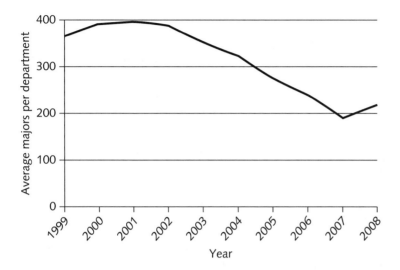

Source: Stuart Zweben, "Computing Degree and Enrollment Trends from the 2007–2008 CRA Taulbee Survey," p. 4, March 2009, Computing Research Association, Washington, DC. Reprinted by permission.

FIGURE 10-1   Average enrollment per U.S. computer science department

Facing a likely long-term shortage of trained and experienced workers, employers are increasingly turning to nontraditional sources to find IT workers with skills that meet their needs; these sources include contingent workers, H-1B workers, and outsourced offshore workers. Employers face ethical decisions about whether to recruit new and more skilled workers from these sources or to develop their own staff to meet the needs of their business.

CONTINGENT WORKERS

The Bureau of Labor Statistics defines **contingent work** as a job situation in which an individual does not have an explicit or implicit contract for long-term employment. The contingent workforce includes independent contractors, temporary workers hired through employment agencies, on-call or day laborers, and on-site workers whose services are provided by contract firms.

A firm is likely to use contingent IT workers if it experiences pronounced fluctuations in its technical staffing needs. Workers are often hired on a contingent basis as consultants on an organizational restructuring project, as technical experts on a product development team, and as supplemental staff for many other short-term projects, such as the design and installation of new information systems.

Typically, these workers join a team of full-time employees and other contingent workers for the life of the project and then move on to their next assignment. Whether they work, when they work, and how much they work depends on the company's need for them. They have neither an explicit nor an implicit contract for continuing employment.

Organizations can obtain contingent workers through temporary staffing firms or employee leasing organizations. Temporary staffing firms recruit, train, and test job seekers in a wide range of job categories and skill levels, and then assign them to clients as needed. Temporary employees are often used to fill in during staff vacations and illnesses, handle seasonal workloads, and help staff special projects. However, they are not considered official employees of the company, nor are they eligible for company benefits such as vacation, sick pay, and medical insurance.

Temporary working arrangements sometimes appeal to people who want maximum flexibility in their work schedule as well as a variety of work experiences. Because temporary workers do not receive additional compensation through company benefits, they are often paid a higher hourly wage than full-time employees doing equivalent work.

In **employee leasing**, a business (called the subscribing firm) transfers all or part of its workforce to another firm (called the leasing firm), which handles all human-resource-related activities and costs, such as payroll, training, and the administration of employee benefits. The subscribing firm leases these workers, but they remain employees of the leasing firm. Employee leasing firms operate with minimal administrative, sales, and marketing staff to keep down overall costs, and they pass the savings on to their clients. Employee leasing is a type of **coemployment relationship**, in which two employers have actual or potential legal rights and duties with respect to the same employee or group of employees. Employee leasing firms are subject to special regulations regarding workers' compensation and unemployment insurance. Because the workers are technically employees of the leasing firm, they may be eligible for some company benefits through the firm.

Organizations can also obtain temporary IT employees by hiring a consulting firm. Consulting organizations maintain a staff of employees with a wide range of skills and experience, up to and including world-renowned industry experts; thus, they can provide the exact skills and expertise that an organization requires. Consulting firms work with their clients on engagements for which there are typically well-defined expected results or deliverables that must be produced (e.g., an IT strategic plan, implementation of an ERP system, or selection of a hardware vendor). The contract with a consulting firm typically specifies the length of the engagement and the rate of pay for each of the consultants, who are directed on the engagement by a senior manager or director from the consulting firm. Table 10-1 shows the world's 10 largest IT consulting firms and the number of consultants that each employs.

TABLE 10-1 The 10 largest IT consulting firms

Firm	Headquarters	Number of employees
IBM Global Business Services	Armonk, New York	398,000[10]
HP (including Electronic Data Systems)	Palo Alto, California	321,000[11]
Accenture	Hamilton, Bermuda	180,000[12]
Deloitte Touche Tohmatsu	New York, New York	165,000[13]
Tata Consultancy Services	Mumbai, India	143,000[14]

TABLE 10-1   The 10 largest IT consulting firms (*continued*)

Firm	Headquarters	Number of employees
Ernst & Young	New York, New York	135,000[15]
KPMG	New York, New York	135,000[16]
Infosys Technologies	Bangalore, India	103,900[17]
Capgemini	Paris, France	90,000[18]
Cognizant Technology Solutions	Teaneck, New Jersey	63,000[19]

Advantages of Using Contingent Workers

When a firm employs a contingent worker, it does not usually have to provide benefits such as insurance, paid time off, and contributions to a retirement plan. A company can easily adjust the number of contingent workers it uses to meet its business needs, and can release contingent workers when they are no longer needed. An organization cannot usually do the same with full-time employees without creating a great deal of ill will and negatively impacting employee morale. Moreover, because many contingent workers are already specialists in a particular task, the firm does not customarily incur training costs. Therefore, the use of contingent workers enables the firm to meet its staffing needs more efficiently, lower its labor costs, and respond more quickly to changing market conditions.

Disadvantages of Using Contingent Workers

One downside to using contingent workers is that they may not feel a strong connection to the company for which they are working. This can result in a low commitment to the company and its projects, along with a high turnover rate. Although contingent workers may already have the necessary technical training for a temporary job, while working for a particular company, many contingent workers gain additional skills and knowledge, which are lost to the company when they depart at the project's completion.

Deciding When to Use Contingent Workers

When an organization decides to use contingent workers for a project, it should recognize the trade-off it is making between completing a single project quickly and cheaply versus developing people within its own organization. If the project requires unique skills that are probably not necessary for future projects, there may be little reason to invest the additional time and costs required to develop those skills in full-time employees.

If a particular project requires only temporary help, and the workers will not be needed for future projects, the use of contingent workers is a good approach. In such a situation, using contingent workers avoids the need to hire new employees and then fire them when staffing needs decrease.

Organizations should carefully consider whether or not to use contingent workers when those workers are likely to learn corporate processes and strategies that are key to the company's success. It is next to impossible to prevent contingent workers from passing on such information to subsequent employers. This can be damaging if the worker's next employer is a major competitor.

Although using contingent workers is often the most flexible and cost-effective way to get a job done, their use can raise ethical and legal issues about the relationships among the staffing firm, its employees, and its customers, including the potential liability of customers for withholding payroll taxes, payment of employee retirement benefits and health insurance premiums, and administration of workers' compensation to the staffing firm's employees.

Depending on how closely workers are supervised and how the job is structured, contingent workers may be viewed as permanent employees by the Internal Revenue Service, the Labor Department, or a state's workers' compensation and unemployment agencies.

For example, in 2001, Microsoft agreed to pay a $97 million settlement to some 10,000 "permatemps"—temporary workers who were employed for an extended length of time as software testers, graphic designers, editors, technical writers, receptionists, and office support staffers. Some had worked at Microsoft for several years. The *Vizcaino v. Microsoft* class action was filed in 1992 by eight former workers who claimed that they and thousands more permatemps had been illegally shut out of a stock purchase plan that allowed employees to buy Microsoft stock at a 15 percent discount. Microsoft shares had skyrocketed in value throughout the 1990s. The sharp appreciation in the stock price meant that had they been eligible, some temporary workers in the lawsuit could have earned more money from stock gains than they received in salary while at Microsoft.[20]

The *Vizcaino v. Microsoft* lawsuit dramatically illustrated the cost of misclassifying employees and violating laws that cover compensation, taxes, unemployment insurance, and overtime. The key lesson of this case is that even if workers sign an agreement indicating that they are contractors and not employees, the deciding factor is not the agreement but the degree of control the company exercises over the employees. The following questions can help determine whether someone is an employee:

- Does the person have the right to control the manner and means of accomplishing the desired result?
- How much work experience does the person have?
- Does the worker provide his own tools and equipment?
- Is the person engaged in a distinct occupation or an independently established business?
- Is the method of payment by the hour or by the job?
- What degree of skill is required to complete the job?
- Does the person hire employees to help?

The Microsoft ruling means that employers must exercise care in their treatment of contingent workers. If a company wants to hire contingent workers through an agency, then the agency must hire and fire the workers, promote and discipline them, do performance reviews, decide wages, and tell them what to do on a daily basis.

Read the manager's checklist in Table 10-2 for questions that pertain to the use of contingent workers. The preferred answer to each question is *yes*.

TABLE 10-2   Manager's checklist for the use of contingent employees

Question	Yes	No
Have you reviewed the definition of an employee in your company's policies and pension plan documents to ensure it is not so broad that it encompasses contingent workers, thus entitling them to benefits?		
Are you careful not to use contingent workers on an extended basis? Do you make sure that the assignments are finite, with break periods in between?		
Do you use contracts designating the worker as a contingent worker?		
Are you aware that the actual circumstances of the working relationship determine whether a worker is considered an employee in various contexts, and that a company's definition of a contingent worker may not be accepted as accurate by a government agency or court?		
Do you avoid telling contingent workers where, when, and how to do their jobs and instead work through the contingent worker's manager to communicate job requirements?		
Do you request that contingent workers use their own equipment and resources, such as computers and e-mail accounts?		
Do you avoid training your contingent workers?		
When leasing employees from an agency, do you let the agency do its job? Do you avoid asking to see résumés and getting involved with compensation, performance feedback, counseling, or day-to-day supervision?		
If you lease employees, do you use a leasing firm that offers its own benefits plan, deducts payroll taxes, and provides required insurance?		

H-1B WORKERS

An **H-1B** is a temporary work visa granted by the U.S. Citizenship and Immigration Services (USCIS) for people who work in specialty occupations—jobs that require at least a four-year bachelor's degree in a specific field, or equivalent experience. Many companies turn to H-1B workers to meet critical business needs or to obtain essential technical skills and knowledge that cannot be readily found in the United States. H-1B workers may also be used when there are temporary shortages of needed skills. Employers often need H-1B professionals to provide special expertise in overseas markets or on projects that enable U.S. businesses to compete globally. A key requirement is that employers must pay H-1B workers the prevailing wage for the work being performed.

A person can work for a U.S. employer as an H-1B employee for a maximum continuous period of six years. After a worker's H-1B visa expires, the foreign worker must remain outside the United States for one year before another H-1B petition will be approved. Table 10-3 shows the employers who received approval for the most H-1B visas in 2008.

TABLE 10-3 Top H-1B visa employers in 2008

Organization	Number of H-1B visas approved in 2008
Infosys Technologies	4,559
Wipro Technologies	2,678
Satyam Computer Services	1,917
Tata Consultancy Services	1,539
Microsoft	1,037
Accenture	731
Cognizant Technology Solutions	467

Source: Patrick Thibodeau, "List of H-1B Visa Employers for 2008," *Computerworld,* June 12, 2008, www.computerworld.com/s/article/9128436/List_of_H_1B_visa_employers_for_2008.

H-1B temporary professionals make up less than 0.1 percent of the U.S. workforce, but nearly 40 percent are employed as computer programmers.[21] The top five outsourcing countries for H-1B workers are India, China, Canada, the United Kingdom, and the Philippines.

Each year the U.S. Congress sets a federal cap on the number of H-1B visas to be granted, although the number of visas issued often varies greatly from this cap. Since 2004, the cap has been set at 65,000 with an additional 20,000 visas available for foreign graduates of U.S. universities with advanced degrees.[22, 23] (Of the 65,000 visas, 5,400 are reserved for nationals of Singapore and 1,400 for nationals of Chile under certain trade agreements between the U.S. and these countries.)[24] The cap only applies to certain IT professionals, such as programmers and engineers at private technology companies. A large number of foreign workers are exempt from the cap, including scientists hired to teach at American universities, work in government research labs, or work for nonprofit organizations.

When considering the use of H-1B visa workers, companies should take into account that even highly skilled and experienced H-1B workers may require help with their English skills. Communication in many business settings is fast paced and full of idiomatic expressions; workers who are not fluent in English may find it difficult and uncomfortable to participate. As a result, some H-1B workers might become isolated. Even worse, H-1B workers who are not comfortable with English may gradually stop trying to acclimate and may create their own cliques, which can hurt a project team's morale and lead to division. Managers and coworkers should make it a priority to assist H-1B workers looking to improve their English skills and to develop beneficial working relationships based on a mutual respect for any cultural differences that may exist. H-1B workers must feel at ease and be able to easily interact to feel like true members of their team.

As concern increases about employment in the IT sector, displaced workers and other critics challenge whether the United States needs to continue importing thousands of H-1B workers every year. Many business managers, however, say such criticisms conceal the real issue, which is the struggle to find qualified people, wherever they are, for increasingly challenging work. Heads of U.S. companies continue to complain that they have trouble finding enough qualified IT employees and have urged the USCIS to loosen the reins on visas for qualified workers. They warn that reducing the number of visas will encourage them to move the work to foreign countries, where they can find the workforce they need. Some human resource managers and educators are concerned that the continued use of H-1B workers may be a symptom of a larger, more fundamental problem—that the United States is not developing a sufficient number of IT employees with the right skills to meet its corporate needs.

Many IT workers in the United States have expressed concern that the use of H-1B workers has a negative impact on their wages. Researchers from New York University and the Wharton School of the University of Pennsylvania examined tens of thousands of résumés as well as demographic and wage data on 156,000 IT workers employed at 7,500 publicly held U.S. firms. Their conclusions were "that H-1B admissions at current levels are associated with a 5% to 6% drop in wages for computer programmers, systems analysts, and software engineers."[25]

H-1B Application Process

Most companies make ethical hiring decisions based on how well an applicant fulfills the job qualifications. Such companies consider the need to obtain an H-1B visa *after* deciding to hire the best available candidate. To receive an H-1B visa, the person must have a job offer from an employer who is also willing to offer sponsorship.

Once a decision has been made to hire a worker who will require an H-1B visa, an employer must begin the application process. There are two application stages: the Labor Condition Application (LCA) and the H-1B visa application. The company files an LCA with the Department of Labor (DOL), stating the job title, the geographic area in which the worker is needed, as well as the salary to be paid. The DOL's Wage and Hour Division reviews the LCA to ensure that the foreign worker's wages will not undercut those of an American worker. After the LCA is certified, the employer may then apply to the USCIS for the H-1B visa, identifying who will fill the position and stating the person's skills and qualifications for the job. A candidate cannot be hired until the USCIS has processed the application, which can take several days or several months.[26]

Companies whose contingent of H1-B workers makes up more than 15 percent of their workforce face hurdles before they can hire any more. To do so, they must prove that they first tried to find U.S. workers—for example, they can show copies of employment ads they placed online or in newspapers. They must also confirm that they are not hiring an H-1B worker after having laid off a similar U.S. worker. Employers must attest to such protections by affirmatively filing with the DOL and maintaining a public file. Failure to comply with DOL regulations can result in an audit and fines in excess of $1,000 per violation, payment of back wages to the employee, and ineligibility to participate in immigration programs.[27]

In order to cut down on the amount of time it can take to hire an H-1B worker, a provision was added to the American Competitiveness in the Twenty-First Century Act of 2000 that allows current H-1B holders to start working for employers as soon as their petitions are filed. Therefore, a company looking to hire a critical person who already has H-1B status can do so in a matter of weeks.

Using H-1B Workers Instead of U.S. Workers

In order to compete in the global economy, U.S. firms must be able to attract the best and brightest workers from all over the world. Most H-1B workers are brought to the United States to fill a legitimate gap that cannot be filled with the existing pool of workers. However, there are some managers who reason that as long as skilled foreign workers can be found to fill critical positions, why spend thousands of dollars and take months to develop their current U.S. workers? Although such logic may appear sound for short-term hiring decisions, it does nothing to develop the strong core of permanent IT workers that the United States will need in the future. Heavy reliance on the use of H-1B workers can lessen the incentive for U.S. companies to educate and develop their own workforces.

Potential Exploitation of H-1B Workers

Even though companies applying for H-1B visas must offer a wage that is not 5 percent less than the average salary for the occupation, some companies use H-1B visas as a way to lower salaries. Because wages in the IT field vary greatly, unethical companies can get around the average salary requirement. Determining an appropriate wage is an imprecise science at best. For example, an H-1B worker may be classified as an entry-level IT employee and yet fill a position of an experienced worker who would make $10,000 to $30,000 more per year. Unethical companies can also find other ways to get around the salary protections included in the H-1B program, as shown in the charges filed in February 2009 against the Vision Systems Group. The company was charged with H-1B visa fraud for allegedly stating that certain H-1B workers in its New Jersey office were actually working in Iowa, where the company had another office and where the prevailing wage is much less.[28] If found guilty of underpayment, an employer must reimburse the underpaid employee for back wages.

Until Congress approved the Visa Reform Act of 2004, there were few investigations into H-1B salary abuses. The act increased the H-1B application fee by $2,000, of which $500 was earmarked for antifraud efforts; the act also defined a modified wage-rate system, allowing for greater variances in pay to visa holders. Investigations are typically triggered by complaints from H-1B holders, but the government can conduct random audits or launch an investigation based on information from third-party sources.

Companies using H-1B workers, as well as the workers themselves, must also consider what will happen at the end of their six-year visa term. The stopgap nature of the visa program can be challenging for both sponsoring companies and applicants. If a worker is not granted a green card, the firm loses the worker without having developed a permanent employee. Many of these foreign workers, finding themselves suddenly unemployed, are forced to uproot their families and return home.

OUTSOURCING

Outsourcing is another approach to meeting staffing needs. **Outsourcing** is a long-term business arrangement in which a company contracts for services with an outside organization that has expertise in providing a specific function. A company may contract with an organization to provide such services as operating a data center, supporting a telecommunications network, or staffing a computer help desk.

Coemployment legal problems with outsourcing are minimal, because the company that contracts for the services does not generally supervise or control the contractor's employees. The primary rationale for outsourcing is to lower costs, but companies also use it to obtain strategic flexibility and to keep their staff focused on the company's core competencies.

In the 1970s, IT executives started the trend toward outsourcing as they began to supplement their IT staff with contractors and consultants. This trend eventually led to outsourcing entire IT business units to such organizations as Accenture, Electronic Data Systems, and IBM, which could take over the operation of a company's data center as well as perform other IT functions.

Offshore Outsourcing

Offshore outsourcing is a form of outsourcing in which the services are provided by an organization whose employees are in a foreign country. Any work done at a relatively high cost in the United States may become a candidate for offshore outsourcing—not just IT work. However, IT professionals in particular can do much of their work anywhere—on a company's premises or thousands of miles away in a foreign country. In addition, companies can reap large financial benefits by reducing labor costs through offshore outsourcing. As a result, and because a large supply of experienced IT professionals is readily available in certain foreign countries, the use of offshore outsourcing in the IT field is increasing. American Express, Aetna, Compaq, General Electric, IBM, Microsoft, Motorola, Shell, Sprint, and 3M are examples of big companies that employ offshore outsourcing for functions such as help-desk support, network management, and information systems development.

As more businesses move their key processes offshore, U.S. IT service providers are forced to lower prices. Many U.S. software firms set up development centers in low-cost foreign countries where they have access to a large pool of well-trained candidates. Intuit—maker of the Quicken tax preparation software—currently has facilities in Canada, Great Britain, and India. Accenture, IBM, and Microsoft maintain large development centers in India. Cognizant Technology Solutions is headquartered in Teaneck, New Jersey, but operates primarily from technology centers in India.

Because of the high salaries earned by application developers in the United States and the ease with which customers and suppliers can communicate, it is now quite common to use offshore outsourcing for major programming projects. According to the Gartner Group, the top four sources of contract programming include India, Ireland, Canada, and Israel. Contract programming is also flourishing in Argentina, Australia, Belarus, Brazil, Bulgaria, China, Malaysia, Malta, Mexico, Nepal, the Philippines, Poland, Russia, Serbia, Singapore, Sri Lanka, and Vietnam.[29] But India, with its rich talent pool (a high percentage of whom speak English) and low labor costs, is widely acknowledged as the best source of programming skills outside Europe and North America.[30]

NXP creates semiconductors, system solutions, and software for mobile phones, TVs, and personal media players. It employs 37,000 people working in 20 countries, with 2007 sales reaching $6.3 billion. In 2008, NXP awarded Tata Consultancy Services a five-year, $100 million outsourcing contract to provide its IT services. These services include data-center operations, networking, and application development out of computing centers located in Bangalore, Bangkok, and Eindhoven in the Netherlands. The purpose of the outsourcing was to provide operational cost savings and to enable NXP's managers and executives to focus on core business activities. The project has gone so well that Tata won the Netherlands Outsourcing Award for 2009.[31]

Table 10-4 shows the leading countries that provide offshore IT services to U.S. firms. Table 10-5 lists several offshore IT outsourcing firms.

TABLE 10-4   Leading countries for providing offshore IT services

Country	Comments
India	Relatively low cost, highly skilled labor pool
Canada	Close to United States, no language barrier, highly skilled labor pool
China	Low cost, large pool of skilled labor, lack of English-language proficiency
Poland	Low overall cost of business operations
Czech Republic	An emerging contender
Russia	Unpredictable political and business climate

Source: Paul McDougall, "India, Canada, and China Are Top Outsourcing Destinations: Study," *Information-Week,* September 21, 2005, www.informationweek.com/news/global-cio/trends/showArticle.jhtml?articleID= 171000615.

TABLE 10-5   Partial list of offshore IT outsourcing firms

Firm	Number of employees	Headquarters location
Tata Consultancy Services	143,000	Mumbai, India[32]
Infosys Technologies	103,900	Bangalore, India[33]
Wipro Technologies	97,810	Bangalore, India[34]
Satyam Computer Services	25,000	Hyderabad, India[35]
HCL Infosystems	5,750	New Delhi, India[36]
EPAM Systems	4,500	Newtown, PA[37]
Luxoft	3,000	Moscow, Russia[38]
Aplana Software	250	Moscow, Russia[39]
Hubport Interactive	NA	Davao City, Philippines
Interxion	192	Amsterdam, the Netherlands[40]

Pros and Cons of Offshore Outsourcing

Wages that an American worker might consider low represent an excellent salary in many other parts of the world, and some companies feel they would be foolish not to exploit such an opportunity. Why pay a U.S. IT worker a six-figure salary, they reason, when they can use offshore outsourcing to hire three India-based workers for the same cost? However, this attitude might represent a short-term point of view—offshore demand is driving up salaries in India by roughly 15 percent per year. Because of this, Indian offshore suppliers have begun to charge more for their services. The cost advantage for offshore outsourcing to India used to be 6:1 or more—you could hire six Indian IT workers for the cost of one U.S. IT worker. The cost advantage has now shrunk to 3:1. Once it shrinks to 1.5:1, the cost savings would no longer be much of an incentive for U.S. offshore outsourcing to India.[41]

Another benefit of offshore outsourcing is its potential to dramatically speed up development efforts. For example, the State of New Mexico contracted the development of a tax system to Syntel, one of the first U.S. firms to successfully launch a global delivery model that enables workers to make progress on a project around the clock. With technical teams working from networked facilities in different time zones, Syntel executes a virtual "24-hour workday" that saves its customers money, speeds projects to completion, and provides continuous support for key software applications.

Although it has its advocates, offshore outsourcing doesn't always pay off. For example, a software developer in Cambridge, Massachusetts, went to India in search of cheap labor but found lots of problems instead. Customs officials charged huge tariffs when the company tried to ship the necessary development software and manuals, and the programmers weren't nearly as experienced as they claimed to be. The code produced in India was inadequate. The company had to send a representative to India for months to work with the programmers there and correct the problems.

While offshore outsourcing can save a company in terms of labor costs, it will also result in some new expenses. In determining how much money and time a company will save with offshore outsourcing, the firm must take into account the additional time that will be required to select an offshore vendor as well as the additional costs that will be incurred for travel and communications. In addition, organizations often find that it can take years of ongoing effort and a large up-front investment to develop a good working relationship with an offshore outsourcing firm. Finding a reputable vendor can be especially difficult for a small or midsized firm that lacks experience in identifying and vetting contractors.

Many of the ethical issues that arise when considering whether to use H-1B and contingent workers also apply to offshore outsourcing. For example, managers must consider the trade-offs between using offshore outsourcing firms and devoting money and time to retain and develop their own staff. Offshore outsourcing tends to upset domestic staff, especially when a company begins to lay off employees in favor of low-wage workers outside the United States. The remaining members of a department can become bitter and nonproductive, and morale can become extremely low.

Cultural and language differences can cause misunderstandings among project members in different countries. Indian programmers, for instance, are known for keeping quiet even when they notice problems. And the difficulty of communicating directly with people over long distances can make offshore outsourcing perilous, especially when key team members speak English as their second language.

The compromising of customer data is yet another potential outsourcing issue. For example, the Australian Broadcasting Corporation uncovered evidence that the personal details of Australian citizens (including names, addresses, telephone and other ID numbers, and employment information) were taken from customer databases in India and offered for sale on the black market.[42] Most countries have laws designed to protect the privacy of their citizens. Firms that outsource must take precautions to protect private data, regardless of where it is stored or processed or who handles it.[43]

Another downside to offshore outsourcing is that a company loses the knowledge and experience gained by outsourced workers when those workers are reassigned after a project's completion. Finally, offshore outsourcing does not advance the development of permanent IT workers in the United States, which increases its dependency on foreign workers to build the IT infrastructure of the future. Many of the jobs that go overseas are entry-level positions that help develop employees for future, more responsible positions.

Strategies for Successful Offshore Outsourcing

Successful projects require day-to-day interaction between software development and business teams, so it is essential for the hiring company to take a hands-on approach to project management. Companies cannot afford to outsource responsibility and accountability.

To improve the chances that an offshore outsourcing project will succeed, a company must carefully evaluate whether an outsourcing firm can provide the following:

1. Employees with the required expertise in the technologies involved in the project
2. A project manager who speaks the employer company's native language
3. A pool of staff large enough to meet the needs of the project
4. A state-of-the-art telecommunications setup
5. High-quality on-site managers and supervisors

To ensure that company data is protected in an outsourcing arrangement, companies can use the Statement on Auditing Standards (SAS) No. 70, Service Organizations, an internationally recognized standard developed by the American Institute of Certified Public Accountants (AICPA). A successful SAS No. 70 audit report demonstrates that an outsourcing firm has effective internal controls in accordance with the Sarbanes-Oxley Act of 2002.

The following list provides several tips for companies that are considering offshore outsourcing:

- Set clear, firm business specifications for the work to be done.
- Assess the probability of political upheavals or factors that might interfere with information flow, and ensure that the risks are acceptable.
- Assess the basic stability and economic soundness of the outsourcing vendor and what might occur if the vendor encounters a severe financial downturn.
- Establish reliable satellite or broadband communications between your site and the outsourcer's location.
- Implement a formal version-control process, coordinated through a quality assurance person.
- Develop and use a dictionary of terms to encourage a common understanding of technical jargon.

- Require vendors to supply project managers at the client site to overcome cultural barriers and facilitate communication with offshore programmers.
- Require a network manager at the vendor site to coordinate the logistics of using several communications providers around the world.
- Obtain advance agreement on the structure and content of documentation to ensure that manuals explain how the system was built, as well as how to maintain it.
- Carefully review a current copy of the outsourcing firm's SAS No. 70 audit report to ascertain its level of control over information technology and related processes.

WHISTLE-BLOWING

Like the subject of contingent workers, whistle-blowing is a significant topic in any discussion of ethics in IT. Both issues raise ethical questions and have social and economic implications. How these issues are addressed can have a long-lasting impact not only on the people and employers involved, but also on the entire IT industry.

As noted above, whistle-blowing is an effort to attract public attention to a negligent, illegal, unethical, abusive, or dangerous act by a company or some other organization. In some cases, whistle-blowers are employees who act as informants on their company, revealing information to enrich themselves or to gain revenge for a perceived wrong. In most cases, however, whistle-blowers act ethically in an attempt to correct what they think is a major wrongdoing, often at great personal risk.

A whistle-blower usually has personal knowledge of what is happening inside the offending organization because of his or her role as an employee of the organization. Sometimes the whistle-blower is not an employee but a person with special knowledge gained from a position as an auditor or business partner.

In going public with the information they have, whistle-blowers often risk their own careers and sometimes even affect the lives of their friends and family. In extreme situations, whistle-blowers must choose between protecting society and remaining silent.

In December 2005, the *New York Times* broke the alarming story that just months after the 9/11 attacks in 2002, President Bush had secretly authorized the National Security Agency (NSA) to listen in on the calls of hundreds, perhaps thousands, of American citizens to obtain evidence of terrorist activity without the court-approved warrants required for domestic spying. (The White House had asked the *New York Times* not to publish the article on the basis that it could jeopardize ongoing investigations and provide an alert to terrorists.)[44] Whistle-blower Russell Tice, who had worked for both the Defense Intelligence Agency and the NSA, later disclosed that he was one of the sources for the story. Tice said that "as far as I'm concerned, as long as I don't say anything that's classified, I'm not worried. We need to clean up the intelligence community. We've had abuses, and they need to be addressed."[45]

Protection for Whistle-Blowers

Whistle-blower protection laws allow employees to alert the proper authorities to employer actions that are unethical, illegal, or unsafe, or that violate specific public policies. Unfortunately, no comprehensive federal law protects all whistle-blowers from retaliatory acts. Instead, numerous laws protect a certain class of specific whistle-blowing acts in various industries. To make things even more complicated, each law has different filing provisions, administrative and judicial remedies, and statutes of limitations (which set time limits for legal action). Thus, the first step in reviewing a whistle-blower's claim of retaliation is for an experienced attorney to analyze the various laws and determine if and how the employee is protected. Once that is known, the attorney can determine what procedures to follow in filing a claim.

From the whistle-blower's perspective, a short statute of limitations is a major weakness of many whistle-blower protection laws. Failure to comply with the statute of limitations is a favorite defense of firms accused of wrongdoing in whistle-blower cases.

The **False Claims Act**, also known as the Lincoln Law, was enacted during the U.S. Civil War to combat fraud by companies that sold supplies to the Union Army. War profiteers sometimes shipped boxes of sawdust instead of guns, for instance, and some swindled the Union Army into purchasing the same cavalry horses several times. When it was enacted, the act's goal was to entice whistle-blowers to come forward by offering them a share of the money recovered.

The **qui tam** ("who sues on behalf of the king as well as for himself") provision of the False Claims Act allows a private citizen to file a suit in the name of the U.S. government, charging fraud by government contractors and other entities who receive or use government funds. In qui tam actions, the government has the right to intervene and join the legal proceedings. If the government declines, the private plaintiff may proceed alone. Some states have passed similar laws concerning fraud in state government contracts.[46]

Qui tam actions can be based on a variety of charges, including mischarging for services, product and service substitution, false certification of entitlement for benefits, and false negotiation to justify an inflated contract. Mischarging is the most common charge in qui tam cases.[47] For example, an IT contractor might overcharge hundreds of hours of programming time as part of a government contract, or a physician might overcharge the government for medical services that a nurse actually performed.

Violators of the False Claims Act are liable for three times the dollar amount for which the government was defrauded. They can also be fined civil penalties of $5,000 to $10,000 for each instance of a false claim. A qui tam plaintiff can receive between 15 and 30 percent of the total recovery from the defendant, depending on how helpful the person was to the success of the case.[48]

Amerigroup, a health insurer that serves people on Medicaid, was found liable in the largest qui tam False Claims Act judgment ever levied—$144 million, plus a possible $190 million in statutory fines. Amerigroup was accused by one of its employees of improperly excluding pregnant women, particularly those in their third trimester, from its Medicaid program in Illinois, thus overcharging the government $48 million. (The government is entitled to collect triple the damages plus additional damages for each false claim.)[49]

A former Oracle vice president filed a whistle-blower lawsuit charging that the company fraudulently billed the federal government for training over a six-year period. In 2005, Oracle agreed to pay $8 million to settle the charges, and the whistle-blower received $1.6 million of the total settlement amount.[50]

The False Claims Act provides strong whistle-blower protection. Any person who is discharged, demoted, harassed, or otherwise discriminated against because of lawful acts of whistle-blowing is entitled to all relief necessary "to make the employee whole." Such relief may include job reinstatement; double back pay; and compensation for any special damages, including litigation costs and reasonable attorney's fees.[51]

The provisions of the False Claims Act are complicated, so it is unwise to pursue a claim without legal counsel. However, because the potential for significant financial recovery is good, attorneys are generally willing to assist.

Whistle-Blowing Protection for Private-Sector Workers

Under state law, an employee could traditionally be terminated for any reason, or no reason, in the absence of an employment contract. However, many states have created laws that prevent workers from being fired because of an employee's participation in "protected" activities. One such activity is the filing of a qui tam lawsuit under the provisions of the False Claims Act. States that recognize the public benefit of such cases offer protection to whistle-blowers; for example, whistle-blowers can file claims against their employers for retaliatory termination and are entitled to jury trials. If successful, they can receive punitive damage awards.

Dealing with a Whistle-Blowing Situation

Each potential whistle-blowing case involves different circumstances, issues, and personalities. Two people working together in the same company may have different values and concerns that cause them to react in different ways to a particular situation—and both reactions might be ethical. It is impossible to outline a definitive step-by-step procedure of how to behave in a whistle-blowing situation. This section provides a general sequence of events, and highlights key issues that a potential whistle-blower should consider.

Assess the Seriousness of the Situation

Before considering whistle-blowing, a person should have specific knowledge that his company or a coworker is acting unethically and that the action represents a *serious* threat to the public interest. The employee should carefully and informally seek trusted resources outside the company and ask for their assessment. Do they also see the situation as serious? Their point of view may help the employee see the situation from a different perspective and alleviate concerns. On the other hand, the outside resources may reinforce the employee's initial suspicions, forcing a series of difficult ethical decisions.

Begin Documentation

An employee who identifies an illegal or unethical practice should begin to compile adequate documentation to establish wrongdoing. The documentation should record all events and facts as well as the employee's insights about the situation. This record helps construct a chronology of events if legal testimony is required in the future. An employee should identify and copy all supporting memos, correspondence, manuals, and other documents *before* taking the next step. Otherwise, records may disappear and become inaccessible. The employee should maintain documentation and keep it up to date throughout the process.

Attempt to Address the Situation Internally

An employee should next attempt to address the problem internally by providing a written summary to the appropriate managers, including a statement that they either responded or clearly chose not to respond. Ideally, the employee can expose the problem and deal with it from inside the organization. The focus should be on disclosing the facts and how the situation affects others. The employee's goal should be to fix the problem, not to place blame.

Given the potential negative impact of whistle-blowing on the employee's future, this step should not be dismissed or taken lightly.

Fortunately, many problems are solved at this point, and further, more drastic actions by the employee are unnecessary. The appropriate managers get involved and resolve the issue that initiated the whistle-blower's action.

On the other hand, managers who are engaged in unethical or illegal behavior might not welcome an employee's questions or concerns. In such cases, the whistle-blower can expect to be strongly discouraged from taking further action. Employee demotion or termination on false or exaggerated claims can occur. Attempts at discrediting the employee can also be expected. As an extreme example, Dr. Jeffrey Wigand, former vice president of research and development at Brown & Williamson, disclosed wrongdoings involving the use of cancer-causing ingredients in the tobacco industry. As a result, he received several anonymous deaths threats; however, none of the threats could be traced back to their source.[52]

Consider Escalating the Situation Within the Company

The employee's initial attempt to deal with a situation internally may be unsuccessful. At this point, the employee may rationalize that he or she has done all that is required by raising the issue. Others may feel so strongly about the situation that they are compelled to take further action. Thus, a determined and conscientious employee may feel forced to choose between escalating the problem and going over the manager's head or going outside the organization to deal with the problem. The employee may feel obligated to sound the alarm on the company because there appears to be no chance to solve the problem internally.

Going over an immediate manager's head can put one's career in jeopardy. Supervisors may retaliate against a challenge to their management, although some organizations may have an effective corporate ethics officer who can be trusted to give the employee a fair and objective hearing. Alternatively, a senior manager with a reputation for fairness and some responsibility for the area of concern might step in. However, in many work environments, the challenger is likely to be fired, demoted, or reassigned to a less desirable position or job location. Such actions send a loud signal throughout an organization that loyalty is highly valued and that challengers will be dealt with harshly. Whether reprisal is ethical depends in large part on the legitimacy of the employee's issue. If the employee is truly overreacting to a minor issue, then the employee may deserve some sort of reprimand for exercising poor judgment.

If senior managers refuse to deal with a legitimate problem, the employee can decide to drop the matter or go outside the organization to try to remedy the situation. Even if a senior manager agrees with the employee's position and overrules the employee's immediate supervisor, the employee may want to request a transfer to avoid working for the same person.

Assess the Implications of Becoming a Whistle-Blower

If the employee feels he has made a strong attempt to resolve the problem internally without results, he must stop and fully assess whether he is prepared to go forward and blow the whistle on the company. Depending on the situation, an employee may incur significant legal fees in order to air or bring charges against an agency or company that may have access to an array of legal resources as well as a lot more money than the individual employee. An employee who chooses to proceed might be accused of having a grievance with the employer or of trying to profit from the accusations. The employee may be fired and may lose the confidence of coworkers, friends, and even family members.

A potential whistle-blower must attempt to answer many ethical questions before making a decision on how to proceed:

- Given the potentially high price, do I really want to proceed?
- Have I exhausted all means of dealing with the problem? Is whistle-blowing all that is left?
- Am I violating an obligation to be loyal to my employer and work for its best interests?
- Will the public exposure of corruption and mismanagement in the organization really correct the underlying cause of these problems and protect others from harm?

From the moment an employee becomes known as a whistle-blower, a public battle may ensue. Whistle-blowers can expect attacks on their personal integrity and character as well as negative publicity in the media. Friends and family members will hear these accusations, and ideally, they should be notified beforehand and consulted for advice before the whistle-blower goes public. This notification helps prevent friends and family members from being surprised at future actions by the whistle-blower or the employer.

The whistle-blower should also consider consulting support groups, elected officials, and professional organizations. For example, the National Whistleblowers Center provides referrals for legal counseling and education about the rights of whistle-blowers.

Use Experienced Resources to Develop an Action Plan

A whistle-blower should consult with competent legal counsel who has experience in these kinds of cases. He or she will determine which statutes and laws apply, depending on the agency, employer, state involved, and nature of the case. Counsel should also know the statute of limitations for reporting the offense, as well as the whistle-blower's protection under the law. Before blowing the whistle, the employee should get an honest assessment of the soundness of his or her legal position and an estimate of the costs of a lawsuit.

Execute the Action Plan

A whistle-blower who chooses to pursue a matter legally should do so based on the research and guidance of legal counsel. If the whistle-blower wants to remain unknown, the safest course of action is to leak information anonymously to the press. The problem with this approach, however, is that anonymous claims are often not taken seriously. In most cases, working directly with appropriate regulatory agencies and legal authorities is more likely to get results, including the imposition of fines, the halting of operations, or other actions that draw the offending organization's immediate attention.

Live with the Consequences

Whistle-blowers must be on guard against retaliation, such as being discredited by coworkers, threatened, or set up; for example, management may attempt to have the whistle-blower transferred, demoted, or fired for breaking some minor rule, such as arriving late to work or leaving early. To justify their actions, management may argue that such behavior has been ongoing. The whistle-blower might need a good strategy and a good attorney to counteract such actions and take recourse under the law.

A massive computer-data breach at TJX (the parent company of T.J. Maxx, Marshalls, and other stores) affecting 94 million Visa and MasterCard accounts occurred in June 2005.[53] Nick Benson was a student at the University of Kansas and an hourly worker at TJX. Although he was not an IT worker, he noticed many computer-related security problems at the firm prior to the data breach. He reported these verbally to TJX managers and also posted information about the breaches on an online security forum, *http://sla.ckers.org*. In the forum, he revealed serious security weaknesses in sufficient detail that the information could be of use to hackers. Benson spoke to store managers and the district loss prevention manager before the data breach occurred, but nothing was done. Eventually Benson was fired over the public disclosures and violation of his nondisclosure agreement.[54] This is a perfect example of how *not* to be a whistle-blower.

GREEN COMPUTING

Many computer manufacturers today are talking about building a "green PC," by which they usually mean one that uses less electricity to run than the standard computer; thus, its carbon footprint on the planet is smaller. However, to manufacturer a truly green PC, hardware companies must also reduce the amount of hazardous materials used and dramatically increase the amount of reusable or recyclable materials used. The manufacturers must also help consumers dispose of their products in an environmentally safe manner at the end of their useful life.

Electronic devices such as personal computers and cell phones contain hundreds or even thousands of components. The components, in turn, are composed of many different materials, including some that are known to be potentially harmful to humans and the environment, including beryllium, cadmium, lead, mercury, brominated flame retardants, selenium, and polyvinyl chloride.[55]

Electronic manufacturing employees and suppliers at all steps along the supply chain and manufacturing process are at risk of unhealthy exposure to these raw materials. Users of these products can also be exposed to these materials when using poorly designed or improperly manufactured devices. Care must also be taken when recycling or destroying these devices to avoid contaminating the environment. The United States has no federal law prohibiting the export of toxic waste, so many used electronic devices intended for recycling are sold to companies in developing countries that try to repair the components or extract valuable metals from them, using methods that release carcinogens and other toxins into the air and the water supply.[56]

EPEAT (Electronic Product Environmental Assessment Tool) is a system that enables purchasers to evaluate, compare, and select electronic products based on a total of 51 environmental criteria. Products are ranked in EPEAT according to three tiers of environmental performance: Bronze (meets all 23 required criteria), Silver (meets all 23 of the required criteria plus at least 50 percent of the optional criteria), and Gold (meets all 23 required criteria plus at least 75 percent of the optional criteria). Individual purchasers of home computers as well as corporate purchasers of thousands of computers can use the EPEAT Web site (*www.epeat.net*) to screen manufacturers and computer models based on certain environmental attributes.

In early 2009, the City of San Francisco upgraded its environmental requirement for purchase of PCs and laptops to EPEAT Gold. Jeff Omelchuck, executive director of the Green Electronics Council, which manages the EPEAT system, states: "By electing to purchase EPEAT Gold products wherever possible, San Francisco is in the forefront of environmental purchasing. This commitment will result in energy and cost savings, as well as significantly reducing the overall environmental impact of the City's IT operations."[57]

The European Union passed the Restriction of Hazardous Substances Directive, which restricts the use of many hazardous materials in computer manufacturing. It also requires that manufacturers use at least 65 percent reusable or recyclable components, implement a plan to manage products at the end of their life cycle in an environmentally safe manner, and reduce or eliminate toxic material in their packaging. The state of California has passed a similar law, called the Electronic Waste Recycling Act. Because of these two acts, some manufacturers—such as Apple and Dell—have promised to remove brominated flame retardants from their PC casings by the end of 2009.[58] The United States has yet to pass a federal law equivalent to the European Union's Restriction of Hazardous Substances Directive.

It is estimated that roughly 20 million computers became obsolete in 1998; by 2007, that number had more than doubled.[59] How should users safely dispose of their obsolete computers? As of 2009, 26 states have established statewide programs for some recycling of obsolete computers. These statutes either impose a fee for each unit sold at retail or require manufacturers to reclaim the equipment at disposal.[60]

Some electronics manufacturers have developed programs to assist their customers in disposing of old equipment. For example, Dell offers a free worldwide recycling program for consumers. It also provides no-charge recycling of any brand of used computer or printer with the purchase of a new Dell computer or printer. This equipment is recycled in an environmentally responsible manner, using Dell's stringent and global recycling guidelines.[61] HP and other manufacturers offer similar programs.

The environmental activist organization Greenpeace issues quarterly ratings of the top manufacturers of personal computers, mobile phones, TVs, and game consoles according to the manufacturers' policies on toxic chemicals, recycling, and climate change. Table 10-6 shows the Greenpeace ratings in July 2009. With 10 being a perfect score, it is clear that these manufacturers have a long way to go in meeting the very high "green" standards of Greenpeace.

TABLE 10-6   Greenpeace ratings of electronics manufacturers

Organization	July 2009 rating
Nokia	7.45
Samsung	7.1
Sony Ericsson	6.5
LG Electronics	5.7
Toshiba	5.5
Motorola	5.5

TABLE 10-6   Greenpeace ratings of electronics manufacturers (*continued*)

Organization	July 2009 rating
Philips	5.3
Sharp	5.3
Acer	4.9
Panasonic	4.9
Apple	4.7
Sony	4.5
Dell	3.9
HP	3.5
Microsoft	2.5
Lenovo	2.5
Fujitsu	2.4
Nintendo	1.0

Source: Greenpeace, "How the Companies Line Up: Guide to Greener Electronics," June 2009, www.greenpeace.org/international/campaigns/toxics/electronics/how-the-companies-line-up#.

ICT INDUSTRY CODE OF CONDUCT

The **Electronic Industry Citizenship Coalition (EICC)** was established to promote a common code of conduct for the electronics and information and communications technology (ICT) industry.[62] The EICC focuses on the areas of worker safety and fairness, environmental responsibility, and business efficiency. Information and communications technology organizations, electronic manufacturers, software firms, and manufacturing service providers may voluntarily join the coalition.

The EICC has established a code of conduct that defines performance, compliance, auditing, and reporting guidelines across five areas of social responsibility: labor, health and safety, environment, management system, and ethics. Adopting organizations apply the code across their entire worldwide supply chain and require their first-tier suppliers to acknowledge and implement it.[63] As of July 2009, the code has been formally adopted by over 38 EICC member organizations, including Adobe, Cisco, Dell, HP, IBM, Intel, Microsoft, Philips, Samsung, and Sony.[64] The following are the five areas of social responsibility and guiding principles covered by the code:

1. **Labor:** "Participants are committed to uphold the human rights of workers, and to treat them with dignity and respect as understood by the international community."[65]

2. **Health and Safety:** "Participants recognize that in addition to minimizing the incidence of work-related injury and illness, a safe and healthy work environment enhances the quality of products and services, consistency of production and worker retention and morale. Participants also recognize that ongoing worker input and education is essential to identifying and solving health and safety issues in the workplace."[66]

3. **Environment:** "Participants recognize that environmental responsibility is integral to producing world class products. In manufacturing operations, adverse effects on the community, environment, and natural resources are to be minimized while safeguarding the health and safety of the public."[67]

4. **Management System:** "Participants shall adopt or establish a management system whose scope is related to the content of this Code. The management system shall be designed to ensure (a) compliance with applicable laws, regulations and customer requirements related to the participant's operations and products; (b) conformance with this Code; and (c) identification and mitigation of operational risks related to this Code. It should also facilitate continual improvement."[68]

5. **Ethics:** "To meet social responsibilities and to achieve success in the marketplace, participants and their agents are to uphold the highest standards of ethics including: business integrity; no improper advantage; disclosure of information; intellectual property; fair business, advertising, and competition; and protection of identity."[69]

Prior to the adoption of the EICC Code of Conduct, many electronic manufacturing companies developed their own codes of conduct and used them to audit their suppliers. Thus, suppliers could be subjected to multiple, independent audits based on different criteria. The adoption of a single, global code of conduct by members of the EICC enables those companies to provide leadership in the area of corporate social responsibility. It also exerts pressure on suppliers to meet a common set of social principles.

The EICC has developed an audit program for member organizations in which audits are conducted by certified, third-party audit firms. EICC members use the audits to measure supplier compliance with the EICC Code of Conduct and to identify areas for improvement.

Summary

- Contingent work is a job situation in which an individual does not have an explicit or implicit contract for long-term employment. The contingent workforce includes independent contractors, temporary workers hired through employment agencies, on-call or day laborers, and on-site workers whose services are provided by contract firms.

- An H-1B is a temporary work visa granted by the U.S. Citizenship and Immigration Services (USCIS) for people who work in specialty occupations—jobs that require at least a four-year bachelor's degree in a specific field, or equivalent experience.

- Employers hire H-1B workers to meet critical business needs or to obtain essential technical skills or knowledge that cannot be readily found in the United States. H-1B workers may also be used when there are temporary shortages of needed skills.

- Some people contend that employers exploit contingent workers, especially H-1B foreign workers, to obtain skilled labor at less than competitive salaries. Others believe that the use of H-1B workers is required to keep the United States competitive.

- Employers must make ethical decisions about whether to recruit new and more skilled workers from these sources or to spend the time and money to develop their current staff to meet the needs of their business.

- Outsourcing is a long-term business arrangement in which a company contracts for services with an outside organization that has expertise in providing a specific function. Offshore outsourcing is a form of outsourcing in which the services are provided by an organization whose employees are in a foreign country.

- Outsourcing and offshore outsourcing are used to meet staffing needs while potentially reducing and speeding up project schedules.

- Many of the same ethical issues that arise when considering whether to hire H-1B and contingent workers apply to outsourcing and offshore outsourcing.

- Whistle-blowing is an effort to attract public attention to a negligent, illegal, unethical, abusive, or dangerous act by a company or some other organization.

- A potential whistle-blower must consider many ethical implications, including whether the high price of whistle-blowing is worth it; whether all other means of dealing with the problem have been exhausted; whether whistle-blowing violates the obligation of loyalty that the employee owes to his or her employer; and whether public exposure of the problem will actually correct its underlying cause and protect others from harm.

- An effective whistle-blowing process includes the following steps: (1) assess the seriousness of the situation, (2) begin documentation, (3) attempt to address the situation internally, (4) consider escalating the situation within the company, (5) assess the implications of becoming a whistle-blower, (6) use experienced resources to develop an action plan, (7) execute the action plan, and (8) live with the consequences.

- Computer companies looking to manufacture green computers are challenged to produce computers that use less electricity, include fewer hazardous materials to harm people or pollute the environment, and contain a high percentage of reusable or recyclable material. These companies should also provide programs to help consumers dispose of their products in an environmentally safe manner at the end of their useful life.

- EPEAT (Electronic Product Environmental Assessment Tool) is a system that enables purchasers to evaluate, compare, and select electronic products based on 51 environmental criteria.

- The European Union passed the Restriction of Hazardous Substances Directive to restrict the use of many hazardous materials in computer manufacturing, require manufacturers to use at least 65 percent reusable or recyclable components, implement a plan to manage products at the end of their life cycle in an environmentally safe manner, and reduce or eliminate toxic material in their packaging.

- The Electronic Industry Citizenship Coalition has established a code of conduct that defines performance, compliance, auditing, and reporting guidelines across five areas of social responsibility: labor, health and safety, environment, management system, and ethics.

- A number of electronics manufacturers have applied this code across their entire worldwide supply chain and also require their first-tier suppliers to acknowledge and implement the code.

Self-Assessment Questions

The answers to the Self-Assessment Questions can be found in Appendix G.

1. The number of declared computer science majors at doctoral-granting computer science departments continued a seven-year decline with the number of students enrolled in the fall of 2007 half of what it was in the fall of 2000. True or False?

2. An employment situation in which two employers have actual or potential legal rights and duties with respect to the same employee is called _____.

3. Which of the following is *not* an advantage for organizations that employ contingent workers?

   a. The firm does not have to offer employee benefits to contingent workers.

   b. Training costs are kept to a minimum.

   c. It provides a way to meet fluctuating staffing needs.

   d. The contingent worker's experience may be useful to the next firm that hires him or her.

4. Depending on how closely workers are supervised and how the job is structured, contingent workers can be viewed as permanent employees by the IRS, the Labor Department, or a state's workers' compensation and unemployment agencies. True or False?

5. A temporary working visa granted by the U.S. Citizenship and Immigration Services for people who work in specialty occupations—jobs that require at least a four-year bachelor's degree in a specific field, or equivalent experience—is called a (an) _____ visa.

6. A business arrangement in which a company contracts for services with an outside organization that has expertise in providing a specific function is called _____.

7. Because of the high cost of U.S.-based application developers and the ease with which customers and suppliers can communicate, it is now quite common to use offshore outsourcing for major software programming projects. True or False?

8. Which of the following is *not* cited as a business advantage of offshore outsourcing?

   a. lower labor costs

   b. the potential to speed up development efforts

   c. the opportunity to learn different languages, cultures, and ways of operating

   d. the ability to tap into a large, well-educated labor pool

9. The cost advantage for Indian workers over U.S. workers has gone from 6:1 to 3:1, so there is now less cost incentive for offshore outsourcing to India. True or False?

10. Which of the following statements about whistle-blowing is true?

   a. From the moment an employee becomes known as a whistle-blower, a public battle may ensue, with negative publicity attacks on the individual's personal integrity.

   b. Whistle-blowing is an effective approach to take in dealing with all work-related matters, from the serious to mundane.

   c. Violators of the False Claim Act are liable for four times the dollar amount that the government is defrauded.

   d. A whistle-blower must be an employee of the company that is the source of the problem.

11. Which of the following are desirable characteristics of a "green computer"?

   a. It runs on less electricity than the typical computer.

   b. It contains a high percentage of reusable or recyclable materials.

   c. Its manufacturer has a program to help consumers dispose of it at the end of its life.

   d. All of the above.

12. Personal computers and cell phones contain hundreds if not thousands of components, which, in turn, are composed of many materials. Many of these materials are potentially harmful to humans and the environment. True or False?

13. Apple, Dell, and HP have earned high marks from Greenpeace for their excellent corporate policies in regards to toxic chemicals, recycling, and climate change. True or False?

14. Products are ranked in EPEAT according to three tiers of environmental performance, with _____ being the highest.

Discussion Questions

1. Briefly discuss the advantages and disadvantages of using H-1B workers. What ethical issues surround the use of these workers?

2. What factors must one consider in deciding whether to employ offshore outsourcing on a project?

3. Which steps of the whistle-blowing process are most important? Why?

4. Apple, Dell, and HP all received low ratings from Greenpeace for their efforts on green computing. Choose one of these companies, visit its Web site, and do research to find out what, if anything, the company is doing to improve its green computing results.

5. You work for an electronics manufacturer that does not belong to the EICC. Present a strong argument for your firm to join. Then present a strong argument for why it makes sense for your firm not to be a member.

6. While labor savings associated with offshore outsourcing may look attractive, what cost increases and other problems can one expect with such projects?

7. Why do companies that make use of a lot of contingent workers fear getting involved in a coemployment situation? What steps should they take to avoid this situation?

8. Make a list of reasons an employer might use to justify paying an H-1B worker less than the average salary for a specific position (e.g., database administrator) in a specific geographic area (e.g., Chicago). Which of these reasons do you think are legitimate and do not violate the spirit of the H-1B wage guidelines?

9. Your company has decided to offshore-outsource a $50 million project to an experienced, reputable firm in India. This is the first offshore outsourcing project of significant size that your company has run. What steps should your company take to minimize the potential for problems?

10. Visit the EPEAT Web site and use the tool to select your next laptop computer. How would you make trade-offs between an expensive machine with a Gold rating and a less expensive machine with the same features and performance but only a Bronze rating?

What Would You Do?

1. You are in the last stages of evaluating laptop vendors for a major hardware upgrade and standardization project for your firm. You will be purchasing a total of 1,200 new laptops to deploy to the worldwide salesforce. One vendor's product carries a Bronze EPEAT rating; the other vendor's product would cost an additional $100,000 but carries a Gold EPEAT rating. The two products are very evenly matched on other key factors, such as performance, features, reliability, and support costs. How would you decide between the two vendors' products?

2. As a relatively new hire within a large multinational firm, you are extremely pleased with the many challenging assignments that have come your way. Now another new hire with whom you have become friends is seeking your input on an important decision that she must make within the next week. She has been challenged to cut costs in her department by outsourcing a large portion of the department's work to an offshore resource firm that has an excellent reputation. Your friend would remain with the firm to oversee the outsourcing work. What advice would you offer your friend?

3. Your firm has just added six H-1B workers to your 50-person department. You have been asked to help get one of the workers "on board." Your manager wants you to introduce him to other team members, provide him with some basic company background and information,

and explain to him how work gets done within your organization. Your manager has also asked you to help your new coworker become familiar with the community, including residential areas, shopping centers, restaurants, and recreational activities. Your goal would be to help the new worker be productive and comfortable with his new surroundings as soon as possible. How would you feel about taking on this responsibility? How would you help the new employee?

4. Dr. Jeffrey Wigand is a whistle-blower who was fired from his position of vice president of research and development at Brown & Williamson Tobacco Corporation in 1993. He was interviewed for a segment of the CBS show *60 Minutes* in August 1995, but the network made a highly controversial decision not to air the interview as initially scheduled. The segment was pulled because CBS management was worried about the possibility of a multibillion-dollar lawsuit for tortuous interference; that is, interfering with Wigand's confidentiality agreement with Brown & Williamson. The interview finally aired on February 4, 1996, after the *Wall Street Journal* published a confidential November 1995 deposition that Wigand gave in a Mississippi case against the tobacco industry, which repeated many of the charges he made to CBS. In the interview, Wigand said that Brown & Williamson had scrapped plans to make a safer cigarette and continued to use a flavoring in pipe tobacco that was known to cause cancer in laboratory animals. Wigand also charged that tobacco industry executives testified untruthfully before Congress about tobacco product safety. Wigand suffered greatly for his actions; he lost his job, his home, his family, and his friends.

Visit Wigand's Web site at *www.jeffreywigand.com* and answer the following questions. (You may also want to watch *The Insider*, a 1999 movie based on Wigand's experience.)

- What motivated Wigand to take an executive position at a tobacco company and then five years later to denounce the industry's efforts to minimize the health and safety issues of tobacco use?
- What whistle-blower actions did Dr. Wigand take?
- If you were in Dr. Wigand's position, what would you have done?

5. Microsoft relies heavily on temporary workers. To minimize potential legal issues, Microsoft sought to ensure that the temporary workers were not mistaken about their place within the company. The temporary agencies provided workers with handbooks that laid out the ground rules in explicit detail. Temporary workers were barred from using company-owned athletic fields—for insurance reasons, the handbooks explained. At some Microsoft facilities, temporary workers were told they could not drive their cars to work because it would create parking problems for regular workers. Instead, they were told to take the bus. Temporary workers were also told not to buy goods at the company store or participate in company social clubs such as chess, tai chi, or rock climbing, which were open to regular employees. They were not permitted to attend parties given for regular employees, a private screening of the latest *Star Wars* film, or company meetings at the Kingdome stadium in Seattle. In addition, their e-mail addresses were to contain an *a-* to indicate their nonpermanent status in the company.

Imagine that you are a senior manager in the Human Resources Department at Microsoft and that you have been asked to respond to temporary workers' complaints about working conditions. How would you handle this?

6. Catalytic Software—a U.S.-based IT outsourcing firm with offices in both Redmond, Washington, and Hyderabad, India—wants to tap India's large supply of engineers as contract software developers for IT projects. However, instead of just outsourcing projects to local Indian software development companies, as is the common practice of U.S. companies, Catalytic has developed a self-contained company community near Hyderabad. Spread over 500 acres, the community of New Oroville is a self-sustaining residential and office community designed to house about 4,000 software developers and their families, as well as 300 support personnel who supply sanitation, police, and fire services.

    The goal of this high-tech city is to knock down barriers that large-scale technology businesses encounter in India. By building a company community, Catalytic is trying to ensure that it has enough qualified employees to staff round-the-clock shifts. The company expects this facility to attract and keep top professionals from all over the world. Building a company town also solved the problem of transportation, which can be challenging for such a large workforce. Because of the terrible state of the local roads, the commute from Hyderabad—25 kilometers from New Oroville—would take almost an hour.

    Catalytic provides private homes with private gardens, all within a short walk of work, school, recreation, shopping, and public facilities. Each house includes cable television, telephones, and a fiber-optic data pipeline that connects to the Internet so that employees can work efficiently even at home. (Employees are awarded bonuses for working overtime.) New Oroville was designed with four indoor recreational complexes, six large retail complexes, and ample green space, including five parks for outdoor exercise and recreation.

    You have just completed a job interview with Catalytic Software for a position as project manager and have been offered a 25 percent raise to join the company. Your position will be based in Redmond and will involve managing U.S.-based projects for customers. The position requires that you spend the first year with Catalytic in the New Oroville facility to learn its methods, culture, and people. (You can take your entire family or accept a pair of three-week company-paid trips back to Redmond.)

    Why do you think the temporary assignment in New Oroville is a requirement? What else would you need to know in considering this position? Would you accept it? Why or why not?

7. A coworker complains to you that he is sick of seeing the company pollute the waters of a nearby stream by dumping runoff water into it from the manufacturing process. He plans to send an anonymous e-mail to the EPA to inform the agency of the situation. What would you do?

Cases

1. The Census Bureau's Outsourcing Debacle

In 2008, just one year before the Census Bureau was scheduled to begin street-canvassing operations for the 2010 decennial census, news broke that the bureau's mobile initiative had flopped.[70] It was revealed that the Field Data Collection Automation (FDCA) system, which was supposed to save taxpayers $1 billion, would now raise the total cost of conducting the census from $11.5 to $14.5 billion.[71] Many reporters quickly described the failure as just one more case of federal IT mismanagement. Yet these reports overlook the bureau's historic role in introducing cutting-edge IT and database development.

The U.S. Constitution requires that a census of the country's population be taken every 10 years. This data is used not only to decide the distribution of congressional seats within the United States but also to determine how to allot federal funds. Census figures are also used by state and local governments as a basis for deciding where to build roads, schools, job-training centers, and so forth. To accommodate both a growing population and the proliferation of uses for census data, the bureau frequently promotes new technology. For example, after spending seven years collecting data for the 1880 census, the bureau became the first organization to use Herman Hollerith's automatic tabulating machine, which performed so successfully that updated versions were used in subsequent censuses as well. (Hollerith's firm, the Tabulating Machine Company, later merged with three other companies to form IBM.) UNIVAC, the first commercial computer, was originally designed for use in the 1950 census.[72]

In the 1980s, the Census Bureau created the Topologically Integrated Geographic Encoding and Referencing (TIGER) database, which provides automated access to and retrieval of relevant geographic information about the United States and its territories.[73] TIGER maps roads as well as state, county, and city boundaries; railroads; and every body of water in the country. For each entity, TIGER lists attributes, such as name, alternative name, longitude, and latitude. For roads, TIGER defines address ranges and associates zip codes with street addresses.[74] This information allows the bureau to mail surveys to each citizen, thus cutting down on the number of workers, called "enumerators," who have to go into the field to locate housing units and conduct interviews. But TIGER's impact extended way beyond the bureau; it jump-started the geographic information system (GIS) industry, facilitating the development of products such as MapQuest, Google Maps, and GPS navigation for automobiles.

After the 1990 census, the bureau created the Master Address File (MAF) database, composed of data collected during the census as well as from the U.S. Postal Service and local, state, and tribal governments.[75] MAF contained a more complete listing of housing units than TIGER did, so the two database systems were used together. Following the 2000 census, the two databases were merged onto an Oracle database platform.

At this time, however, the Census Bureau was already embracing outsourcing to meet its mounting IT needs. "By the 1990s, we recognized that we needed to move away from Census home-grown technologies. The capabilities of a robust, nascent IT industry had by then exceeded our internal abilities," reported Census Bureau director Charles Kincannon at a 2007 Congressional hearing.[76]

The bureau relies on technology to support its wide range of operations, enhancing legacy systems and acquiring new IT systems when necessary. For each decennial census, the bureau must first identify where to count by collecting all known addresses of all people living within the United States. While the MAF/TIGER system provides much of this data, street canvassing must be carried out to update the information in the database. The second step in this operation is to collect information from all households. Once the addresses are identified, the bureau mails a survey to each household. If the survey is not returned, the bureau must send out enumerators to the address and interview the resident. The data assembled from the surveys and the reports of the enumerators is integrated by the Decennial Response Integration System (DRIS). The final step of tabulating and summarizing the results is carried out through the Data Access and Dissemination System (DADS II).[77]

Following the 2000 census, the bureau elected to develop a mobile system that would allow enumerators working in the field to collect and transmit data back to the offices. In 2006, the Census Bureau outsourced the work and awarded a $600 million contract to Harris Corporation to create the Field Data Collection Automation (FDCA) system to automate the collection of field data. Harris Corporation is a communications and IT company with $5 billion in annual revenue and over 15,000 employees. The company had already helped the bureau integrate MAF and

TIGER. Through the new contract, Harris was supposed to provide the IT system and the mobile hardware—handheld computers—used by enumerators. The bureau planned to use this system not only to collect and transmit data during the canvassing and the interviews, but also to manage the field operations.

On May 1, 2008, the Census Bureau conducted a test of the new FDCA system. The results revealed that the mobile handheld computers were slow, sometimes froze up, and did not transmit data consistently.[78] In addition, the test showed problems with the program designed to manage operations in the field. The bureau announced that it would need to push back the development schedule and allocate considerably more funds to complete the project. In August 2009, it announced that enumerators would drop the use of the mobile devices during the interview process and rely exclusively on paper-based operations. The new contract will implement just the handheld-driven update of addresses, and even with this greatly reduced scope, the cost is $200 million more than the original budget for the entire system.[79]

The question that both reporters and members of the federal government asked was: What went wrong with this outsourcing project? The U.S. Government Accountability Office (GAO) had been carefully monitoring preparations for the 2010 census and identified the following project management shortcomings:

- Failure to identify key project deliverables and milestones
- Failure to gain stakeholder buy-in on the project plan, including key project parameters such as estimated costs and schedule
- Failure to validate key project requirements
- Failure to assign responsibility for risks and to prepare mitigation plans
- Failure to define key metrics for contract tracking and executive oversight[80]

In a March 2009 report, the Commerce Department's inspector general stated that a root cause of the project failure was "the failure of senior Census Bureau managers in place at the time to anticipate the complex IT requirements involved in automating the census." Indeed, the Census Bureau kept changing the requirements, with each change adding to the cost and further delaying the project. In addition, the Census Bureau set up a cost-plus contract with Harris Corporation instead of a fixed-cost contract. With a cost-plus contract, Harris could increase the cost each time the Census Bureau changed its mind on what it wanted.

While the enumerators will have to use pencil and paper to collect data from the interviews, the bureau is planning to use the FDCA system to conduct street canvassing to identify and correct the locations of housing units. So the millions spent on this project will not be entirely for naught. And perhaps the bureau will learn from these mistakes and develop procedures to manage future outsourced IT projects more effectively.

Discussion Questions

1. After many years of conducting successful IT projects, why did the Census Bureau decide to outsource creation of the FDCA system to automate the collection of field data?

2. Go to the Harris Corporation Web site (*www.harris.com*) to gain an understanding of the broad range of projects that Harris is working on for the U.S. government. Do you think it is appropriate that the government continues to spend so much money with this firm based on the Census Bureau experience? Why or why not?

3. Make a list of three key principles for the successful outsourcing of IT projects, based on the Census Bureau's experience.

2. American Engineer Blows the Whistle on Airbus' Superjumbo A380

The Airbus A380, the largest passenger airliner ever built, flew its maiden voyage on April 27, 2005. This superjumbo double-decker surpasses Boeing's 747 in size, and orders have poured in, although the project is behind schedule and over budget. *Popular Science* named the Airbus A380 one of its 2005 Grand Award winners of the "Best of What's New Award."[81]

Indeed, the aircraft is a titanic project—complete with one lethal flaw, according to Joseph Mangan, an aerospace engineer from Kansas City, Kansas.

In February 2004, Mangan moved to Vienna to work as an aerospace manager for TTTech Computertechnik AG, an Austrian subcontractor that supplies the computer chips that control cabin pressure in the A380. Mangan says that by March 2004 he had confronted his employers about problems with the documentation submitted to the U.S. Federal Aviation Administration and the European Aviation Safety Agency (EASA), the two organizations that needed to certify the chips. By the summer, Mangan insists that he had discovered serious defects in the software and repeatedly requested that TTTech correct the software before continuing with the certification process. Finally, in September 2004, Mangan revealed the design flaw to Airbus and EASA during an official audit. Mangan was fired a few days later.[82]

Mangan believes the flaw in the software could lead to rapid loss of cabin pressure, leaving pilots, crew, and passengers with little time to don their air masks. Experts believe that such depressurization may have contributed to the 2005 crash of a Boeing 737 in Greece, which killed 121 people, and the 2002 crash of a China Airlines Boeing 747, which killed 225 people.[83]

A typical passenger jet has two outflow valves that control cabin pressure. To achieve redundancy and ensure safety, manufacturers install three separate motors, each with a different chip, to operate each outflow valve. For example, the Boeing 777 uses chips manufactured by Advanced Micro Devices, Motorola, and Intel. In addition, pilots can manually override this system on most jets.[84]

To reduce the superjumbo A380's weight, Airbus decided to install four outflow valves—each operated by only one motor, and each motor driven by a TTTech controller chip. Mangan says he discovered that these chips were executing unpredictable commands when fed certain data. If one chip fails, he contends, all four will fail. So, while Airbus claims to have achieved redundancy by installing four outflow valves, Mangan vehemently rejects this assertion. His nightmare is that the failure of all four valves would cause the 555-seat jetliner to crash.[85]

Mangan says he felt a moral obligation to warn the public. In the event of a crash, Mangan could also be held legally accountable. After determining that a metal strip from a Continental DC-10 sliced the tires of a Concord jet, causing a crash in 2000, French prosecutors went after the American mechanic who installed the strip.[86]

TTTech insists that Mangan was fired for poor job performance and is exacting revenge by trying to destroy the company. The company has filed both civil and criminal charges against Mangan, claiming that the information he released to the media was proprietary and that he has damaged the company's reputation.[87]

Austria, unlike the United States, offers little protection for whistle-blowers. TTTech was able to obtain a gag order from an Austrian judge. Mangan is currently facing jail time for failing to pay a $185,000 fine for violating that gag order.[88]

In the meantime, an EASA investigation concluded that TTTech's chip was unacceptable and would have to be fixed before Airbus received certification for the A380. TTTech's CEO Stefan Poledna says the company never received any indication from EASA that its chip was noncompliant. TTTech did identify and fix a glitch, Poledna admits, but only as part of the routine software development and review process.[89]

In late 2005, the European Aeronautic Defence and Space Company—manufacturer of the A380—issued a press release stating that all parties had "ensured through the most varied control channels that there is no safety deficit with regards to the scenario as described by Mangan."[90] By late 2006, the Airbus A380—including the microprocessors that control the cabin pressure—was certified by EASA and the FAA.[91] As of July 2009, the A380 has not been involved in any accident related to faulty cabin pressure control systems.

Discussion Questions

1. Discuss the importance of redundancy to such safety-critical systems as the Airbus A380's cabin.

2. If TTTech were located in Kansas City rather than Vienna, what protection would Joseph Mangan receive as a whistle-blower?

3. Mangan risked a year in jail and went bankrupt trying to convince people of a problem. If he were simply a disgruntled employee, there are many ways he could do damage without putting himself under such pressure. Is it possible that there was a problem and that Mangan's actions forced TTTech to address and fix it before EASA and the FAA discovered it themselves?

3. Manufacturers Compete on Green Computing

Dell, Apple, and HP have long competed on the basis of price and performance. Recently, the three companies have begun to focus on green computing as a way to differentiate themselves to consumers who have come to see green computing as an excellent opportunity to save money through reduced power consumption and to lessen their negative impact on the environment.[92]

In 2007, Dell announced that it had set a goal for itself of becoming the "greenest technology company on Earth," and the company frequently touts its strong recycling program. Apple claims to have the "greenest family of notebooks" and emphasizes the progress it is making in removing toxic materials from all of its computers. HP highlights its efforts to develop more environmentally friendly packaging as well as its long tradition of environmentalism.[93] "Power to Change" is HP's latest green initiative campaign, which urges PC users around the world to shut down their computers at the end of the day to save energy and reduce carbon emissions.[94]

The public relations battle over which computer manufacturer is greener is heating up. Apple's claim that it produces the "world's greenest family of notebooks" was disputed by Dell and investigated by the National Advertising Division (NAD) of the Council of Better Business Bureaus in 2009. The basis for Dell's complaint was that other computer manufacturers meet the same standards and that Apple's ads implied superiority. Apple was asked by NAD to change its claim to clarify that the basis of comparison is between all MacBooks and all notebooks made by a given competitor. Rather than drop its "world's greenest" claim, Apple changed the wording to "world's greenest lineup of notebooks."[95]

While companies differ in their environmental standards, consumers now have a number of new factors to evaluate when purchasing a computer based on its "green credentials."[96] Green computing proponents are glad to see the innovation and competition in this arena.

Discussion Questions

1. How have green computing efforts lowered the total cost of computer ownership?

2. Which approach can yield greater benefits—building greener computers or implementing programs that change users' behavior so that they operate their computers in a more responsible manner? Explain your response.

3. Do research at the EPEAT Web site and determine which computer manufacturer currently has the best green computing ratings.

End Notes

1. "Jason Gooljar, "Chinese Factory That Supplies IBM, Microsoft, Dell, Lenovo and Hewlett-Packard to Be Investigated," February 15, 2009, www.jasongooljar.com/?tag=meitai-plastic-and-electronics.

2. Tom Espiner, "Tech Coalition Launches Sweatshop Probe," *CNET*, February 14, 2009, http://news.cnet.com/8301-1001_3-10164325-92.html.

3. National Labor Committee, "High Tech Misery in China: The Dehumanization of Young Workers Producing Our Computer Keyboards," February 2009, www.nlcnet.org/article.php?id=613.

4. Joe Bavier, "Congo War-Driven Crisis Kills 45,000 A Month—Study," *Reuters*, January 22, 2008, www.alertnet.org/thenews/newsdesk/L22802012.htm.

5. The Enough Project, "Electronics Companies Respond to the Enough Project," www.raisehopeforcongo.org/responses.

6. Elizabeth Murphy, "Computer Science Major Sees Enrollment Decline," *Daily Collegian Online*, March 21, 2008, www.collegian.psu.edu/archive/2008/03/21/computer_science_major_sees_en.aspx; Jay Vegso, "Enrollments and Degree Production at US CS Departments Drop Further in 2006–07," Computing Research Association, March 2008, www.cra.org/CRN/articles/march08/jvegso_enrollments.html.

7. John Timmer, "Computer Science Degrees Rebound from dotcom Bust," *Ars Technica*, March 17, 2009, http://arstechnica.com/science/news/2009/03/computer-science-degrees-rebound-from-dotcom-bust.ars.

8. Bureau of Labor Statistics, "Occupational Outlook Handbook, 2008–09 Edition," May 13, 2009, www.bls.gov/oco/oco2003.htm.

9. Bureau of Labor Statistics, "Occupational Outlook Handbook, 2008–09 Edition," May 13, 2009, www.bls.gov/oco/oco2003.htm.

10. IBM, "About IBM," www.ibm.com/ibm/us/en.

11. Hewlett-Packard Company, "HP Employees," www.hp.com/hpinfo/globalcitizenship/gcreport/employees.html.

12. Accenture, "IT Transformation – The Journey to High Performance," www.accenture.com/Global/Services/CIO/TheServices.htm.

13. Deloitte Touche Tohmatsu, "About Deloitte," www.deloitte.com/view/en_US/us/About/index.htm.

14 Tata Consultancy Services, "Corporate Facts," www.tcs.com/about/corp_facts/Pages/default.aspx.

15 Ernst & Young, "People," www.ey.com/US/en/Careers/Experienced/Life-at-Ernst---Young/People.

16 KPMG, Home page, www.kpmg.com/Global/Pages/default.aspx.

17 Infosys Technologies Ltd., "About Us: What We Do," www.infosys.com/about/what-we-do/default.asp.

18 Capgemini, "Who We Are," www.us.capgemini.com/about.

19 Cognizant Technology Solutions, "About Us," www.cognizant.com/html/aboutus/landingpage.asp.

20 Bill Virgin, "Microsoft Settles 'Permatemp' Suits," *Seattle-Post Intelligencer*, December 13, 2000, www.seattlepi.com/business/micr13.shtml.

21 W. David Gardner, "U.S. Expects 20,000 H1B Visas To Go Quickly," *ChannelWeb*, May 11, 2005, www.crn.com/it-channel/163101217.

22 Patrick Thibodeau, "Update: H-1B Visa Cap Reached; IT groups May Press for More," *Computerworld*, August 12, 2005, www.computerworld.com/s/article/103883/Update_H_1B_visa_cap_reached_IT_groups_may_press_for_more.

23 U.S. Citizenship and Immigration Services, "USCIS-Cap Count for H-1B and H-2B Workers for Fiscal Year 2010," www.uscis.gov/portal/site/uscis/menuitem.5af9bb95919f35e66f6141765 43f6d1a/?vgnextoid=138b6138f898d010VgnVCM10000048f3d6a1RCRD.

24 Ivener & Fullmer, "H-1B Visas for Professionals of Singapore & Chile," www.usworkvisa.com/visas/H1B-singapore-chile.html.

25 Patrick Thibodeau, "Hiring H-1B Visa Workers Trims U.S. Tech Workers' Wages," *PC World*, April 19, 2009, www.pcworld.com/article/163383/hiring_h1b_visa_workers_trims_us_tech_workers_wages.html.

26 United States Department of Labor, Employment Law Guide, "Workers in Professional and Specialty Occupations (H-1B and H-1B1 Visas)," www.dol.gov/compliance/guide/h1b.htm#who.

27 United States Department of Labor, Employment Law Guide, "Workers in Professional and Specialty Occupations (H-1B and H-1B1 Visas)," www.dol.gov/compliance/guide/h1b.htm#who.

28 Don Sears, "Careers," *eWeek.com*, May 29, 2009.

29 "Where Is Your Software From?," *TechByter Worldwide*, January 11, 2009, www.techbyter.com/2009/20090111.html.

30 Cognizant Technology Solutions, "10K SEC Filing," March 7, 2000, http://sec.edgar-online.com/cognizant-technology-solutions-corp/10-k-annual-report/2000/03/07/Section2.aspx.

31 Tata Consultancy Services, "Best IT Outsourcing Project Award for TCS in the Netherlands," June 15, 2009, www.tcs.com/news_events/press_releases/Pages/Best-IT-Outsourcing-Project-Award-Netherlands-TCS.aspx.

32 Tata Consultancy Services, "Corporate Facts," www.tcs.com/about/corp_facts/Pages/default.aspx.

33 Infosys Technologies Ltd., "What We Do," www.infosys.com/about/what-we-do/default.asp.

34 Wipro Limited, "Annual Report 2008-2009," pg. 16.

35 Satyam Computer Services, Ltd., "Satyam Allots Shares to Venturbay," July 10, 2009, www.mahindrasatyam.net/media/pr2Jul09.asp.

36 HCL Infosystems Ltd., "Annual Report 2007-2008," www.hclinfosystems.in/annual_rep_07-08.pdf.

37 EPAM Systems, "Fact Sheet," www.epam.com/software-development-company-facts.htm.

38 Luxoft, "Fact Sheet," www.luxoft.com/about/glance/fact_sheet.html.

39 RUSSOFT Association, "Aplana Software Company Profile," www.russoft.org/directory/?profile=106.

40 Interxion, "At a Glance," www.interxion.com/About-Interxion/At-a-Glance-Fact-Sheet.

41 Computer Merchant, *The Navigator* 12 (April 2009), www.tcml.com/about/news/articles/Navigator_2009_April.pdf.

42 Julian Bajkowski, "Report: Black Market Growing for Offshore Data," *Computerworld*, August 16, 2005, www.computerworld.com/s/article/103962/Report_Black_market_growing_for_offshore_data.

43 Dinesh C. Sharma, "Indian Call Center Under Suspicion of ID Theft," *CNET News*, August 16, 2005.

44 James Risen and Eric Lichtblau, "Bush Lets U.S. Spy on Callers Without Courts," *New York Times*, December 16, 2005, www.nytimes.com/2005/12/16/politics/16program.html.

45 Brian Ross, "NSA Whistleblower Alleges Illegal Spying, *ABC News*, January 10, 2006, http://abcnews.go.com/WNT/Investigation/story?id=1491889.

46 "False Claims," Cornell University Law School, www.law.cornell.edu/uscode/31/usc_sec_31_00003729----000-.html.

47 The Law Offices of Jason S. Coomer, "Texas Whistleblower Claim, Government Contractor Corruption, and Qui Tam Claim Lawyer," www.texaslawyers.com/coomer/governmentcorruptionlawyer.htm.

48 "False Claims," Cornell University Law School, www.law.cornell.edu/uscode/31/usc_sec_31_00003729----000-.html.

49 Melissa Davis, "Medicaid Suit Mangles Amerigroup," *TheStreet.com*, October 31, 2006, www.thestreet.com/newsanalysis/healthcare/10318892.html.

50 Matt Hines, "Oracle Settles Whistleblower Suit for $8 Million," *CNET*, May 13, 2005, http://news.cnet.com/Oracle-settles-whistleblower-suit-for-8-million/2100-1012_3-5706531.html.

51 "False Claims," Cornell University Law School, www.law.cornell.edu/uscode/31/usc_sec_31_00003729----000-.html.

52 Federal Accountability Initiative for Reform, "The Whistleblower's Ordeal," http://fairwhistleblower.ca/wbers/wb_ordeal.html.

53 Jaikumar Vijayan, "Scope of TJX Data Breach Doubles: 94 Million Cards Now Said to Be Affected," *Computerworld*, October 24, 2007.

54 Steve Ragan, "TJX Fires Whistleblower—Was It Justified Action or Something Else?," *Tech Herald*, May 26, 2008, www.thetechherald.com/article.php/200821/1070/TJX-fires-whistleblower-%E2%80%93-was-it-justified-action-or-something-else.

55 Brad Wells, "What Truly Makes a Computer 'Green'?," *OnEarth*, September 8, 2008, www.onearth.org/node/658.

56 Elizabeth Royte, "E-Gad! Americans Discard More Than 100 Million Computers, Cellphones and Other Electronic Devices Each Year. As 'E-Waste' Piles Up, So Does Concern about This Growing Threat to the Environment," *Smithsonian*, August 2005, www.smithsonianmag.com/arts-culture/e-gad.html.

57 "San Francisco Goes for EPEAT Gold," EPEAT Press Release, March 10, 2009, www.epeat.net/NewsDocuments/San%20Francisco.GOLD.09-03.pdf.

58 Apple, "A Greener Apple," www.apple.com/hotnews/agreenerapple.

59 Senate Fiscal Agency, "Electronic Waste Recycling," Michigan Legislature, November 21, 2008, www.legislature.mi.gov/documents/2007-2008/billanalysis/Senate/pdf/2007-SFA-0897-B.pdf.

60 "State Legislation: States Are Passing E-Waste Legislation," Electronics TakeBack Coalition, March 20, 2008, www.electronicstakeback.com/legislation/state_legislation.htm.

61 Dell, "Dell Recycling," www.dell.com/content/topics/segtopic.aspx/dell_recycling?c=us&l=en&cs=19.

62 Electronic Industry Citizenship Coalition, "History," www.eicc.info/ABOUT.htm.

63 Electronic Industry Citizenship Coalition, "Electronic Industry Code of Conduct" (Version 3.01, 1 June 2009), www.eicc.info/EICC%20CODE.htm.

64 Electronic Industry Citizenship Coalition, "Membership," www.eicc.info/MEMBERSHIP.htm.

65 Electronic Industry Citizenship Coalition, "Electronic Industry Code of Conduct" (Version 3.01, 1 June 2009), www.eicc.info/EICC%20CODE.htm.

66 Electronic Industry Citizenship Coalition, "Electronic Industry Code of Conduct" (Version 3.01, 1 June 2009), www.eicc.info/EICC%20CODE.htm.

67 Electronic Industry Citizenship Coalition, "Electronic Industry Code of Conduct" (Version 3.01, 1 June 2009), www.eicc.info/EICC%20CODE.htm.

68 Electronic Industry Citizenship Coalition, "Electronic Industry Code of Conduct" (Version 3.01, 1 June 2009), www.eicc.info/EICC%20CODE.htm.

69 Electronic Industry Citizenship Coalition, "Electronic Industry Code of Conduct" (Version 3.01, 1 June 2009), www.eicc.info/EICC%20CODE.htm.

70 Harris Corporation, "Harris Corporation Demonstrates Functionality of Handheld Mobile Computing Device for 2010 Decennial Census," January 3, 2007, www.harris.com/view_pressrelease.asp?act=lookup&pr_id=2031.

71 Jean Thilmany, "Behind the Census Bureau's Mobile SNAFU," *CIO Insight*, May 20, 2008, www.cioinsight.com/c/a/Case-Studies/Census-Mobile-SNAFU.

72 "Prepared Statement of Charles Louis Kincannon, Director US Census Bureau, Hearing to Examine Issues Relating to the Census Bureau's Risk Management of Key 2010 Information Technology Acquisitions, Before the Subcommittee on Information Policy, Census, and

National Archives, U.S. House of Representatives," December 11, 2007, www.ogc.doc.gov/ogc/legreg/testimon/110f/Kincannon121107.pdf.

73 U.S. Census Bureau, "TIGER® Overview," August 31, 2005, www.census.gov/geo/www/tiger/overview.html.

74 Jacob S. Siegel, *Applied Demography* (San Diego, CA. Academic Press, 2002), 201–3, http://books.google.com/books?id=a5Ax1oRbkDMC&pg=PA201&lpg=PA201&dq=TIGER+Census+Bureau+1983+DIME&source=bl&ots=8fCjHu7ILF&sig=pECg3Pz6WJc7QUGw09ztwrYV1GA&hl=en&ei=1-4jSvnYJMyntgezhcSvBg&sa=X&oi=book_result&ct=result&resnum=4#PPA202,M1.

75 Shawana P. Johnson and J. Edward Kunz, "Private Sector Makes Census Bureau's TIGER Roar," *GPS World*, May 1, 2005, http://www.gpsworld.com/gis/local-government/private-sector-makes-census-bureau039s-tiger-roar-5360.

76 "Prepared Statement of Charles Louis Kincannon, Director, US Census Bureau, Hearing to Examine Issues Relating to the Census Bureau's Risk Management of Key 2010 Information Technology Acquisitions, Before the Subcommittee on Information Policy, Census, and National Archives, U.S. House of Representatives," December 11, 2007, http://informationpolicy.oversight.house.gov/documents/20080318161001.pdf.

77 GAO-09-262, "Information Technology: Census Bureau Testing of 2010 Decennial Systems Can Be Strengthened," Government Accountability Office, March 5, 2009, www.gao.gov/htext/d09262.html.

78 GAO-09-262, "Information Technology: Census Bureau Testing of 2010 Decennial Systems Can Be Strengthened," Government Accountability Office, March 5, 2009, www.gao.gov/htext/d09262.html.

79 Brian Friel, "The Right Stuff," *MyTwoCensus.com*, May 20, 2009, www.mytwocensus.com/2009/05/20/investigative-series-spotlight-on-harris-corp-part-1.

80 Michael Krigsman, "Billion-Dollar IT Failure at Census Bureau," *ZDNet*, March 20, 2008, http://blogs.zdnet.com/projectfailures/?cat=22&paged=2&paged=3.

81 "Popular Science Names Top Tech Innovations of 2005," *PRNewswire*, November 7, 2005, www.prnewswire.com/cgi-bin/stories.pl?ACCT=104&STORY=/www/story/11-07-2005/0004209487&EDATE=.

82 Daniel Michaels and Matthew Karnitschnig, "Airbus Feud Lands in Court," *Wall Street Journal*, April 28, 2005.

83 "Decompression May Have Caused Crash," *FoxNews.com*, August 14, 2005, http://origin.foxnews.com/story/0,2933,165686,00.html.

84 Doug Merrill, "This Just Looks Bad," *A Fistful of Euros*, October 17, 2005, http://fistfulofeuros.net/afoe/political-issues/this-just-looks-bad.

85 Peter Pae, "A380 Jet Flawed, Fired Worker Alleges," *Los Angeles Times*, October 2, 2005.

86 "The Buzz: Guilt by Association," *FoxNews.com*, July 16, 2001.

87 Peter Pae, "A380 Jet Flawed, Fired Worker Alleges," *Los Angeles Times*, October 2, 2005

[88] "Airbus Whistleblower Faces Prison," Telegraph.co.uk, October 15, 2005, www.telegraph.co.uk/finance/2923861/Airbus-whistleblower-faces-prison.html.

[89] Peter Pae, "A380 Jet Flawed, Fired Worker Alleges," *Los Angeles Times*, October 2, 2005.

[90] "Statement Regarding Claims by a Former A380 Component Supplier Employee About Safety Concerns," October 5, 2005, European Aeronautic Defence and Space Company, www.eads.com/1024/en/pressdb/archiv/2005/2005/en_20051006_stellungnahme.html.

[91] "Airbus A380 Gets Type Certification," *Airwise News*, December 12, 2006, http://news.airwise.com/story/view/1165958573.html.

[92] Reuters, "Computer Makers Dell, HP and Apple Push Green," *Channel Insider*, June 15, 2009, www.channelinsider.com/c/a/News/Computer-Makers-Dell-HP-and-Apple-Push-Green-281835.

[93] Reuters, "Computer Makers Dell, HP and Apple Push Green," *Channel Insider*, June 15, 2009, www.channelinsider.com/c/a/News/Computer-Makers-Dell-HP-and-Apple-Push-Green-281835.

[94] "Leo Burnett Launches HP's Green-Computing Campaign," HP Press Release, June 6, 2009, http://www.domain-b.com/companies/companies_h/hewlett_packard/20090606_green-computing_campaign.html.

[95] Antone Gonsalves, "Apple Maintains 'World's Greenest' Laptop Claims," *InformationWeek*, June 19, 2009, www.informationweek.com/news/hardware/mac/showArticle.jhtml?articleID=218100403.

[96] Jackie Ammons, "A Green Marketing Ploy? NGOs Scrutinize Apple and Other Computers," *Global Governance Watch*, July 24, 2009, www.globalgovernancewatch.org/ngo_watch/a-green-marketing-ploy-ngos-scrutinize-apple-and-other-computers-2.

A BRIEF INTRODUCTION TO MORALITY

By Clancy Martin, Assistant Professor of Philosophy, University of Missouri—Kansas City

INTRODUCTION

This appendix offers a quick survey of various attempts by Western civilization to make sense of the ethical question "What is the good?" As you will recall from Chapter 1, *ethics* is the discipline dealing with what is good and bad and with moral duty and obligation. How should we live our lives? How should we act? Which goals are worth pursuing and which are not? What do we owe to ourselves and to others? These are all ethical questions.

The answers to these questions are provided in what we call *moralities* or *moral codes*. The Judeo-Christian morality, for example, attempts to tell us how we should live our lives, the difference between right and wrong, how we ought to act toward others, and so on. If you ask a question like "Is it wrong to lie?," the Judeo-Christian morality has a ready answer: "Yes, it is wrong to lie; it is right to tell the truth." Speaking loosely, we could also say that, according to Judeo-Christian morality, it is *immoral* to lie and *moral* to tell the truth.

Moralities, or moral codes, differ by time and place. According to some people— 8th-century BC Greeks, for example—it is not always wrong to lie, and it is not always right to tell the truth. So we are confronted with the *ethical* problem of choosing between different *moralities*. Some moralities may be better than others. It may even be true—as many thinkers have argued—that only *one* system of morality is ultimately acceptable. Thinking about ethics means thinking about the strengths and weaknesses of moralities, understanding why we might endorse one morality and reject another, and searching for better systems of morality or even "the best" morality. Especially in our own day, when globalization and accelerating advances in communication have created a cultural blending (and cultural conflicts) like never before, our ability to understand different moralities is crucial.

This appendix introduces you to the way various Western philosophers have answered the ethical question "What is the good?" Because the Western tradition is complicated enough, we have not addressed Eastern moralities and the ethical thinking of many fascinating Eastern philosophers. One of the interesting things about studying ethics is the enormous variety of moralities that humans have created and the many similarities between competing moralities. Unlike the rest of your textbook, this appendix is not specifically focused on the ethical problems created by technology. But as you read through the various moralities in the appendix, ask yourself how you would deal with the moral dilemmas you have studied and confronted in your own life.

THE KNOTTY QUESTION OF GOODNESS

Achilles kills Hector outside the gates of Troy. He binds Hector's corpse by the ankles, ties the ankles to the back of his chariot, and drags the body around the city walls. The treatment of the fallen Trojan hero by his victorious Greek enemy is so outrageous that not only Trojans, but most of Achilles' Greek allies and even the Gods, are shocked. But what is wrong with Achilles' action?

To an ancient Greek of the time, the answer would not have been obvious. When the poet **Homer (8th century BC)** tells this story in his epic *The Iliad*, his purpose is to illustrate a failure in the morality of his own day. Among Greeks of Homer's day, the prevailing moral code was: "Help to friends and harm to enemies." That code may sound naïve or ridiculously simplistic today. But for the collection of small and largely independent city-states that was ancient Greece, it was a moral code that had worked reasonably well for centuries. Yet Homer saw that different times were on the way. When the Greeks banded together, as they did to combat the Trojans, the old morality looked barbaric. There was nothing heroic about the lone Achilles dragging his vanquished enemy behind him. On the contrary, he seemed like a savage.

When a society is passing from an old moral code to a new one, or when two different cultures clash in their moral codes, the extraordinarily difficult question of which moral code is correct inevitably appears. *Ethics*, the systematic study of moral codes, is the attempt to answer that question. Almost every philosopher and most thinking people will agree that some moral codes are better than others; many philosophers and others will argue that a particular moral code is the best.

Perhaps the most famous philosopher of all time, **Socrates (470–399 BC)**, argued that there was only one true moral code, and it was simple: "No person should ever willingly do evil." Socrates thought that no harm could come to a person who always sought the good, because what truly counted in life was the caretaking of one's self or soul. But Socrates also acknowledged that identifying the good was rarely easy, and his method of constantly interrogating his friends and fellow citizens—what came to be called Socratic questioning, or the Socratic dialectic—tried to improve everyone's thinking about what one ought and ought not do.

Socrates never wrote down any of his philosophy. But his student **Plato (427–347 BC)** made Socrates the hero of almost all of his many philosophical dialogues. Plato was the first "professional" philosopher in the West: he established a school of philosophy called the Academy (where we get the word *academic*), published a great number of books both for general readers and his own students, and formed arguments on virtually every subject in philosophy (not only morality). In fact, Plato possessed such breadth that the 20th-century philosopher Lord Alfred North Whitehead wrote that "all subsequent philosophy is only a footnote to Plato."

In many of his dialogues Plato raises the question: "What is the good?" Like Homer (who was one of Plato's favorite writers), Plato lived in a time when great political, social, and cultural changes were occurring. Athens had lost the first major war in its history, trade was accelerating across the Mediterranean, and people were traveling deeper into Asia and Africa and discovering new cultures, religions, and values. Many candidates for "the good" were being offered by different thinkers: some thought that "pleasure" was the highest good, others argued that "peace" (both personal and social) and what contributed to it was the best, others argued for "flourishing" and material wealth and power, while still others endorsed "honor and fame." But Plato responded that, while all of these things might be examples of goodness, they were not good itself. What is it that makes them good? What is the nature of the property "goodness" that they all share? And because we recognize that most "goods" may also mislead us into badness—the good of pleasure is an obvious example—how shall we sort the good from the bad?

Plato's idea is that we cannot reliably say what is good and what is not until we know what goodness is. Once we have identified goodness itself, we can discriminate among particular goods and particular activities that are designed to seek the good. We will judge what is "good" and "better" by comparing it with what is "best": the truly and wholly good. And the truly and wholly good ought always and everywhere to be good. Could we say that something was truly, wholly good if it was good only in some countries and not others, during some times and not others? So, if we can identify goodness as such, Plato said, we can solve every problem posed by the clash between good and bad; that is, we can solve every problem of morality.

One way to think about Plato's insight is to see the moral importance of *standards*. We have standards for good hamburgers, for good businesses, and for good hammers, so why not have standards for good people and good actions? A standard is one way of providing a *justification* for an evaluation. Suppose Rebecca insists, "It is always wrong to kill an innocent human being." And Thomas replies, "But why?" Rebecca may justify her evaluation by appealing to a standard of rightness and wrongness. Of course, identifying that standard may prove more difficult than appealing to it, and the history of ethics, again, may be seen as the struggle to provide such a standard. The philosophers you will read about in the following sections attempted to answer Plato's knotty questions in their own ways.

RELATIVISM: WHY "COMMON SENSE" WON'T WORK

What about simply using common sense to find the good? Some 20th-century philosophers argued for what they called moral "intuitions": a kind of "consult your conscience" approach to morality. This view is initially compelling for most people; it holds that the standard for goodness demanded by Plato is accessible to all of us if we simply think through our moral decisions carefully enough. (Socrates may have been arguing for the same view.) There is a "voice" in our heads that tells us what is morally right and wrong, and if you honestly and thoroughly interrogate yourself about what you ought to do, that "voice" will praise the right action and warn you against the wrong one. Someone who says "Do the right thing!" is invoking this common-sense notion. We all know what the right thing is, a moral intuitionist argues, if we use our common sense and are tough on ourselves. The difficulty is that we don't always want to use common sense or ask ourselves tough questions. Therefore, the problem of right and wrong is not so much that of moral knowledge as it is weakness of will. We *know* what we ought to do, but it is hard to make ourselves *do* it.

A crippling difficulty with this view is called the problem of relativism. *Cultural relativism* is the simple observation that different cultures employ different norms (or standards). Implicit in this view is that it is morally legitimate for different cultures to create and embrace different norms. So, for example, among the Greeks of Homer's day, lying was considered to be a virtue. Odysseus was praised specifically for his ability to lie well. In 18th-century Germany, on the other hand, lying was widely considered as morally reprehensible as theft. Some philosophers even argued that lying was just as morally foul as murder. For the relativist, lying is neither right nor wrong; rather, it can be right at a certain time and place and wrong in another. Another example is bribery. Although people in many nations condemn bribery, it is perfectly acceptable in other countries, particularly in Latin America. The relativist would say: "Bribery itself is not right or wrong. Rather, some people at some times and in some places say it is wrong, and other people say it is right, depending on the circumstances. Bribery is therefore wrong for some people, right for others."

You have probably encountered this relativism with something as simple as e-mail. The conventions that govern e-mail etiquette vary dramatically from user to user, group to group, and culture to culture. The emoticon-laced e-mail you send to a friend would be wholly inappropriate if sent to a professor. The kind of language you use in an e-mail to a college admissions officer is not what you would use to e-mail your parents or an e-mail pal in India. A practical platitude that embodies this idea is: "When in Rome, do as Romans." What is *appropriate* and what counts as a "good" e-mail (as opposed to a "bad" or offensive e-mail) depends on the conventions within its cultural context. Even e-mails have *norms*.

Moral relativists argue that all norms and values are relative to the cultures in which they are created and expressed. For the moral relativist, it makes no sense to say that there are any transcultural or transhistorical values, and that any attempt to construct them would still be informed by the particular cultural values of a person or group. All you can talk about are the values "on the ground": the values that particular cultures embrace. And common sense may be one of the best tools for discovering those values. Common sense may be the psychological embodiment of the complex structure of rules, standards, and values that are the substance of every robust culture.

But moral relativists run into trouble, because there are some moral claims they cannot consistently make. Moral relativists can say "slavery is wrong in my society" or "slavery is wrong in the 20th century," but they cannot say that slavery is always wrong. Furthermore, because they cannot appeal to transcultural standards for morality, they cannot speak of *moral progress*. Moral values (like all other values) change over time for the relativist, but they do not improve or degenerate. Yet, most of us would agree that the growing worldwide prohibition against slavery and torture, for example, is not merely a change, it is moral progress. And if we believe in moral progress, we cannot be relativists.

Egoism vs. Altruism

Throughout this book we have seen that ethics deals with the question of how we should treat one another. But some thinkers would say we have already misconstrued the question when we ask "How should we treat others?" For an *egoist*, the salient moral question is "How do I best benefit myself?," and the answer to Plato's question "What is the good?" is simply "The good is whatever is pleasing to *me*."

Egoism is usually divided into two types. *Psychological egoism* is the thesis that people always act from selfish motives, whether they should or not. *Ethical egoism* is the more controversial thesis that, whether people always act from selfish motives, they should if they want to be moral.

There is a superficial plausibility to psychological egoism, because it might appear that most of us make many of our choices for self-interested reasons. You probably decided that you wanted to go to college rather than immediately finding a job. You might respond: "No, I went to college because my parents wanted me to!" But the psychological egoist would reply: "That simply means that, for you, pleasing your parents is more important than other things that would have kept you out of college."

However, some of the problems with psychological egoism already are glaringly apparent. First, though we may make many decisions based on our own interests, it is far from obvious that *all* of our decisions are motivated by self-interest. We make many decisions, including decidedly uncomfortable ones, because we are thinking of the interests of others. It is silly to suppose that our own interests must always and implicitly conflict with those of others, as a psychological egoist believes. Why did you go to college? Because you wanted to, and your parents, teachers, and friends wanted you to. Everyone's interests happily coincided, and it is oversimplifying your complex choice to say, as a psychological egoist would, "I did it because *I* wanted to."

While considering ethical egoism, we should also look at its opposite: *altruism*. The altruist argues that the morally correct action always best serves the interest of others. Wouldn't the world be a better place, the altruist asks, if we worried about ourselves less and tried to help other people?

No one will deny that everyone benefits from altruism, but problems arise if we try to adopt altruism as a moral code. Practically speaking, it is sometimes difficult to know what best serves the interest of another, beyond helping people with the basic necessities of life. For example, a devout Southern Baptist might sincerely believe that his neighbors are condemned to hell unless they accept his religious views, and might feel an altruistic urge to convert them, despite their hesitation. Another more famous example involves a boat full of altruists lost at sea. They can only survive if one of them volunteers to be eaten, but if the only moral action is to serve the interests of others, how can any of the adrift altruists be truly moral when one of them has to die to save the rest?

Problems like these help to motivate advocates of ethical egoism. We do not reliably know the interests of others, the ethical egoist says, but we certainly know our own. And, unlike altruists, whose satisfaction is in helping others, ethical egoists try to create a happy and moral world by seeking good for themselves. The hacker who thinks she can morally break the rules because she has the smarts to do so is both a psychological egoist ("you would break the rules too, if you could") and an ethical egoist ("everyone who can break the rules to help themselves should do so"). Given the choice between self-interest and altruism, the ethical egoist takes the former.

Of course, the only choice is not between ethical egoism and altruism. Most moral codes and most people recognize the importance of both self-interest and the interest of others. The more telling objection to ethical egoism is that it does not respect our deepest intuitions about moral goodness. If an ethical egoist can serve his own interest by performing some horrific act against another human being, and be guaranteed that the act will not interfere with his self-interest, he is morally permitted to perform that act. In fact, if he finds that he can *only* serve his interest by performing the horrific act and getting away with it, he is morally *required* to do so. An employer who could benefit from spying on her employee's e-mail would be morally required to do it if it served her long-term interest. But for most of us, such examples are sufficient to defeat ethical egoism. Moral codes are plausible only if they accommodate basic intuitions about our sense of right and wrong, and ethical egoism fails on that ground.

DEONTOLOGY, OR THE ETHICS OF LOGICAL CONSISTENCY AND DUTY

Most people find they cannot accept relativism as a moral code because of their moral intuitions that some things are *always* wrong (like slavery or the torture of innocents). For this reason, they must also abandon a "common sense" approach to morality, which relies on embedded knowledge of cultural norms. The problems with egoism and altruism are even more glaring. But don't despair—there are lots more moral theories to consider. The rest of this appendix reviews several modern attempts to articulate a consistent morality.

Immanuel Kant (1724–1804) is generally considered the most important philosopher since Aristotle. Kant's moral theory is an attempt to refine and provide a sound philosophical foundation for the strict Judeo-Christian morality of his own day. Most people, when they begin thinking about ethics in a philosophical way, find that they are some brand of Kantian. Kant's theory is called *deontology*, from the Greek word *deon*, meaning *duty*. For Kant, to do what is morally right is to do one's duty.

Understanding what one's duty requires is the difficult part, of course. Kant begins with the idea that the only thing in the world that is wholly good, without any qualification, is good will. Most good things may be turned to evil or undesirable ends, or are mixed with bad qualities. Human beings do not seem wholly good: they are a mix of good and bad. Money is a good that most of us seek, while "love of money is the root of all evil." But the will to do good—the desire or intention—must be wholly good. If we think through what we mean by "moral goodness," Kant argues, we realize that the notion of moral goodness is just another name for this will to goodness. Kant recognizes that, as the old saying goes, "the road to Hell is paved with good intentions"; he is not saying that good will must always have good consequences. (In general, Kant is suspicious of the moral worth of consequences.) But the intention to do good, before it gets tangled up in the difficulties of the world, must itself be purely good.

Morality, therefore, comes from our ability to intend that certain things happen: that is, from our ability to choose. The good choice will come from a good will. But how do we sort the good choice from the bad? Kant, following the ancient Greek philosopher Aristotle, believed that the property that makes human beings unique, and that propels us into the moral sphere, is the faculty of *reason*. Kant saw human beings as constantly torn between their passions, drives, and desires (what he called "inclinations") and the rational ability to make good choices on the basis of good and defensible reasons. For Kant, with his dim view of human nature, what we *want* to do is very rarely what we *ought* to do. But we can recognize what we ought to do by the application of reason.

Kant's derivation of the *categorical imperative*, which he argued is the fundamental principle of all morality, is notoriously complex. But the key idea is simple: reason demands consistency and rejects contradiction. Accordingly, Kant argued that the moral principle we should follow must preserve consistency in all cases and prevent any possibility of contradiction. This moral principle might be expressed as: "Act only on that maxim such that the maxim of your action can be willed to be a universal law." (Although Kant offered several different formulations of the categorical imperative, this is the most famous and most basic formulation.) Kant's prose is dense and confusing, and the categorical imperative is no exception. What does Kant mean?

Kant observed that we make choices according to rules. We tell the truth even when it is inconvenient or embarrassing because we have a rule in our heads that tells us to do so. This is an example of what Kant calls a "subjective principle of action" or a *maxim*. Other examples of maxims are "don't steal" and "keep your promises." Our heads are full of rules that we use to guide our choices. When we worry about *moral* choices, Kant tells us in the categorical imperative that we should act only on choices that "can be willed to be a universal law." That is, before acting on a maxim that informs a moral choice, one must ask: "Could this rule (this maxim) be applied to everyone, everywhere, for all time?" Kant argues that, by *universalizing* a maxim, one can see whether it generates a contradiction. If it generates a contradiction, it cannot be rational, and so it is not a legitimate expression of a good will. If it does not generate a contradiction, it looks morally permissible. When we follow the categorical imperative, Kant thinks, we are doing our (moral) duty.

Take a couple of examples. Suppose you decide to borrow money without intending to pay it back. Your maxim might be: "If I need to borrow money I should do so, even though I know I will never pay it back." Now universalize this maxim according to the categorical imperative. Suppose everyone, everywhere, always borrowed money without the intention of paying it back? Obviously no one would lend money and the very possibility of borrowing would be eliminated. It is rationally contradictory to choose to borrow money without intending to pay it back.

Or, suppose you are caught cheating and try to lie your way out of it. Your maxim is: "When caught cheating, I should lie to get out of trouble." But suppose everyone, everywhere, always lied to get out of trouble when caught cheating? To lie you must hide the truth, and in this situation, were it universalized, it would be impossible to hide the truth. Lies depend on being exceptions to the rule of truthful communication; if lies are no longer the exception but the rule, there is no more truthful communication, and a lie becomes impossible. Again, this is a rational contradiction, and we see that the lie is immoral.

Suppose, however, that you try a maxim like "Thou shalt not kill." What if everyone, everywhere, always avoided killing others? No contradiction is generated. There may be many impractical consequences of universal not-killing, but there are no logical problems with it. If you try a maxim of "Thou shalt kill," on the other hand, you see how quickly it falls apart.

It is not difficult to generate objections to this theory. If one makes maxims specific enough, it is easy to justify apparently immoral actions while following the rule of universal maxims. For example, one can easily universalize a maxim like "a woman with no money whose children are dying of pneumonia should steal penicillin if necessary to save her children's lives," yet Kant would maintain that theft is always wrong and irrational.

Kant also maintains that it is always irrational and wrong to lie, even in the attempt to save an innocent life. But to most of us that sounds absurd. Should a mother never lie, even if it means saving the life of her child? Should the Danes who lied to the Nazis about whether they were protecting Jews have told the truth? Surely not.

Perhaps the most controversial aspect of Kant's moral theory is his distinction between moral duty and happiness. Kant argues that choosing freely on the basis of what we rationally see is right—following the categorical imperative, acting from duty—is the only way we can choose *morally*. But suppose we are acting a certain way solely because it makes us happy, even though those actions happen to agree with what would otherwise be our duty. For Kant, actions motivated by inclination (with the result of happiness) are not motivated by duty, and so we should not consider them *moral* actions. For example, a suicidal person who does not shoot herself because she recognizes that it would be irrational (and thus contrary to her duty) is acting morally. However, another person who fleetingly considers shooting himself but then declines because he loves his life is not acting morally; he is merely inclining toward his happiness.

But if moral duty and happiness are opposed, it seems that only miserable people can be moral. Wouldn't it be nicer if we could have both moral worth in our actions and happy lives? This leads us to *utilitarianism*, the theory of morality that responds specifically to deontology by insisting that morality and happiness are not opposites, but the very same thing.

HAPPY CONSEQUENCES, OR UTILITARIANISM

Hedonism is the notion, first advocated by the Greek philosopher **Epicurus (342–270 BC)**, that pleasure is the greatest good for human beings. (Epicurus is the source of the word *epicurean*.) To be moral is to live the life that produces the most pleasure and avoids pain. But we should not suppose that Epicurus was arguing for a life of debauchery. Drinking too much wine, for example, though fun while it lasts, produces more pain than pleasure in the end, so Epicurus sorted pleasures into categories:

- Natural and necessary, like sleeping and moderate eating
- Natural but unnecessary, like drinking wine or playing chess
- Unnatural and unnecessary, which hurt one's body (for example, smoking cigarettes)
- Unnatural but necessary (but there are no such pleasures)

Epicurus said that we should cultivate natural and necessary pleasures, enjoy natural but unnecessary pleasures in moderation, and avoid all other sorts. The true hedonist does not seek what is immediately pleasurable, but looks for pleasures that will guarantee a long, healthy life full of them. For this reason, *friendship* is Epicurus' favorite example of a pleasure that everyone should cultivate; friendship was consistently considered one of the highest human goods among ancient Greeks.

Jeremy Bentham (1748–1832) adopted Epicurus' basic principles when he developed the theory that later became known as *utilitarianism*. In response to Plato's question "What is the good?," Bentham argued that it is easy to see what humans consider good because they are always seeking it: pleasure. But Bentham was not an egoist, and he argued that the highest good would result from a maximum of pleasure for all people concerned in any moral decision. Decisions that promote *utility* are those that create the most pleasure (the words *utility* and *pleasure* were virtually interchangeable to Bentham, though later utilitarians would ascribe many different meanings to *utility*). Whenever making a decision, the person who desires a moral result should weigh all possible outcomes, and choose the action that produces the most pleasure for everyone concerned. Bentham called this weighing of outcomes a "utilitarian calculus."

Bentham's new moral theory enjoyed enormous popularity, but brought inevitable objections. Some philosophers argued that such a theory made people look no better than swine (because they were just pursuing pleasure). Others objected that people would surely frame their moral decisions to enable them to do whatever they pleased. **John Stuart Mill (1806–1873)** responded to these objections and gave us the form of utilitarianism that, in its fundamentals, is the same moral theory that so many philosophers and economists still endorse today.

Mill argued that the good that human beings seek is not so much pleasure as happiness, and that the basic principle of utilitarianism was what he called the "Greatest Happiness Principle": that action is good which creates the greatest happiness, and the least unhappiness, for the greatest number. He also insisted that people who used this principle must adopt a disinterested view when deciding what would create the greatest happiness. He called this the perspective of "the perfectly disinterested benevolent spectator."

When making a moral decision, then, people will consider the various outcomes and make the choice that produces the most happiness for themselves and everyone else. This is not the same as asking which choice will produce the most pleasure. Accepting a job selling computer software for $55,000 a year might produce more short-term pleasure than going to graduate school, but it might not produce the most happiness. You might be broke and hungry in graduate school, but still very happy because you are progressing toward a goal and finding intellectual stimulation along the way.

The utilitarian must also ask: does this decision produce the most happiness for everyone else, and am I evaluating their happiness fairly and reasonably? Suppose that the recent graduate is again deliberating whether to go to graduate school. Her mother and her father, both attorneys, very much want her to go into the law. But she is fed up with school and will be miserable sitting in a classroom all day. She is sick of eating Ramen noodles and having roommates, and would like to drink a nice bottle of wine once in a while and buy a new car. It is true that her parents' happiness is relevant to the decision, but she must try to weigh the happiness of everyone involved. How unhappy will her parents be if she takes a few years off? How unhappy will she be back in a lecture hall? Utilitarians admit that finding the good is not always easy, but they insist that they offer a practical method for finding the good that anyone can use to solve a moral dilemma.

Utilitarianism is a kind of *consequentialism*, because we evaluate the morality of actions on the basis of their probable outcomes or consequences. For this reason utilitarianism is also what we call a *teleological* theory. Coming from the Greek word *telos*, meaning *purpose* or *end*, teleology refers to the notion that some things and processes are best understood by considering their goals. For utilitarians the goal of life is happiness, and thus they argue that the good (moral) life for humans is the happy life.

Utilitarianism is probably the most popular moral theory of the last hundred years. It is widely used by economists, because one easy way of measuring utility is by assigning dollar signs to outcomes. Today's most famous advocate of animal rights, Peter Singer, is also a well-known utilitarian. Many different versions of utilitarianism have been advanced. In *rule-utilitarianism*, we first rationally determine the general rules that will produce good outcomes, and then follow those rules. In *preference-utilitarianism*, we solve the difficult problem of what will create the most happiness for others by simply asking every person involved for their preference.

But there are many strong objections to utilitarianism. One was raised by the German philosopher **Friedrich Nietzsche (1844–1900)** in his masterpiece *Thus Spoke Zarathustra*. At the end of the book, Zarathustra asks himself if his efforts to find the good for human beings and for himself have increased his personal happiness. He responds to himself: "Happiness? Why should I strive for happiness? I strive for my work!" Nietzsche's point is that many profound and praiseworthy human goals are unquestionably moral, and yet they cannot be said to contribute to the happiness of the person who has those goals, and perhaps not even to the happiness of the greater number. It is true that Van Gogh's paintings, though they destroyed him, created a greater happiness for the rest of us. But that did not count for him as a reason to paint them—he had no idea of his own legacy. For a utilitarian, such self-sacrifice is not only confused but immoral. And yet if our moral theory has difficulty accounting for the value of Van Gogh sacrificing his happiness and everything he loved to his art, we might be in trouble.

Perhaps the most telling objection to utilitarianism is that it could be used to morally sanction a "tyranny of the majority." Suppose you could solve all of the suffering of the world and create universal happiness by flipping a switch on a black box. But, in order to power the box, you had to place one person inside it, who would suffer unspeakably painful torture. None of us would be willing to flip that switch, and yet for a utilitarian such an action would not only be permissible, it would be morally demanded.

A related objection comes from the British philosopher Bernard Williams. Suppose you are an explorer in the Amazon basin and you stumble on a tribe that is about to slaughter 20 captured warriors from another tribe. You interrupt the gruesome execution, and the tribal chief offers to release 19 prisoners in your honor, on the condition that you accept the ceremonial role of choosing one victim and killing him yourself. A utilitarian would be morally required to accept, but most of us would be morally appalled at the idea of killing a complete stranger who presented no threat to us.

Utilitarians have responded to such objections by introducing the notion of certain irreducible human *rights* into utilitarianism. The discussion now turns to these rights and their origins in *social contract theory*.

Promises and Contracts

Although there are good reasons for being suspicious of egoism of any stamp, **Thomas Hobbes (1588–1679)** was a psychological egoist, which was essential to his moral and political theory. Hobbes argued that there are two fundamental facts about human beings: (1) we are all selfish, and (2) we can only survive by banding together. You may have heard Hobbes' famous dictum that human life outside of a society—that is, in his imagined "state of nature"—is "solitary, poor, nasty, brutish, and short." We form groups for self-interested reasons because we need one another to survive and prosper. But the fact that we band together as selfish beings inevitably results in tension between people. Because resources are always scarce, there is competition, and competition creates conflict. Accordingly, if we are to survive as a group, we need rules that everyone promises to follow. These rules, which may be simple at first but become enormously complex, are an exchange of protections for freedoms. "I promise not to punch you in the nose as long as you promise not to punch me in the nose" is precisely such an exchange. You trade the freedom to throw your fists wherever you please for the protection of not being punched yourself. These rules of mutual agreement are, of course, called *laws*, and they guarantee our protections or *rights*. The system of laws and rights that make up the society is called the *social contract*.

Social contract theory builds on the Greek notion that good people are most likely encouraged by a good society. Few social contract theorists would argue that morality can be reduced to societal laws. But most would insist that it is extremely difficult to be a good person unless you are in a good society with good laws. Hobbes argued that the habit of exchanging liberties for protections would extend itself into all dimensions of a good citizen's behavior. The law is an expression of the reciprocity expressed in the Golden Rule—"do unto others as you would have them do unto you"—and so through repeated obedience to the law we would develop the habit, Hobbes thought, of treating others as we would like to be treated.

The most famous American social contract theorist was **John Rawls (1921–2002)**. Rawls argued that "justice is fairness," and for him the morally praiseworthy society distributes its goods in a way that helps the least advantaged of its members. Rawls asked us to imagine what rules we would propose for a society if, when we thought about the rules, we imagined that we had no idea what our own role in that society would be. What rules would we want for our society if we did not know whether we would be poor or rich, African-American or Native Indian, man or woman, or a teacher, plumber, or famous actor? Rawls imagined that this thought experiment—which he called standing behind "the veil of ignorance"—would guarantee fairness in the formulation of the social contract. Existing social rules and laws that did not pass this test—that no rational person would endorse if standing behind the veil of ignorance—were obviously unfair and should be changed or discarded.

Strictly speaking, social contract theory is not a moral code. But because so many of our moral decisions are made in the context of laws and rights, we should understand that the foundation of those laws and rights is a system of promises that have been made, either implicitly or explicitly, by every citizen who freely chooses to live in and benefit from a commonwealth.

A RETURN TO THE GREEKS: THE GOOD LIFE OF VIRTUE

In the 20th century, many philosophers grew increasingly suspicious of the possibility of founding a workable moral system upon rules or principles. The problem with moral rules or principles is that they self-consciously ignore the particulars of the situations in which people actually make moral decisions. For the dominant moralities of the 20th century, deontology and utilitarianism, what is moral for one person is moral for another, regardless of the many differences that undoubtedly exist between their lives, personalities, and stations. This serious weakness in prevailing moral systems caused philosophers to turn once more to the ancient Greeks for help.

Aristotle (384–307 BC) argued that it does not make sense to speak of good actions unless one recognizes that good actions are performed by good people. But good people deliberate over their actions in particular situations, each of which may differ importantly from other situations in which a person has to make a moral choice. But what is a good person?

Aristotle would have responded to this question with his famous "function argument," which posits that the goodness of anything is expressed in its proper function. A good hammer is good because it pounds nails well. A good ship is good if it sails securely across the sea. A bad ship, on the other hand, will take on water and drift aimlessly across the waves. Moreover, we can recognize the function of a thing by identifying what makes it different from other things. The difference between a door and a curtain lies fundamentally in the way they do their jobs. Human function, the particular ability that makes mankind different from all other species, is the ability to reason. The good life is the life of the mind: to be a good person is to actively think.

But to pursue the life of the mind, we need many things. We need health; we need the protection and services of a good society; we need friends for conversation. We need leisure time and enough money to satisfy our physical needs (but not so much as to distract or worry us); we need education, books, art, music, culture, and pleasant distractions to relax the mind.

This does sound like the good *life*. But how does the thinking person *act*? Presumably, Aristotle's happy citizen will encounter moral conflicts and dilemmas like the rest of us. How do we resolve these dilemmas? What guides our choices?

Aristotle did not believe that human beings confront each choice as though it were the first they ever made. Rather, he thought, we develop habits that guide our choices. There are good habits and bad habits. Good habits contribute to our flourishing and are called *virtues* (Aristotle's word, *arête*, may also be translated as *excellence*). Bad habits diminish our happiness and are called *vices*. And happily, for Aristotle, the thinking person will see that there is a practical method for sorting between virtues and vices built into the nature of human beings. Aristotle insisted that human beings are animals, like any other warm-blooded creature on the earth; just as a tiger can act in ways that cause it to flourish or fail, so human beings have a natural guide to their betterment. This has come to be called Aristotle's "golden mean": the notion that our good lies between the extremes of the deficiency of an activity and its excess. Healthy virtue lies in moderation.

An example will help. Suppose you are sitting in the classroom with your professor and fellow students when a wild buffalo storms into the room. The buffalo is enraged and ready to gore all comers. What do you do? An excessive action would be to attack the buffalo with your bare hands: this would, for Aristotle, show the vice of rashness. A deficient action would be to cower behind your desk and shriek for help: this would show the vice of cowardice. But a moderate action would be to make a loud noise to frighten the buffalo, or perhaps to distract it so that others could make for the door, or to do whatever might reasonably reduce the danger to others and yourself. This moderate course of action exemplifies the virtue of courage. Notice, however, that the courageous course of action would change if an enraged tomcat came spitting into the room. Then the moderate and virtuous choice might be to trap the feline with a handy trash basket.

Aristotle's list of virtues includes courage, temperance, justice, liberality or generosity, magnificence (living well), pride, high-mindedness, aspiration, gentleness, truthfulness, friendliness, modesty, righteous indignation, and wittiness. But one could write many such lists, depending on one's own society and way of life. Aristotle would doubtless argue that at least some of these virtues are virtuous for any human being in any place or time, but a strength of his theory is that others' virtues depend on the when, where, and how of differing human practices and communities. One appeal of virtue ethics is that it insists on the context of our moral deliberations.

But is human goodness fully expressed by moderation? Or by being a good citizen? And what about people who lack Aristotle's material requirements of health, friends, and a little property? Aristotle is committed to the idea that such people cannot live fully moral lives, but can that be right? As powerful as it is, one weakness of Aristotle's virtue ethics is that it seems to overemphasize the importance of "fitting" into one's society. The rebel, the outcast, or the romantic chasing an iconoclastic ideal has no place. And Aristotle's theory may sanction some gross moral injustices—such as slavery—if they contribute to the flourishing of society as a whole. Aristotle himself would have had no problem with this: his theory was explicitly designed for the aristocratic way of life. But today we would insist that the good life, if it is to be truly *good* for any of us, must at least in principle be available to every member of our society.

Feminism and the Ethics of Care

Psychologist and philosopher **Carol Gilligan** discovered that moral concepts develop differently in young children. Boys tend to emphasize reasons, rules, and justifications; girls tend to emphasize relationships, the good of the group, and mutual nurturing. From these empirical studies Gilligan developed what came to be called the "ethics of care": the idea that morality might be better grounded on the kind of mutual nurturing and love that takes place in close friendships and family groups. The ethical ideal, according to Gilligan, is a good mother.

Gilligan's ethics of care is compelling because it seems to reflect how many of us make our daily moral decisions. Consider the moral decisions you face in a typical day: telling the truth or lying to a parent or sibling, skipping a party to take care of a heartsick friend or going to see that cute guy, keeping a promise to another student to copy your notes or saying "oops, I forgot." We often confront the moral difficulties of being a good son, sister, friend, or colleague. Generally speaking, we do not settle these moral issues on the basis of impersonal moral principles—we wonder whether it would even be appropriate to do so, given that we are personally involved in these decisions. Should you treat your best friend in precisely the same way you treat a stranger on the street? Some moralists would say, "Of course!" Yet, many of us would consider such behavior odd or psychologically impossible.

The feminist attack on traditional ethics does not accuse one Western morality or another, but indicts its whole history. Western morality has insisted on rationality at the expense of emotions, on impartiality at the expense of relationships, on punishment at the expense of forgiveness, and on "universal principles" at the expense of real, concrete moral problems. In a phrase, morality has been male at the expense of the female. Thus, the feminist argues, a radical rethinking of the entire history of morality is necessary.

As a negative attack on traditional morality, it is hard to disagree with feminism. Our moral tradition does have a suspiciously masculine cast; it is not surprising that virtually every philosopher mentioned in this appendix was a man. But feminism has struggled to develop a positive ethics of its own. Many consider Gilligan's ethics of care to be the best attempt so far, and it works well in family contexts. But when we try to extend the ethics of care into larger spheres, we run into trouble. Gilligan insists on the moral urgency of partiality (as a mother is partial to her children, and even among children). But you would object if you were a defendant in a lawsuit and saw the plaintiff enter, wave genially to the judge, and say, "Hi Mom!" The point, of course, is that in many situations we insist on *impartiality*, and for good reasons. And we all agree that people we have never met may still exercise moral demands upon us. We believe that a man rotting in prison on the other side of the world ought not be tortured, and maybe that we should do something about it if he is (if only by donating money to Amnesty International). Everyone deserves protection from torture for reasons that apply equally to all of us.

PLURALISM

When the German philosopher Nietzsche famously proclaimed that "God is dead," he was not proposing that the nature of the universe had changed. Rather, he was proposing that a change had taken place in the way we view ourselves in the universe. He meant that the Judeo-Christian tradition that has informed all of our values in the West can no longer do the job for us that it used to do. Part of that tradition, Nietzsche thought, was the unfortunate Platonic idea that there is an answer to the question "What is the good?" There is no more one "good" than there is one "God" or one "truth": there are, Nietzsche insisted, many goods, like there are many truths. Nietzsche argued the moral position that we now call *pluralism*.

Pluralism is the idea that there are many goods and many sources of value. Pluralism is explicitly opposed to Plato's insistence that all good things and actions must share some quality that makes them all good. But does this make the pluralist a relativist? No, because the pluralist argues for the moral significance of two ideas that the relativist rejects: (1) that some aspects of human nature are transcultural and transhistorical, and (2) that some methods of inquiry reveal transcultural and transhistorical human values.

When we look at human history, we see goods that repeatedly contribute to human flourishing and evils that interfere with it. War is almost always viewed as an evil in history that has consistently interfered with human flourishing; health, on the other hand, is almost always viewed as a good (with the exception of aberrant religious practices like asceticism). "Avoid war and seek health" is not a moral code—although it might go further than we think—but it does provide an example of what a pluralist is looking for. The pluralist wants concrete goods and practices that actually enrich human life. For the pluralist, the choice between Plato's absolutism and moral relativism is a false dichotomy. Just because there is no absolute "good" does not mean that all goods or values are relative to the time, place, and culture in which we find them. Some things and practices are usually bad for humans, others are usually good, and the discovery and encouragement of the good things and practices is the game the smart ethicist plays.

For this reason, pluralists emphasize the importance of investigating and questioning. Is our present culture enhancing or diminishing us as human beings? Is the American attitude toward sexuality, say, improving the human condition or interfering with it? (And before we can answer that question, what *is* the American attitude toward sexuality? Or are there many attitudes?) The ethical contribution to the history of philosophy made by the fascinating 20th-century movement called *existentialism* is its insistence on this kind of vigorous, ruthlessly honest interrogation of oneself and one's culture. The danger of hypocrisy and self-deception, or what the leading existentialist **Jean-Paul Sartre (1912–1984)** called *bad faith*, is rampant in every culture: challenging our values is uncomfortable. It is much easier for us, like the subjects of the nude ruler in H. C. Andersen's fable *The Emperor's New Clothes*, to collectively pretend that something is good (even if we know there is really nothing there at all). Thus, the project of becoming a good person becomes not just a matter of following the rules, doing one's duty, seeking happiness, becoming virtuous, or caring for others. It is also the lifelong project of discovering *if, when, and why* the apparently good things we seek are what we ought to pursue.

SUMMARY

After reading this appendix, a reasonable student might ask: "But which of these moralities is the *right* one?" Admittedly, philosophers are better at posing problems than solving them. But the lesson was not in demonstrating that one or another morality is the one a person ought to follow. Rather, this appendix has attempted to show you how different people have struggled with the enormously difficult questions of ethics. Many people think they simply know the difference between right and wrong, or unreflectively accept the definitions of right and wrong offered by their parents, churches, communities, or societies. This appendix tried to show that there is nothing simple about ethics. To understand ethics means to think, to challenge, to question, and to reflect. Accordingly, being a good person might mean attempting your own struggle with, and attempting to find your own answer to, what we called Plato's knotty question of goodness.

APPENDIX **B**

ACM CODE OF ETHICS AND PROFESSIONAL CONDUCT

Adopted by ACM Council 10/16/92.

Preamble

Commitment to ethical professional conduct is expected of every member (voting members, associate members, and student members) of the Association for Computing Machinery (ACM).

This Code, consisting of 24 imperatives formulated as statements of personal responsibility, identifies the elements of such a commitment. It contains many, but not all, issues professionals are likely to face. Section 1 outlines fundamental ethical considerations, while Section 2 addresses additional, more specific considerations of professional conduct. Statements in Section 3 pertain more specifically to individuals who have a leadership role, whether in the workplace or in a volunteer capacity such as with organizations like ACM. Principles involving compliance with this Code are given in Section 4.

The Code shall be supplemented by a set of Guidelines, which provide explanation to assist members in dealing with the various issues contained in the Code. It is expected that the Guidelines will be changed more frequently than the Code.

The Code and its supplemented Guidelines are intended to serve as a basis for ethical decision making in the conduct of professional work. Secondarily, they may serve as a basis for judging the merit of a formal complaint pertaining to violation of professional ethical standards.

It should be noted that although computing is not mentioned in the imperatives of Section 1, the Code is concerned with how these fundamental imperatives apply to one's conduct as a computing professional. These imperatives are expressed in a general form to emphasize that ethical principles which apply to computer ethics are derived from more general ethical principles.

It is understood that some words and phrases in a code of ethics are subject to varying interpretations, and that any ethical principle may conflict with other ethical principles in specific situations. Questions related to ethical conflicts can best be answered by thoughtful consideration of fundamental principles, rather than reliance on detailed regulations.

1. GENERAL MORAL IMPERATIVES.

As an ACM member I will ....

1.1 Contribute to society and human well-being.

This principle concerning the quality of life of all people affirms an obligation to protect fundamental human rights and to respect the diversity of all cultures. An essential aim of computing professionals is to minimize negative consequences of computing systems, including threats to health and safety. When designing or implementing systems, computing professionals must attempt to ensure that the products of their efforts will be used in socially responsible ways, will meet social needs, and will avoid harmful effects to health and welfare.

In addition to a safe social environment, human well-being includes a safe natural environment. Therefore, computing professionals who design and develop systems must be alert to, and make others aware of, any potential damage to the local or global environment.

1.2 Avoid harm to others.

"Harm" means injury or negative consequences, such as undesirable loss of information, loss of property, property damage, or unwanted environmental impacts. This principle prohibits use of computing technology in ways that result in harm to any of the following: users, the general public, employees, employers. Harmful actions include intentional destruction or modification of files and programs leading to serious loss of resources or unnecessary expenditure of human resources such as the time and effort required to purge systems of "computer viruses."

Well-intended actions, including those that accomplish assigned duties, may lead to harm unexpectedly. In such an event the responsible person or persons are obligated to undo or mitigate the negative consequences as much as possible. One way to avoid unintentional harm is to carefully consider potential impacts on all those affected by decisions made during design and implementation.

To minimize the possibility of indirectly harming others, computing professionals must minimize malfunctions by following generally accepted standards for system design and testing. Furthermore, it is often necessary to assess the social consequences of systems to project the likelihood of any serious harm to others. If system features are misrepresented to users, coworkers, or supervisors, the individual computing professional is responsible for any resulting injury.

In the work environment the computing professional has the additional obligation to report any signs of system dangers that might result in serious personal or social damage. If one's superiors do not act to curtail or mitigate such dangers, it may be necessary to "blow the whistle" to help correct the problem or reduce the risk. However, capricious or misguided reporting of violations can, itself, be harmful. Before reporting violations, all relevant aspects of the incident must be thoroughly assessed. In particular, the assessment of risk and responsibility must be credible. It is suggested that advice be sought from other computing professionals. See principle 2.5 regarding thorough evaluations.

1.3 Be honest and trustworthy.

Honesty is an essential component of trust. Without trust an organization cannot function effectively. The honest computing professional will not make deliberately false or deceptive claims about a system or system design, but will instead provide full disclosure of all pertinent system limitations and problems.

A computer professional has a duty to be honest about his or her own qualifications, and about any circumstances that might lead to conflicts of interest.

Membership in volunteer organizations such as ACM may at times place individuals in situations where their statements or actions could be interpreted as carrying the "weight" of a larger group of professionals. An ACM member will exercise care to not misrepresent ACM or positions and policies of ACM or any ACM units.

1.4 Be fair and take action not to discriminate.

The values of equality, tolerance, respect for others, and the principles of equal justice govern this imperative. Discrimination on the basis of race, sex, religion, age, disability, national origin, or other such factors is an explicit violation of ACM policy and will not be tolerated.

Inequities between different groups of people may result from the use or misuse of information and technology. In a fair society, all individuals would have equal opportunity to participate in, or benefit from, the use of computer resources regardless of race, sex, religion, age, disability, national origin or other such similar factors. However, these ideals do not justify unauthorized use of computer resources nor do they provide an adequate basis for violation of any other ethical imperatives of this code.

1.5 Honor property rights including copyrights and patent.

Violation of copyrights, patents, trade secrets and the terms of license agreements is prohibited by law in most circumstances. Even when software is not so protected, such violations are contrary to professional behavior. Copies of software should be made only with proper authorization. Unauthorized duplication of materials must not be condoned.

1.6 Give proper credit for intellectual property.

Computing professionals are obligated to protect the integrity of intellectual property. Specifically, one must not take credit for other's ideas or work, even in cases where the work has not been explicitly protected by copyright, patent, etc.

1.7 Respect the privacy of others.

Computing and communication technology enables the collection and exchange of personal information on a scale unprecedented in the history of civilization. Thus there is increased potential for violating the privacy of individuals and groups. It is the responsibility of professionals to maintain the privacy and integrity of data describing individuals. This includes taking precautions to ensure the accuracy of data, as well as protecting it from unauthorized access or accidental disclosure to inappropriate individuals. Furthermore, procedures must be established to allow individuals to review their records and correct inaccuracies.

This imperative implies that only the necessary amount of personal information be collected in a system, that retention and disposal periods for that information be clearly defined and enforced, and that personal information gathered for a specific purpose not be used for other purposes without consent of the individual(s). These principles apply to electronic communications, including electronic mail, and prohibit procedures that capture or monitor electronic user data, including messages, without the permission of users or bona fide authorization related to system operation and maintenance. User data observed during the normal duties of system operation and maintenance must be treated with strictest confidentiality, except in cases where it is evidence for the violation of law, organizational regulations, or this Code. In these cases, the nature or contents of that information must be disclosed only to proper authorities.

1.8 Honor confidentiality.

The principle of honesty extends to issues of confidentiality of information whenever one has made an explicit promise to honor confidentiality or, implicitly, when private information not directly related to the performance of one's duties becomes available. The ethical concern is to respect all obligations of confidentiality to employers, clients, and users unless discharged from such obligations by requirements of the law or other principles of this Code.

2. MORE SPECIFIC PROFESSIONAL RESPONSIBILITIES.

As an ACM computing professional I will ....

2.1 Strive to achieve the highest quality, effectiveness and dignity in both the process and products of professional work.

Excellence is perhaps the most important obligation of a professional. The computing professional must strive to achieve quality and to be cognizant of the serious negative consequences that may result from poor quality in a system.

2.2 Acquire and maintain professional competence.

Excellence depends on individuals who take responsibility for acquiring and maintaining professional competence. A professional must participate in setting standards for appropriate levels of competence, and strive to achieve those standards. Upgrading technical knowledge and competence can be achieved in several ways: doing independent study; attending seminars, conferences, or courses; and being involved in professional organizations.

2.3 Know and respect existing laws pertaining to professional work.

ACM members must obey existing local, state, province, national, and international laws unless there is a compelling ethical basis not to do so. Policies and procedures of the organizations in which one participates must also be obeyed. But compliance must be balanced with the recognition that sometimes existing laws and rules may be immoral or inappropriate and, therefore, must be challenged. Violation of a law or regulation may be ethical when that law or rule has inadequate moral basis or when it conflicts with another law judged to be more important. If one decides to violate a law or rule because it is viewed as unethical, or for any other reason, one must fully accept responsibility for one's actions and for the consequences.

2.4 Accept and provide appropriate professional review.

Quality professional work, especially in the computing profession, depends on professional reviewing and critiquing. Whenever appropriate, individual members should seek and utilize peer review as well as provide critical review of the work of others.

2.5 Give comprehensive and thorough evaluations of computer systems and their impacts, including analysis of possible risks.

Computer professionals must strive to be perceptive, thorough, and objective when evaluating, recommending, and presenting system descriptions and alternatives. Computer professionals are in a position of special trust, and therefore have a special responsibility to provide objective, credible evaluations to employers, clients, users, and the public. When providing evaluations the professional must also identify any relevant conflicts of interest, as stated in imperative 1.3.

As noted in the discussion of principle 1.2 on avoiding harm, any signs of danger from systems must be reported to those who have opportunity and/or responsibility to resolve them. See the guidelines for imperative 1.2 for more details concerning harm, including the reporting of professional violations.

2.6 Honor contracts, agreements, and assigned responsibilities.

Honoring one's commitments is a matter of integrity and honesty. For the computer professional this includes ensuring that system elements perform as intended. Also, when one contracts for work with another party, one has an obligation to keep that party properly informed about progress toward completing that work.

A computing professional has a responsibility to request a change in any assignment that he or she feels cannot be completed as defined. Only after serious consideration and with full disclosure of risks and concerns to the employer or client, should one accept the assignment. The major underlying principle here is the obligation to accept personal accountability for professional work. On some occasions other ethical principles may take greater priority.

A judgment that a specific assignment should not be performed may not be accepted. Having clearly identified one's concerns and reasons for that judgment, but failing to procure a change in that assignment, one may yet be obligated, by contract or by law, to proceed as directed. The computing professional's ethical judgment should be the final guide in deciding whether or not to proceed. Regardless of the decision, one must accept the responsibility for the consequences.

However, performing assignments "against one's own judgment" does not relieve the professional of responsibility for any negative consequences.

2.7 Improve public understanding of computing and its consequences.

Computing professionals have a responsibility to share technical knowledge with the public by encouraging understanding of computing, including the impacts of computer systems and their limitations. This imperative implies an obligation to counter any false views related to computing.

2.8 Access computing and communication resources only when authorized to do so.

Theft or destruction of tangible and electronic property is prohibited by imperative 1.2 - "Avoid harm to others." Trespassing and unauthorized use of a computer or communication system is addressed by this imperative. Trespassing includes accessing communication networks and computer systems, or accounts and/or files associated with those systems, without explicit authorization to do so. Individuals and organizations have the right to restrict access to their systems so long as they do not violate the discrimination principle (see 1.4). No one should enter or use another's computer system, software, or data files without permission. One must always have appropriate approval before using system resources, including communication ports, file space, other system peripherals, and computer time.

3. ORGANIZATIONAL LEADERSHIP IMPERATIVES.

As an ACM member and an organizational leader, I will ....

BACKGROUND NOTE: This section draws extensively from the draft IFIP Code of Ethics, especially its sections on organizational ethics and international concerns. The ethical obligations of organizations tend to be neglected in most codes of professional conduct, perhaps because these codes are written from the perspective of the individual member. This dilemma is addressed by stating these imperatives from the perspective of the organizational leader. In this context "leader" is viewed as any organizational member who has leadership or educational responsibilities. These imperatives generally may apply to organizations as well as their leaders. In this context "organizations" are corporations, government agencies, and other "employers," as well as volunteer professional organizations.

3.1 Articulate social responsibilities of members of an organizational unit and encourage full acceptance of those responsibilities.

Because organizations of all kinds have impacts on the public, they must accept responsibilities to society. Organizational procedures and attitudes oriented toward quality and the welfare of society will reduce harm to members of the public, thereby serving public interest and fulfilling social responsibility. Therefore, organizational leaders must encourage full participation in meeting social responsibilities as well as quality performance.

3.2 Manage personnel and resources to design and build information systems that enhance the quality of working life.

Organizational leaders are responsible for ensuring that computer systems enhance, not degrade, the quality of working life. When implementing a computer system, organizations must consider the personal and professional development, physical safety, and human dignity of all workers. Appropriate human-computer ergonomic standards should be considered in system design and in the workplace.

3.3 Acknowledge and support proper and authorized uses of an organization's computing and communication resources.

Because computer systems can become tools to harm as well as to benefit an organization, the leadership has the responsibility to clearly define appropriate and inappropriate uses of organizational computing resources. While the number and scope of such rules should be minimal, they should be fully enforced when established.

3.4 Ensure that users and those who will be affected by a system have their needs clearly articulated during the assessment and design of requirements; later the system must be validated to meet requirements.

Current system users, potential users and other persons whose lives may be affected by a system must have their needs assessed and incorporated in the statement of requirements. System validation should ensure compliance with those requirements.

3.5 Articulate and support policies that protect the dignity of users and others affected by a computing system.

Designing or implementing systems that deliberately or inadvertently demean individuals or groups is ethically unacceptable. Computer professionals who are in decision making positions should verify that systems are designed and implemented to protect personal privacy and enhance personal dignity.

3.6 Create opportunities for members of the organization to learn the principles and limitations of computer systems.

This complements the imperative on public understanding (2.7). Educational opportunities are essential to facilitate optimal participation of all organizational members. Opportunities must be available to all members to help them improve their knowledge and skills in computing, including courses that familiarize them with the consequences and limitations of particular types of systems. In particular, professionals must be made aware of the dangers of building systems around oversimplified models, the improbability of anticipating and designing for every possible operating condition, and other issues related to the complexity of this profession.

4. COMPLIANCE WITH THE CODE.

As an ACM member I will ....

4.1 Uphold and promote the principles of this code.

The future of the computing profession depends on both technical and ethical excellence. Not only is it important for ACM computing professionals to adhere to the principles expressed in this Code, each member should encourage and support adherence by other members.

4.2 Treat violations of this code as inconsistent with membership in the ACM.

Adherence of professionals to a code of ethics is largely a voluntary matter. However, if a member does not follow this code by engaging in gross misconduct, membership in ACM may be terminated.

This Code and the supplemental Guidelines were developed by the Task Force for the Revision of the ACM Code of Ethics and Professional Conduct: Ronald E. Anderson, Chair, Gerald Engel, Donald Gotterbarn, Grace C. Hertlein, Alex Hoffman, Bruce Jawer, Deborah

G. Johnson, Doris K. Lidtke, Joyce Currie Little, Dianne Martin, Donn B. Parker, Judith A. Perrolle, and Richard S. Rosenberg. The Task Force was organized by ACM/SIGCAS and funding was provided by the ACM SIG Discretionary Fund. This Code and the supplemental Guidelines were adopted by the ACM Council on October 16, 1992.

This Code may be published without permission as long as it is not changed in any way and it carries the copyright notice. Copyright ©1997, Association for Computing Machinery, Inc.

ASSOCIATION OF INFORMATION TECHNOLOGY PROFESSIONALS (AITP) CODE OF ETHICS AND STANDARD OF CONDUCT

CODE OF ETHICS

I acknowledge:

That I have an obligation to management, therefore, I shall promote the understanding of information processing methods and procedures to management using every resource at my command.

That I have an obligation to my fellow members, therefore, I shall uphold the high ideals of AITP as outlined in the Association Bylaws. Further, I shall cooperate with my fellow members and shall treat them with honesty and respect at all times.

That I have an obligation to society and will participate to the best of my ability in the dissemination of knowledge pertaining to the general development and understanding of information processing. Further, I shall not use knowledge of a confidential nature to further my personal interest, nor shall I violate the privacy and confidentiality of information entrusted to me or to which I may gain access.

That I have an obligation to my College or University, therefore, I shall uphold its ethical and moral principles.

That I have an obligation to my employer whose trust I hold, therefore, I shall endeavor to discharge this obligation to the best of my ability, to guard my employer's interests, and to advise him or her wisely and honestly.

That I have an obligation to my country, therefore, in my personal, business, and social contacts, I shall uphold my nation and shall honor the chosen way of life of my fellow citizens.

I accept these obligations as a personal responsibility and as a member of this Association. I shall actively discharge these obligations and I dedicate myself to that end.

STANDARD OF CONDUCT

These standards expand on the Code of Ethics by providing specific statements of behavior in support of each element of the Code. They are not objectives to be strived for, they are rules that no true professional will violate. It is first of all expected that an information processing professional will abide by the appropriate laws of their country and community. The following standards address tenets that apply to the profession.

In recognition of my obligation to management I shall:

- Keep my personal knowledge up-to-date and insure that proper expertise is available when needed.
- Share my knowledge with others and present factual and objective information to management to the best of my ability.
- Accept full responsibility for work that I perform.
- Not misuse the authority entrusted to me.
- Not misrepresent or withhold information concerning the capabilities of equipment, software or systems.
- Not take advantage of the lack of knowledge or inexperience on the part of others.

In recognition of my obligation to my fellow members and the profession I shall:

- Be honest in all my professional relationships.
- Take appropriate action in regard to any illegal or unethical practices that come to my attention. However, I will bring charges against any person only when I have reasonable basis for believing in the truth of the allegations and without any regard to personal interest.
- Endeavor to share my special knowledge.
- Cooperate with others in achieving understanding and in identifying problems.
- Not use or take credit for the work of others without specific acknowledgement and authorization.
- Not take advantage of the lack of knowledge or inexperience on the part of others for personal gain.

In recognition of my obligation to society I shall:

- Protect the privacy and confidentiality of all information entrusted to me.
- Use my skill and knowledge to inform the public in all areas of my expertise.
- To the best of my ability, insure that the products of my work are used in a socially responsible way.
- Support, respect, and abide by the appropriate local, state, provincial, and federal laws.
- Never misrepresent or withhold information that is germane to a problem or situation of public concern nor will I allow any such known information to remain unchallenged.

- Not use knowledge of a confidential or personal nature in any unauthorized manner or to achieve personal gain.

In recognition of my obligation to my employer I shall:

- Make every effort to ensure that I have the most current knowledge and that the proper expertise is available when needed.
- Avoid conflict of interest and insure that my employer is aware of any potential conflicts.
- Present a fair, honest, and objective viewpoint.
- Protect the proper interests of my employer at all times.
- Protect the privacy and confidentiality of all information entrusted to me.
- Not misrepresent or withhold information that is germane to the situation.
- Not attempt to use the resources of my employer for personal gain or for any purpose without proper approval.
- Not exploit the weakness of a computer system for personal gain or personal satisfaction.

INSTITUTE OF ELECTRICAL AND ELECTRONICS ENGINEERS COMPUTER SOCIETY CODE

SOFTWARE ENGINEERING CODE OF ETHICS AND PROFESSIONAL PRACTICE

(Version 5.2) as recommended by the IEEE-CS/ACM Joint Task Force on Software Engineering Ethics and Professional Practices and Jointly approved by the ACM and the IEEE-CS as the standard for teaching and practicing software engineering.

Short Version

Preamble

The short version of the code summarizes aspirations at a high level of the abstraction; the clauses that are included in the full version give examples and details of how these aspirations change the way we act as software engineering professionals. Without the aspirations, the details can become legalistic and tedious; without the details, the aspirations can become high sounding but empty; together, the aspirations and the details form a cohesive code.

Software engineers shall commit themselves to making the analysis, specification, design, development, testing and maintenance of software a beneficial and respected profession. In accordance with their commitment to the health, safety and welfare of the public, software engineers shall adhere to the following Eight Principles:

1. PUBLIC - Software engineers shall act consistently with the public interest.
2. CLIENT AND EMPLOYER - Software engineers shall act in a manner that is in the best interests of their client and employer consistent with the public interest.
3. PRODUCT - Software engineers shall ensure that their products and related modifications meet the highest professional standards possible.
4. JUDGMENT - Software engineers shall maintain integrity and independence in their professional judgment.

5. MANAGEMENT - Software engineering managers and leaders shall subscribe to and promote an ethical approach to the management of software development and maintenance.

6. PROFESSION - Software engineers shall advance the integrity and reputation of the profession consistent with the public interest.

7. COLLEAGUES - Software engineers shall be fair to and supportive of their colleagues.

8. SELF - Software engineers shall participate in lifelong learning regarding the practice of their profession and shall promote an ethical approach to the practice of the profession.

Full Version

Preamble

Computers have a central and growing role in commerce, industry, government, medicine, education, entertainment and society at large. Software engineers are those who contribute by direct participation or by teaching, to the analysis, specification, design, development, certification, maintenance and testing of software systems. Because of their roles in developing software systems, software engineers have significant opportunities to do good or cause harm, to enable others to do good or cause harm, or to influence others to do good or cause harm. To ensure, as much as possible, that their efforts will be used for good, software engineers must commit themselves to making software engineering a beneficial and respected profession. In accordance with that commitment, software engineers shall adhere to the following Code of Ethics and Professional Practice.

The Code contains eight Principles related to the behavior of and decisions made by professional software engineers, including practitioners, educators, managers, supervisors and policy makers, as well as trainees and students of the profession. The Principles identify the ethically responsible relationships in which individuals, groups, and organizations participate and the primary obligations within these relationships. The Clauses of each Principle are illustrations of some of the obligations included in these relationships. These obligations are founded in the software engineer's humanity, in special care owed to people affected by the work of software engineers, and in the unique elements of the practice of software engineering. The Code prescribes these as obligations of anyone claiming to be or aspiring to be a software engineer.

It is not intended that the individual parts of the Code be used in isolation to justify errors of omission or commission. The list of Principles and Clauses is not exhaustive. The Clauses should not be read as separating the acceptable from the unacceptable in professional conduct in all practical situations. The Code is not a simple ethical algorithm that generates ethical decisions. In some situations, standards may be in tension with each other or with standards from other sources. These situations require the software engineer to use ethical judgment to act in a manner which is most consistent with the spirit of the Code of Ethics and Professional Practice, given the circumstances.

Ethical tensions can best be addressed by thoughtful consideration of fundamental principles, rather than blind reliance on detailed regulations. These Principles should influence software engineers to consider broadly who is affected by their work; to examine if they and their colleagues are treating other human beings with due respect; to consider how the public, if reasonably well informed, would view their decisions; to analyze how the least empowered will be affected by their decisions; and to consider whether their acts would be judged worthy of the ideal professional working as a software engineer. In all these judgments concern for the health, safety and welfare of the public is primary; that is, the "Public Interest" is central to this Code.

The dynamic and demanding context of software engineering requires a code that is adaptable and relevant to new situations as they occur. However, even in this generality, the Code provides support for software engineers and managers of software engineers who need to take positive action in a specific case by documenting the ethical stance of the profession. The Code provides an ethical foundation to which individuals within teams and the team as a whole can appeal. The Code helps to define those actions that are ethically improper to request of a software engineer or teams of software engineers.

The Code is not simply for adjudicating the nature of questionable acts; it also has an important educational function. As this Code expresses the consensus of the profession on ethical issues, it is a means to educate both the public and aspiring professionals about the ethical obligations of all software engineers.

Principles

Principle 1 PUBLIC

Software engineers shall act consistently with the public interest. In particular, software engineers shall, as appropriate:

1.01. Accept full responsibility for their own work.

1.02. Moderate the interests of the software engineer, the employer, the client and the users with the public good.

1.03. Approve software only if they have a well-founded belief that it is safe, meets specifications, passes appropriate tests, and does not diminish quality of life, diminish privacy or harm the environment. The ultimate effect of the work should be to the public good.

1.04. Disclose to appropriate persons or authorities any actual or potential danger to the user, the public, or the environment, that they reasonably believe to be associated with software or related documents.

1.05. Cooperate in efforts to address matters of grave public concern caused by software, its installation, maintenance, support or documentation.

1.06. Be fair and avoid deception in all statements, particularly public ones, concerning software or related documents, methods and tools.

1.07. Consider issues of physical disabilities, allocation of resources, economic disadvantage and other factors that can diminish access to the benefits of software.

1.08. Be encouraged to volunteer professional skills to good causes and to contribute to public education concerning the discipline.

Principle 2 CLIENT AND EMPLOYER

Software engineers shall act in a manner that is in the best interests of their client and employer, consistent with the public interest. In particular, software engineers shall, as appropriate:

2.01. Provide service in their areas of competence, being honest and forthright about any limitations of their experience and education.

2.02. Not knowingly use software that is obtained or retained either illegally or unethically.

2.03. Use the property of a client or employer only in ways properly authorized, and with the client's or employer's knowledge and consent.

2.04. Ensure that any document upon which they rely has been approved, when required, by someone authorized to approve it.

2.05. Keep private any confidential information gained in their professional work, where such confidentiality is consistent with the public interest and consistent with the law.

2.06. Identify, document, collect evidence and report to the client or the employer promptly if, in their opinion, a project is likely to fail, to prove too expensive, to violate intellectual property law, or otherwise to be problematic.

2.07. Identify, document, and report significant issues of social concern, of which they are aware, in software or related documents, to the employer or the client.

2.08. Accept no outside work detrimental to the work they perform for their primary employer.

2.09. Promote no interest adverse to their employer or client, unless a higher ethical concern is being compromised; in that case, inform the employer or another appropriate authority of the ethical concern.

Principle 3 PRODUCT

Software engineers shall ensure that their products and related modifications meet the highest professional standards possible. In particular, software engineers shall, as appropriate:

3.01. Strive for high quality, acceptable cost, and a reasonable schedule, ensuring significant tradeoffs are clear to and accepted by the employer and the client, and are available for consideration by the user and the public.

3.02. Ensure proper and achievable goals and objectives for any project on which they work or propose.

3.03. Identify, define and address ethical, economic, cultural, legal and environmental issues related to work projects.

3.04. Ensure that they are qualified for any project on which they work or propose to work, by an appropriate combination of education, training, and experience.

3.05. Ensure that an appropriate method is used for any project on which they work or propose to work.

3.06. Work to follow professional standards, when available, that are most appropriate for the task at hand, departing from these only when ethically or technically justified.

3.07. Strive to fully understand the specifications for software on which they work.

3.08. Ensure that specifications for software on which they work have been well documented, satisfy the users' requirements and have the appropriate approvals.

3.09. Ensure realistic quantitative estimates of cost, scheduling, personnel, quality and outcomes on any project on which they work or propose to work and provide an uncertainty assessment of these estimates.

3.10. Ensure adequate testing, debugging, and review of software and related documents on which they work.

3.11. Ensure adequate documentation, including significant problems discovered and solutions adopted, for any project on which they work.

3.12. Work to develop software and related documents that respect the privacy of those who will be affected by that software.

3.13. Be careful to use only accurate data derived by ethical and lawful means, and use it only in ways properly authorized.

3.14. Maintain the integrity of data, being sensitive to outdated or flawed occurrences.

3.15. Treat all forms of software maintenance with the same professionalism as new development.

Principle 4 JUDGMENT

Software engineers shall maintain integrity and independence in their professional judgment. In particular, software engineers shall, as appropriate:

4.01. Temper all technical judgments by the need to support and maintain human values.

4.02. Only endorse documents either prepared under their supervision or within their areas of competence and with which they are in agreement.

4.03. Maintain professional objectivity with respect to any software or related documents they are asked to evaluate.

4.04. Not engage in deceptive financial practices such as bribery, double billing, or other improper financial practices.

4.05. Disclose to all concerned parties those conflicts of interest that cannot reasonably be avoided or escaped.

4.06. Refuse to participate, as members or advisors, in a private, governmental or professional body concerned with software related issues, in which they, their employers or their clients have undisclosed potential conflicts of interest.

Principle 5 MANAGEMENT

Software engineering managers and leaders shall subscribe to and promote an ethical approach to the management of software development and maintenance. In particular, those managing or leading software engineers shall, as appropriate:

5.01. Ensure good management for any project on which they work, including effective procedures for promotion of quality and reduction of risk.

5.02. Ensure that software engineers are informed of standards before being held to them.

5.03. Ensure that software engineers know the employer's policies and procedures for protecting passwords, files and information that is confidential to the employer or confidential to others.

5.04. Assign work only after taking into account appropriate contributions of education and experience tempered with a desire to further that education and experience.

5.05. Ensure realistic quantitative estimates of cost, scheduling, personnel, quality and outcomes on any project on which they work or propose to work, and provide an uncertainty assessment of these estimates.

5.06. Attract potential software engineers only by full and accurate description of the conditions of employment.

5.07. Offer fair and just remuneration.

5.08. Not unjustly prevent someone from taking a position for which that person is suitably qualified.

5.09. Ensure that there is a fair agreement concerning ownership of any software, processes, research, writing, or other intellectual property to which a software engineer has contributed.

5.10. Provide for due process in hearing charges of violation of an employer's policy or of this Code.

5.11. Not ask a software engineer to do anything inconsistent with this Code.

5.12. Not punish anyone for expressing ethical concerns about a project.

Principle 6 PROFESSION

Software engineers shall advance the integrity and reputation of the profession consistent with the public interest. In particular, software engineers shall, as appropriate:

6.01. Help develop an organizational environment favorable to acting ethically.

6.02. Promote public knowledge of software engineering.

6.03. Extend software engineering knowledge by appropriate participation in professional organizations, meetings and publications.

6.04. Support, as members of a profession, other software engineers striving to follow this Code.

6.05. Not promote their own interest at the expense of the profession, client or employer.

6.06. Obey all laws governing their work, unless, in exceptional circumstances, such compliance is inconsistent with the public interest.

6.07. Be accurate in stating the characteristics of software on which they work, avoiding not only false claims but also claims that might reasonably be supposed to be speculative, vacuous, deceptive, misleading, or doubtful.

6.08. Take responsibility for detecting, correcting, and reporting errors in software and associated documents on which they work.

6.09. Ensure that clients, employers, and supervisors know of the software engineer's commitment to this Code of ethics, and the subsequent ramifications of such commitment.

6.10. Avoid associations with businesses and organizations which are in conflict with this code.

6.11. Recognize that violations of this Code are inconsistent with being a professional software engineer.

6.12. Express concerns to the people involved when significant violations of this Code are detected unless this is impossible, counter-productive, or dangerous.

6.13. Report significant violations of this Code to appropriate authorities when it is clear that consultation with people involved in these significant violations is impossible, counter-productive or dangerous.

Principle 7 COLLEAGUES

Software engineers shall be fair to and supportive of their colleagues. In particular, software engineers shall, as appropriate:

7.01. Encourage colleagues to adhere to this Code.

7.02. Assist colleagues in professional development.

7.03. Credit fully the work of others and refrain from taking undue credit.

7.04. Review the work of others in an objective, candid, and properly-documented way.

7.05. Give a fair hearing to the opinions, concerns, or complaints of a colleague.

7.06. Assist colleagues in being fully aware of current standard work practices including policies and procedures for protecting passwords, files and other confidential information, and security measures in general.

7.07. Not unfairly intervene in the career of any colleague; however, concern for the employer, the client or public interest may compel software engineers, in good faith, to question the competence of a colleague.

7.08. In situations outside of their own areas of competence, call upon the opinions of other professionals who have competence in that area.

Principle 8 SELF

Software engineers shall participate in lifelong learning regarding the practice of their profession and shall promote an ethical approach to the practice of the profession. In particular, software engineers shall continually endeavor to:

8.01. Further their knowledge of developments in the analysis, specification, design, development, maintenance and testing of software and related documents, together with the management of the development process.

8.02. Improve their ability to create safe, reliable, and useful quality software at reasonable cost and within a reasonable time.

8.03. Improve their ability to produce accurate, informative, and well-written documentation.

8.04. Improve their understanding of the software and related documents on which they work and of the environment in which they will be used.

8.05. Improve their knowledge of relevant standards and the law governing the software and related documents on which they work.

8.06. Improve their knowledge of this Code, its interpretation, and its application to their work.

8.07. Not give unfair treatment to anyone because of any irrelevant prejudices.

8.08. Not influence others to undertake any action that involves a breach of this Code.

8.09. Recognize that personal violations of this Code are inconsistent with being a professional software engineer.

This Code was developed by the IEEE-CS/ACM joint task force on Software Engineering Ethics and Professional Practices (SEEPP):

Executive Committee: Donald Gotterbarn (Chair), Keith Miller and Simon Rogerson;

Members: Steve Barber, Peter Barnes, Ilene Burnstein, Michael Davis, Amr El-Kadi, N. Ben Fairweather, Milton Fulghum, N. Jayaram, Tom Jewett, Mark Kanko, Ernie Kallman, Duncan Langford, Joyce Currie Little, Ed Mechler, Manuel J. Norman, Douglas Phillips, Peter Ron Prinzivalli, Patrick Sullivan, John Weckert, Vivian Weil, S. Weisband and Laurie Honour Werth.

PROJECT MANAGEMENT INSTITUTE CODE OF ETHICS AND PROFESSIONAL CONDUCT

CHAPTER 1. VISION AND APPLICABILITY

1.1 Vision and Purpose

As <u>practitioners</u> of project management, we are committed to doing what is right and honorable. We set high standards for ourselves and we aspire to meet these standards in all aspects of our lives—at work, at home, and in service to our profession.

This Code of Ethics and Professional Conduct describes the expectations that we have of ourselves and our fellow practitioners in the global project management community. It articulates the ideals to which we aspire as well as the behaviors that are mandatory in our professional and volunteer roles.

The purpose of this Code is to instill confidence in the project management profession and to help an individual become a better practitioner. We do this by establishing a profession-wide understanding of appropriate behavior. We believe that the credibility and reputation of the project management profession is shaped by the collective conduct of individual practitioners.

We believe that we can advance our profession, both individually and collectively, by embracing this Code of Ethics and Professional Conduct. We also believe that this Code will assist us in making wise decisions, particularly when faced with difficult situations where we may be asked to compromise our integrity or our values.

Our hope that this Code of Ethics and Professional Conduct will serve as a catalyst for others to study, deliberate, and write about ethics and values. Further, we hope that this Code will ultimately be used to build upon and evolve our profession.

1.2 Persons to Whom the Code Applies

The Code of Ethics and Professional Conduct applies to:

1.2.1 All <u>PMI members</u>

1.2.2 Individuals who are not members of PMI but meet one or more of the following criteria:

.1 Non-members who hold a PMI certification

.2 Non-members who apply to commence a PMI certification process

.3 Non-members who serve PMI in a volunteer capacity.

Comment: *Those holding a Project Management Institute (PMI®) credential (whether members or not) were previously held accountable to the Project Management Professional (PMP®) or Certified Associate in Project Management (CAPM®) Code of Professional Conduct and continue to be held accountable to the PMI Code of Ethics and Professional Conduct. In the past, PMI also had separate ethics standards for members and for credentialed individuals. Stakeholders who contributed input to develop this Code concluded that having multiple codes was undesirable and that everyone should be held to one high standard. Therefore, this Code is applicable to both <u>PMI members</u> and individuals who have applied for or received a credential from PMI, regardless of their membership in <u>PMI</u>*

1.3 Structure of the Code

The Code of Ethics and Professional Conduct is divided into sections that contain standards of conduct which are aligned with the four values that were identified as most important to the project management community. Some sections of this Code include comments. Comments are not mandatory parts of the Code, but provide examples and other clarification. Finally, a glossary can be found at the end of the standard. The glossary defines words and phrases used in the Code. For convenience, those terms defined in the glossary are underlined in the text of the Code.

1.4 Values that Support this Code

<u>Practitioners</u> from the global project management community were asked to identify the values that formed the basis of their decision making and guided their actions. The values that the global project management community defined as most important were: responsibility, respect, fairness, and honesty. This Code affirms these four values as its foundation.

1.5 Aspirational and Mandatory Conduct

Each section of the Code of Ethics and Professional Conduct includes both aspirational standards and mandatory standards. The aspirational standards describe the conduct that we strive to uphold as <u>practitioners</u>. Although adherence to the aspirational standards is not easily measured, conducting ourselves in accordance with these is an expectation that we have of ourselves as professionals—it is not optional.

The mandatory standards establish firm requirements, and in some cases, limit or prohibit practitioner behavior. Practitioners who do not conduct themselves in accordance with these standards will be subject to disciplinary procedures before PMI's Ethics Review Committee.

Comment: *The conduct covered under the aspirational standards and conduct covered under the mandatory standards are not mutually exclusive; that is, one specific act or omission could violate both aspirational and mandatory standards.*

CHAPTER 2. RESPONSIBILITY

2.1 Description of Responsibility

Responsibility is our duty to take ownership for the decisions we make or fail to make, the actions we take or fail to take, and the consequences that result.

2.2 Responsibility: Aspirational Standards

As <u>practitioners</u> in the global project management community:

2.2.1 We make decisions and take actions based on the best interests of society, public safety, and the environment.

2.2.2 We accept only those assignments that are consistent with our background, experience, skills, and qualifications.

Comment: *Where developmental or stretch assignments are being considered, we ensure that key stakeholders receive timely and complete information regarding the gaps in our qualifications so that they may make informed decisions regarding our suitability for a particular assignment.*

In the case of a contracting arrangement, we only bid on work that our organization is qualified to perform and we assign only qualified individuals to perform the work.

2.2.3 We fulfill the commitments that we undertake – we do what we say we will do.

2.2.4 When we make errors or omissions, we take ownership and make corrections promptly. When we discover errors or omissions caused by others, we communicate them to the appropriate body as soon as they are discovered. We accept accountability for any issues resulting from our errors or omissions and any resulting consequences.

2.2.5 We protect proprietary or confidential information that has been entrusted to us.

2.2.6 We uphold this Code and hold each other accountable to it.

2.3 Responsibility: Mandatory Standards

As <u>practitioners</u> in the global project management community, we require the following of ourselves and our fellow practitioners:

Regulations and Legal Requirements

2.3.1 We inform ourselves and uphold the policies, rules, regulations and laws that govern our work, professional, and volunteer activities.

2.3.2 We report unethical or illegal conduct to appropriate management and, if necessary, to those affected by the conduct.

Comment: *These provisions have several implications. Specifically, we do not engage in any illegal behavior, including but not limited to: theft, fraud, corruption,* *embezzlement, or bribery. Further, we do not take or abuse the property of others, including intellectual property, nor do we engage in slander or libel. In focus groups conducted with practitioners around the globe, these types of illegal behaviors were mentioned as being problematic.*

As practitioners and representatives of our profession, we do not condone or assist others in engaging in illegal behavior. We report any illegal or unethical conduct. Reporting is not easy and we recognize that it may have negative consequences. Since recent corporate scandals, many organizations have adopted policies to protect employees who reveal the truth about illegal or unethical activities. Some governments have also adopted legislation to protect employees who come forward with the truth.

Ethics Complaints

2.3.3 We bring violations of this Code to the attention of the appropriate body for resolution.

2.3.4 We only file ethics complaints when they are substantiated by facts.

Comment: *These provisions have several implications. We cooperate with PMI concerning ethics violations and the collection of related information whether we are a complainant or a respondent. We also abstain from accusing others of ethical misconduct when we do not have all the facts. Further, we pursue disciplinary action against individuals who knowingly make false allegations against others.*

2.3.5 We pursue disciplinary action against an individual who retaliates against a person raising ethics concerns.

CHAPTER 3. RESPECT

3.1 Description of Respect

Respect is our duty to show a high regard for ourselves, others, and the resources entrusted to us. Resources entrusted to us may include people, money, reputation, the safety of others, and natural or environmental resources.

An environment of respect engenders trust, confidence, and performance excellence by fostering mutual cooperation — an environment where diverse perspectives and views are encouraged and valued.

3.2 Respect: Aspirational Standards

As <u>practitioners</u> in the global project management community:

3.2.1 We inform ourselves about the norms and customs of others and avoid engaging in behaviors they might consider disrespectful.

3.2.2 We listen to others' points of view, seeking to understand them.

3.2.3 We approach directly those persons with whom we have a conflict or disagreement.

3.2.4 We conduct ourselves in a professional manner, even when it is not reciprocated.

Comment: *An implication of these provisions is that we avoid engaging in gossip and avoid making negative remarks to undermine another person's reputation. We also have a duty under this Code to confront others who engage in these types of behaviors.*

3.3 Respect: Mandatory Standards

As <u>practitioners</u> in the global project management community, we require the following of ourselves and our fellow practitioners:

3.3.1 We negotiate in good faith.

3.3.2 We do not exercise the power of our expertise or position to influence the decisions or actions of others in order to benefit personally at their expense.

3.3.3 We do not act in an <u>abusive manner</u> toward others.

3.3.4 We respect the property rights of others.

CHAPTER 4. FAIRNESS

4.1 Description of Fairness

Fairness is our duty to make decisions and act impartially and objectively. Our conduct must be free from competing self interest, prejudice, and favoritism.

4.2 Fairness: Aspirational Standards

As <u>practitioners</u> in the global project management community:

4.2.1 We demonstrate transparency in our decision-making process.

4.2.2 We constantly reexamine our impartiality and objectivity, taking corrective action as appropriate.

Comment: *Research with practitioners indicated that the subject of conflicts of interest is one of the most challenging faced by our profession. One of the biggest problems practitioners report is not recognizing when we have conflicted loyalties and recognizing when we are inadvertently placing ourselves or others in a conflict-of-interest situation. We as practitioners must proactively search for potential conflicts and help each other by highlighting each other's potential conflicts of interest and insisting that they be resolved.*

4.2.3 We provide equal access to information to those who are authorized to have that information.

4.2.4 We make opportunities equally available to qualified candidates.

Comment: *An implication of these provisions is, in the case of a contracting arrangement, we provide equal access to information during the bidding process.*

4.3 Fairness: Mandatory Standards

As practitioners in the global project management community, we require the following of ourselves and our fellow practitioners:

<u>Conflict of Interest</u> Situations

4.3.1 We proactively and fully disclose any real or potential conflicts of interest to the appropriate stakeholders.

4.3.2 When we realize that we have a real or potential conflict of interest, we refrain from engaging in the decision-making process or otherwise attempting to influence outcomes, unless or until: we have made full disclosure to the affected stakeholders; we have an approved mitigation plan; and we have obtained the consent of the stakeholders to proceed.

Comment: *A conflict of interest occurs when we are in a position to influence decisions or other outcomes on behalf of one party when such decisions or outcomes could affect one or more other parties with which we have competing loyalties. For example, when we are acting as an employee, we have a duty of loyalty to our employer. When we are acting as a PMI volunteer, we have a duty of loyalty to the Project Management Institute. We must recognize these divergent interests and refrain from influencing decisions when we have a conflict of interest.*

Further, even if we believe that we can set aside our divided loyalties and make decisions impartially, we treat the appearance of a conflict of interest as a conflict of interest and follow the provisions described in the Code.

Favoritism and Discrimination

4.3.3 We do not hire or fire, reward or punish, or award or deny contracts based on personal considerations, including but not limited to, favoritism, nepotism, or bribery.

4.3.4 We do not discriminate against others based on, but not limited to, gender, race, age, religion, disability, nationality, or sexual orientation.

4.3.5 We apply the rules of the organization (employer, Project Management Institute, or other group) without favoritism or prejudice.

CHAPTER 5. HONESTY

5.1 Description of Honesty

Honesty is our duty to understand the truth and act in a truthful manner both in our communications and in our conduct.

5.2 Honesty: Aspirational Standards

As practitioners in the global project management community:

5.2.1 We earnestly seek to understand the truth.

5.2.2 We are truthful in our communications and in our conduct.

5.2.3 We provide accurate information in a timely manner.

Comment: *An implication of these provisions is that we take appropriate steps to ensure that the information we are basing our decisions upon or providing to others is accurate, reliable, and timely.*

This includes having the courage to share bad news even when it may be poorly received. Also, when outcomes are negative, we avoid burying information or shifting blame to others. When outcomes are positive, we avoid taking credit for the achievements of others. These provisions reinforce our commitment to be both honest and responsible.

5.2.4 We make commitments and promises, implied or explicit, in good faith.

5.2.5 We strive to create an environment in which others feel safe to tell the truth.

5.3 Honesty: Mandatory Standards

As practitioners in the global project management community, we require the following of ourselves and our fellow practitioners:

5.3.1 We do not engage in or condone behavior that is designed to deceive others, including but not limited to, making misleading or false statements, stating half-truths, providing information out of context or withholding information that, if known, would render our statements as misleading or incomplete.

5.3.2 We do not engage in dishonest behavior with the intention of personal gain or at the expense of another.

Comment: *The aspirational standards exhort us to be truthful. Half-truths and non-disclosures intended to mislead stakeholders are as unprofessional as affirmatively making misrepresentations. We develop credibility by providing complete and accurate information.*

APPENDIX A

A.1 History of this Standard

PMI's vision of project management as an independent profession drove our early work in ethics. In 1981, the PMI Board of Directors formed an Ethics, Standards and Accreditation Group. One task required the group to deliberate on the need for a code of ethics for the profession. The team's report contained the first documented PMI discussion of ethics for the project management profession. This report was submitted to the PMI Board of Directors in August 1982 and published as a supplement to the August 1983 *Project Management Quarterly.*

In the late 1980's, this standard evolved to become the Ethics Standard for the Project Management Professional [PMP®]. In 1997, the PMI Board determined the need for a member code of ethics. The PMI Board formed the Ethics Policy Documentation Committee to draft and publish an ethics standard for PMI's membership. The Board approved the new Member Code of Ethics in October 1998. This was followed by Board approval of the Member Case Procedures in January 1999, which provided a process for the submission of an ethics complaint and a determination as to whether a violation had occurred.

Since the 1998 Code was adopted, many dramatic changes have occurred within PMI and the business world. PMI membership has grown significantly. A great deal of growth has also occurred in regions outside North America. In the business world, ethics scandals have caused the downfall of global corporations and non-profits, causing public outrage and sparking increased government regulations. Globalization has brought economies closer together but has caused a realization that our practice of ethics may differ from culture to culture. The rapid, continuing pace of technological change has provided new opportunities, but has also introduced new challenges, including new ethical dilemmas.

For these reasons, in 2003 the PMI Board of Directors called for the reexamination of our codes of ethics. In 2004, the PMI Board commissioned the Ethics Standards Review Committee [ESRC] to review the codes of ethics and develop a process for revising the codes. The ESRC developed processes that would encourage active participation by the global project management community. In 2005, the PMI Board approved the processes for revising the code, agreeing that global participation by the project management community was paramount. In 2005, the Board also commissioned the Ethics Standards Development Committee to carry out the Board-approved process and deliver the revised code by the end of 2006. This Code of Ethics and Professional Development was approved by the PMI Board of Directors in October 2006.

A.2 Process Used to Create this Standard

The first step by the Ethics Standards Development Committee [ESDC] in the development of this Code was to understand the ethical issues facing the project management community and to understand the values and viewpoints of practitioners from all regions of the globe. This was accomplished by a variety of mechanisms including focus group discussions and two internet surveys involving practitioners, members, volunteers, and people holding a PMI certification. Additionally, the team analyzed the ethics codes of 24 non-profit associations from various regions of the world, researched best practices in the development of ethics standards, and explored the ethics-related tenets of PMI's strategic plan.

This extensive research conducted by the ESDC provided the backdrop for developing the exposure draft of the PMI Code of Ethics and Professional Conduct. The exposure draft was circulated to the global project management community for comment. The rigorous, standards development processes established by the American National Standards Institute were followed during the development of the Code because these processes were used for PMI technical standard development projects and were deemed to represent the best practices for obtaining and adjudicating stakeholder feedback to the exposure draft.

The result of this effort is a Code of Ethics and Professional Conduct that not only describes the ethical values to which the global project management community aspires, but also addresses the specific conduct that is mandatory for every individual bound by this Code. Violations of the PMI Code of Ethics and Professional Conduct may result in sanctions by PMI under the ethics Case Procedures.

The ESDC learned that as practitioners of project management, our community takes its commitment to ethics very seriously and we hold ourselves and our peers in the global project management community accountable to conduct ourselves in accordance with the provisions of this Code.

APPENDIX B

B.1 Glossary

Abusive Manner. Conduct that results in physical harm or creates intense feelings of fear, humiliation, manipulation, or exploitation in another person.

Conflict of Interest. A situation that arises when a practitioner of project management is faced with making a decision or doing some act that will benefit the practitioner or another person or organization to which the practitioner owes a <u>duty of loyalty</u> and at the same time will harm another person or organization to which the practitioner owes a similar <u>duty of loyalty</u>. The only way practitioners can resolve conflicting duties is to disclose the conflict to those affected and allow them to make the decision about how the practitioner should proceed.

Duty of Loyalty. A person's responsibility, legal or moral, to promote the best interest of an organization or other person with whom they are affiliated.

Project Management Institute [PMI]. The totality of the Project Management Institute, including its committees, groups, and chartered components such as chapters, colleges, and specific interest groups.

PMI Member. A person who has joined the Project Management Institute as a member.

PMI-Sponsored Activities. Activities that include, but are not limited to, participation on a PMI Member Advisory Group, PMI standard development team, or another PMI working group or committee. This also includes activities engaged in under the auspices of a chartered PMI component organization—whether it is in a leadership role in the component or another type of component educational activity or event.

Practitioner. A person engaged in an activity that contributes to the management of a project, portfolio, or program, as part of the project management profession.

PMI Volunteer. A person who participates in PMI-sponsored activities, whether a member of the Project Management Institute or not.

417

SYSADM, AUDIT, NETWORK, SECURITY (SANS) IT CODE OF ETHICS

I will strive to know myself and be honest about my capability.

- I will strive for technical excellence in the IT profession by maintaining and enhancing my own knowledge and skills. I acknowledge that there are many free resources available on the Internet and affordable books and that the lack of my employer's training budget is not an excuse nor limits my ability to stay current in IT.
- When possible I will demonstrate my performance capability with my skills via projects, leadership, and/or accredited educational programs and will encourage others to do so as well.
- I will not hesitate to seek assistance or guidance when faced with a task beyond my abilities or experience. I will embrace other professionals' advice and learn from their experiences and mistakes. I will treat this as an opportunity to learn new techniques and approaches. When the situation arises that my assistance is called upon, I will respond willingly to share my knowledge with others.
- I will strive to convey any knowledge (specialist or otherwise) that I have gained to others so everyone gains the benefit of each other's knowledge.
- I will teach the willing and empower others with Industry Best Practices (IBP). I will offer my knowledge to show others how to become security professionals in their own right. I will strive to be perceived as and be an honest and trustworthy employee.
- I will not advance private interests at the expense of end users, colleagues, or my employer.
- I will not abuse my power. I will use my technical knowledge, user rights, and permissions only to fulfill my responsibilities to my employer.
- I will avoid and be alert to any circumstances or actions that might lead to conflicts of interest or the perception of conflicts of interest. If such circumstance occurs, I will notify my employer or business partners.

- I will not steal property, time or resources.
- I will reject bribery or kickbacks and will report such illegal activity.
- I will report on the illegal activities of myself and others without respect to the punishments involved. I will not tolerate those who lie, steal, or cheat as a means of success in IT.

I will conduct my business in a manner that assures the IT profession is considered one of integrity and professionalism.

- I will not injure others, their property, reputation, or employment by false or malicious action.
- I will not use availability and access to information for personal gains through corporate espionage.
- I distinguish between advocacy and engineering. I will not present analysis and opinion as fact.
- I will adhere to Industry Best Practices (IBP) for system design, rollout, hardening and testing.
- I am obligated to report all system vulnerabilities that might result in significant damage.
- I respect intellectual property and will be careful to give credit for other's work. I will never steal or misuse copyrighted, patented material, trade secrets or any other intangible asset.
- I will accurately document my setup procedures and any modifications I have done to equipment. This will ensure that others will be informed of procedures and changes I've made.

I respect privacy and confidentiality.

- I respect the privacy of my co-workers' information. I will not peruse or examine their information including data, files, records, or network traffic except as defined by the appointed roles, the organization's acceptable use policy, as approved by Human Resources, and without the permission of the end user.
- I will obtain permission before probing systems on a network for vulnerabilities.
- I respect the right to confidentiality with my employers, clients, and users except as dictated by applicable law. I respect human dignity.
- I treasure and will defend equality, justice and respect for others.
- I will not participate in any form of discrimination, whether due to race, color, national origin, ancestry, sex, sexual orientation, gender/sexual identity or expression, marital status, creed, religion, age, disability, veteran's status, or political ideology.

APPENDIX **G**

ANSWERS TO SELF-ASSESSMENT QUESTIONS

Chapter 1 answers: 1. moral code; 2. virtue; 3. goodwill; 4. consistency; 5. respondeat superior or "let the master answer"; 6. reputation; 7. vision and leadership; 8. board of directors; 9. Section 406 of the Sarbanes-Oxley Act; 10. renew investor's trust in the content and preparation of disclosure documents by public companies; 11. code of ethics; 12. social audit; 13. formal ethics training; 14. problem definition; 15. utilitarian; 16. brainstorming

Chapter 2 answers: 1. d.; 2. IT staff; 3. stop the unauthorized copying of software produced by its members; 4. False; 5. Fraud; 6. False; 7. d.; 8. Bribery; 9. b.; 10. True; 11. Negligence; 12. code of ethics

Chapter 3 answers: 1. d.; 2. True; 3. exploit; 4. botnet; 5. c.; 6. spam; 7. CAPTCHA; 8. False; 9. False; 10. malicious insider; 11. True; 12. trustworthy computing; 13. risk assessment; 14. security policy; 15. False; 16. intrusion prevention system

Chapter 4 answers: 1. Bill of Rights (particularly the Fourth Amendment); 2. information privacy; 3. True; 4. b.; 5. opt-out; 6. d.; 7. Communications Act of 1934; 8. Katz; 9. d.; 10. a.; 11. USA PATRIOT Act; 12. National Security Letter; 13. True; 14. d.; 15. Personalization software; 16. True

Chapter 5 answers: 1. c.; 2. True; 3. Miller vs. California; 4. c.; 5. Reno vs. ACLU; 6. True; 7. a.; 8. d.; 9. False; 10. John Doe lawsuit; 11. anonymous Internet speakers; 12. True; 13. a.; 14. CAN-SPAM

Chapter 6 answers: 1. d.; 2. True; 3. fair use doctrine; 4. Digital Millennium Copyright Act; 5. patent; 6. trade mark; 7. patent infringements; 8. c.; 9. True; 10. Trade-Related Aspects of Intellectual Property Rights (TRIPS) Agreement; 11. False; 12. reverse engineering; 13. b.; 14. True; 15. True

Chapter 7 answers: 1. True; 2. Software defect; 3. d.; 4. process control system; 5. False; 6. software development methodology; 7. True; 8. software quality assurance; 9. d.; 10. c.; 11. True; 12. b.; 13. Failure Mode and Effects Analysis (FEMA); 14. ISO 9000 standards; 15. negligence

Chapter 8 answers: 1. b.; 2. productivity; 3. True; 4. five to seven; 5. digital divide; 6. True; 7. c.; 8. d.; 9. Ed-Tech; 10. a low income; 11. $75; 12. d.; 13. shielding; 14. True; 15. 98,000; 16. 2; 17. False

Chapter 9 answers: 1. b.; 2. False; 3. LinkedIn; 4. True; 5. a.; 6. viral marketing; 7. True; 8. b.; 9. MySpace; 10. False; 11. physical assault; 12. d.; 13. False; 14. sharing with retailers data about their members' likes and dislikes

Chapter 10 answers: 1. True; 2. coemployment; 3. d.; 4. True; 5. H-1B; 6. outsourcing; 7. True; 8. c.; 9. True; 10. a.; 11. d.; 12. True; 13. False; 14. gold

affiliated Web sites   A group of Web sites served by a single advertising network.

Agreement on Trade-Related Aspects of Intellectual Property Rights (TRIPS Agreement)   An agreement of the World Trade Organization that requires member governments to ensure that intellectual property rights can be enforced under their laws and that penalties for infringement are tough enough to deter further violations.

anonymous expression   The expression of opinions by people who do not reveal their identity.

anonymous remailer   A company that provides a service in which an originating IP number (physical address) is stripped from an e-mail message before the message is sent on to its destination.

antivirus software   Software that regularly scans a computer's memory and disk drives for viruses.

avatar   A virtual world visitor's representation of him- or herself—usually in the form of a human but sometimes in some other form.

black-box testing   A form of dynamic testing that involves viewing the software unit as a device that has expected input and output behaviors but whose internal workings are unknown (a black box). If the unit demonstrates the expected behaviors for all the input data in the test suite, it passes the test.

body of knowledge   An agreed-upon set of skills and abilities that all licensed professionals in a particular type of profession must possess.

botnet   A large group of computers controlled centrally from one or more remote locations by hackers, without the knowledge or consent of their owners.

breach of contract   The failure of one party to meet the terms of a contract.

breach of the duty of care   The failure to act as a reasonable person would act.

breach of warranty   The failure of a product to meet the terms of its warranty.

bribery   The act of providing money, property, or favors to someone in business or government to obtain a business advantage.

business information system   A set of interrelated components—including hardware, software, databases, networks, people, and procedures—that collects data, processes it, and disseminates the output.

Business Software Alliance (BSA)   Trade group that represents the world's largest software and hardware manufacturers; its mission is to stop the unauthorized copying of software produced by its members.

Capability Maturity Model Integration (CMMI)   A process improvement approach developed by the Software Engineering Institute at Carnegie Mellon that defines the essential elements of effective processes.

certification   A recognition that a professional possesses a particular set of skills, knowledge, or abilities—in the opinion of the certifying organization.

chief privacy officer (CPO)   A senior manager within an organization whose role is to both ensure that the organization does not violate government regulations and reassure customers that their privacy will be protected.

Child Online Protection Act (COPA) A 1998 law that was intended to protect children from online pornography while preserving the rights of adults; it was eventually ruled unconstitutional.

Children's Internet Protection Act (CIPA) A 2000 law that required federally financed schools and libraries to use some form of technological protection (such as an Internet filter) to block computer access to obscene material, pornography, and anything else considered harmful to minors.

Children's Online Privacy Protection Act (COPPA) A 1998 law that requires Web sites that cater to children to offer comprehensive privacy policies, notify parents or guardians about their data-collection practices, and receive parental consent before collecting any personal information from children under 13 years of age.

click-stream data Information gathered by monitoring a consumer's online activity through the use of cookies.

CMMI-Development (CMMI-DEV) An application of CMMI, frequently used to assess and improve software development practices.

code of ethics A statement that highlights an organization's key ethical issues and identifies the overarching values and principles that are important to the organization and its decision making.

coemployment relationship A employment situation in which two employers have actual or potential legal rights and duties with respect to the same employee or group of employees.

collusion Cooperation between two or more people, often an employee and a company outsider, to commit fraud.

common good approach An approach to ethical decision making based on a vision of society as a community whose members work together to achieve a common set of values and goals.

Communications Act of 1934 The law that established the Federal Communications Commission and gave it responsibility for regulating all non-federal-government use of radio and television broadcasting and all interstate telecommunications—including wire, satellite, and cable—as well as all international communications that originate or terminate in the United States.

Communications Assistance for Law Enforcement Act (CALEA) A 1994 law that amended both the Wiretap Act and ECPA; it required the telecommunications industry to build tools into its products that federal investigators could use—after obtaining a court order—to eavesdrop on conversations and intercept electronic communications.

Communications Decency Act (CDA) A part of the 1996 Telecommunications Act directed at protecting children from online pornography; it was eventually ruled unconstitutional.

competitive intelligence Legally obtained information gathered to help a company gain an advantage over its rivals.

compiler A language translator that converts computer program statements expressed in a source language (such as COBOL, Pascal, or C) into a machine language (a series of binary codes of 0s and 1s) that the computer can execute.

Completely Automated Public Turing Test to Tell Computers and Humans Apart (CAPTCHA) Software that generates and grades tests that humans can pass but all but the most sophisticated computer programs cannot.

contingent work A job situation in which an individual does not have an explicit or implicit contract for long-term employment.

contributory negligence A defense in a negligence case in which the defendant argues that the plaintiffs' own actions contributed to their injuries.

Controlling the Assault of Non-Solicited Pornography and Marketing (CAN-SPAM) Act A 2004 law that specifies requirements that commercial e-mailers must follow when sending out messages that advertise or promote a commercial product or service.

cookie A text file that a Web site downloads to visitors' hard drives so that it can identify them on subsequent visits.

copyright The exclusive right to distribute, display, perform, or reproduce an original work in copies or to prepare derivative works based on the work; granted to creators of original works of authorship.

copyright infringement A violation of the rights secured by the owner of a copyright; occurs when someone copies a substantial and material part of another's copyrighted work without permission.

corporate compliance officer *See also* corporate ethics officer.

corporate ethics officer A senior-level manager who provides an organization with vision and leadership in the area of business conduct.

cracker Someone who breaks into other people's networks and systems to cause harm.

cyberbullying The harassment, torment, humiliation, or threatening of one minor by another minor or group of minors via the Internet or cell phone.

cybercriminal An individual, motivated by the potential for monetary gain, who hacks into corporate computers to steal, often by transferring money from one account to another to another.

cybersquatter A person or company that registers domain names for famous trademarks or company names to which they have no connection, with the hope that the trademark's owner will buy the domain name for a large sum of money.

cyberstalking Threatening behavior or unwanted advances directed at an adult using the Internet or other forms of online and electronic communications; it is the adult version of cyberbullying.

cyberterrorist An individual who launches computer-based attacks against other computers or networks in an attempt to intimidate or coerce a government in order to advance certain political or social objectives.

decision support system (DSS) A type of business information system used to improve decision making in a variety of industries.

decompiler Software that can read the machine language version of software and produce the source code.

defamation Making either an oral or a written statement of alleged fact that is false and harms another person.

deliverables The products of a software development process, such as statements of requirements, flowcharts, and user documentation.

digital divide The gulf between those who do and those who do not have access to modern information and communications technology such as cell phones, personal computers, and the Internet.

distributed denial-of-service attack (DDoS) An attack in which a malicious hacker takes over computers on the Internet and causes them to flood a target site with demands for data and other small tasks.

duty of care   The obligation to protect people against any unreasonable harm or risk.

dynamic testing   An approach to software QA in which the code for a completed unit of software is tested by entering test data and comparing the actual results to the expected results.

Education Rate (E-Rate) program   A program of the Telecommunications Act of 1996 whose primary goal is to help schools and libraries obtain access to state-of-the-art services and technologies at discounted rates.

Electronic Communications Privacy Act of 1986 (ECPA)   A law focusing on three main issues: (1) the protection of communications while in transfer from sender to receiver; (2) the protection of communications held in electronic storage; and (3) the prohibition of devices to record dialing, routing, addressing, and signaling information without a search warrant.

electronic health record (EHR)   A summary of health information generated by each patient encounter in any healthcare delivery setting. It can include patient demographics, medical history, immunization records, laboratory data, problems, progress notes, medications, vital signs, and radiology reports.

Electronic Industry Citizenship Coalition (EICC)   An industry organization established to promote a common code of conduct for the electronics and information and communications technology (ICT) industry.

e-mail spam   The abuse of e-mail systems to send unsolicited e-mail to large numbers of people.

employee leasing   A business arrangement in which an organization (called the subscribing firm) transfers all or part of its workforce to another firm (called the leasing firm), which handles all human resource-related activities and costs, such as payroll, training, and the administration of employee benefits. The subscribing firm leases these workers back but as employees of the leasing firm.

Enhancing Education Through Technology (Ed-Tech) program   A federal program with the following goals: (1) improve student academic achievement through the use of technology in schools; (2) assist children in crossing the digital divide by ensuring that every student is technologically literate by the end of eighth grade; and (3) encourage the effective integration of technology with teacher training and curriculum.

ethics   A set of beliefs about right and wrong behavior within a society.

European Union Data Protection Directive   A directive passed by the European Union in 1998 that requires any company doing business within the borders of 15 western European nations to implement a set of privacy directives on the fair and appropriate use of information; it also bars the export of data to countries that do not have comparable data privacy protection standards.

exploit   An attack on an information system that takes advantage of a particular system vulnerability.

failure mode and effects analysis (FMEA)   A technique used to develop ISO 9000-compliant quality systems by both evaluating reliability and determining the effects of system and equipment failures.

Fair Credit Reporting Act   A law passed in 1970 that regulates the operations of credit-reporting bureaus, including how they collect, store, and use credit information.

Fair Information Practices A set of eight principles created by the Organisation for Economic Co-operation and Development that provides guidelines for the ethical treatment of consumer data.

fair use doctrine A legal doctrine that allows portions of copyrighted materials to be used without permission under certain circumstances. Title 17, section 107, of the U.S. Code established four factors that courts should consider when deciding whether a particular use of copyrighted property is fair and can be allowed without penalty: (1) the purpose and character of the use (such as commercial use or non-profit, educational purposes); (2) the nature of the copyrighted work; (3) the portion of the copyrighted work used in relation to the work as a whole; and (4) the effect of the use on the value of the copyrighted work.

fairness approach An approach to ethical decision making that focuses on how fairly actions and policies distribute benefits and burdens among people affected by the decision.

False Claims Act A law enacted during the U.S. Civil War to combat fraud by companies that sold supplies to the Union Army; also known as the Lincoln Law.

firewall A hardware or software device that serves as a barrier between an organization's network and the Internet; a firewall also limits access to the company's network based on the organization's Internet usage policy.

foreign intelligence Information relating to the capabilities, intentions, or activities of foreign governments, agents of foreign governments, or foreign organizations.

Foreign Intelligence Surveillance Act (FISA) An act passed in 1978 that describes procedures for the electronic surveillance and collection of foreign intelligence information in communications between foreign powers and agents of foreign powers.

Foreign Intelligence Surveillance Amendments Act An act passed in 2008 that both revised many of the FISA procedures for gathering foreign intelligence and implemented legal protections for electronic communications service providers who previously provided consumer data to the NSA and the CIA.

fraud The crime of obtaining goods, services, or property through deception or trickery.

Freedom of Information Act (FOIA) A law passed in 1966 and amended in 1974 that grants citizens the right to access certain information and records of the federal government upon request.

government license A government-issued permission to engage in an activity or to operate a business; it is generally administered at the state level and often requires that the recipient pass a test of some kind.

Gramm-Leach-Bliley Act (GLBA) A 1999 bank deregulation law, also known as the Financial Services Modernization Act, which granted banks the right to offer investment, commercial banking, and insurance services through a single entity.

green computing Efforts directed toward the efficient design, manufacture, operation, and disposal of IT-related products, including personal computers, laptops, servers, printers, and printer supplies.

H-1B A temporary work visa granted by the U.S. Citizenship and Immigration Services (USCIS) for people who work in specialty occupations—jobs that require a four-year bachelor's degree in a specific field, or equivalent experience.

hacker Someone who tests the limitations of information systems out of intellectual curiosity—to see if he or she can gain access.

hacktivism   Hacking to achieve a political or social goal.

Health Insurance Portability and Accountability Act of 1996 (HIPAA)   A law designed to improve the portability and continuity of health insurance coverage; to reduce fraud, waste, and abuse in health insurance and healthcare delivery; and to simplify the administration of health insurance.

identity theft   The act of stealing key pieces of personal information to impersonate a person.

industrial espionage   The use of illegal means to obtain business information not available to the general public.

industrial spy   Someone who uses illegal means to obtain trade secrets from competitors of their firm.

information privacy   The combination of communications privacy (the ability to communicate with others without those communications being monitored by other persons or organizations) and data privacy (the ability to limit access to one's personal data by other individuals and organizations in order to exercise a substantial degree of control over that data and its use).

integration testing   A form of software testing in which individual software units are combined into an integrated subsystem that undergoes rigorous testing to ensure that the linkages among the various subsystems work successfully.

integrity   Adherence to a personal code of principles.

intellectual property   Works of the mind—such as art, books, films, formulas, inventions, music, and processes—that are distinct, and owned or created by a single person or group. Intellectual property is protected through copyright, patent, trade secret, and trademark laws.

intentional misrepresentation   Fraud that occurs when a seller or lessor either misrepresents the quality of a product or conceals a defect in it.

Internet filter   Software that can be used to block access to certain Web sites that contain material deemed inappropriate or offensive.

intrusion detection system   Software and/or hardware that monitors system and network resources and activities, and notifies network security personnel when it identifies possible intrusions from outside the organization or misuse from within the organization.

intrusion prevention system (IPS)   A network security device that prevents an attack by blocking viruses, malformed packets, and other threats from getting into the protected network.

ISO 9000 standard   A standard that serves as a guide to quality products, services, and management.

IT user   A person for whom a hardware or software product is designed.

John Doe lawsuit   A lawsuit in which the identity of the defendant is temporarily unknown, typically because the defendant is communicating anonymously or using a pseudonym.

lamer   A technically inept hacker. *See also* script kiddie.

law   A system of rules that tells us what we can and cannot do. Laws are enforced by a set of institutions.

libel   A written defamatory statement.

live telemedicine   A form of telemedicine in which patients and healthcare providers are present at the same time; often involves a videoconference link between the two sites.

logic bomb A type of Trojan horse that executes when it is triggered by a specific event.

material breach of contract The failure of one party to perform certain express or implied obligations, which impairs or destroys the essence of the contract.

Miller v. California The 1973 Supreme Court case that established a test to determine if material is obscene and therefore not protected by the First Amendment.

misrepresentation The misstatement or incomplete statement of a material fact.

moral code A set of rules that establishes the boundaries of generally accepted behavior within a society.

morality Social conventions about right and wrong that are widely shared throughout a society.

morals One's personal beliefs about right and wrong.

negligence The failure to do what a reasonable person would do, or doing something that a reasonable person would not do.

noncompete agreement Terms of an employment contract that prohibit an employee from working for any competitors for a period of time, often one to two years.

nondisclosure clause Terms of an employment contract that prohibit an employee from revealing secrets.

N-version programming A form of redundancy in which two computer systems execute a series of program instructions simultaneously.

offshore outsourcing A form of outsourcing in which the services are provided by an organization whose employees are in a foreign country.

One Laptop per Child (OLPC) A nonprofit organization whose goal is to provide children around the world with low-cost (less than $100) laptop computers to aid in their education.

online virtual world A computer-simulated world in which a visitor can move in three-dimensional space, communicate and interact with other visitors, and manipulate elements of the simulated world.

open source code Any program whose source code is made available for use or modification, as users or other developers see fit.

opt in To agree (either implicitly or by default) to allow an organization to collect and share one's personal data with other institutions.

opt out To refuse to give an organization the right to collect and share one's personal data with unaffiliated parties.

outsourcing A long-term business arrangement in which a company contracts for services with an outside organization that has expertise in providing a specific function.

patent A grant of a property right issued by the U.S. Patent and Trademark Office to an inventor; permits its owner to exclude the public from making, using, or selling a protected invention, and allows for legal action against violators.

patent farming An unethical strategy of influencing a standards organization to make use of a patented item without revealing the existence of a patent; later, the patent holder might demand royalties from all implementers of the standard.

patent infringement A violation of the rights secured by the owner of a patent; occurs when someone makes unauthorized use of another's patent.

patent troll A firm that acquires patents for the purpose of licensing the patents to others rather than manufacturing anything itself.

pen register A device that records electronic impulses to identify the numbers dialed for outgoing calls.

personalization software Software used by online marketers to optimize the number, frequency, and mixture of their ad placements as well as to evaluate how visitors react to new ads.

phishing The act of using e-mail fraudulently to try to get the recipient to reveal personal data.

plagiarism The act of stealing someone's ideas or words and passing them off as one's own.

Platform for Privacy Preferences (P3P) Screening software that shields users from sites that do not provide the level of privacy protection they desire.

pretexting The use of false pretenses to gain information.

prior art The existing body of knowledge that is available to a person of ordinary skill in the art.

Privacy Act of 1974 A law decreeing that no agency of the U.S. government can conceal the existence of any personal data record-keeping system; under this law, any agency that maintains such a system must publicly describe both the kinds of information in it and the manner in which the information will be used.

problem statement A clear, concise description of the issue that needs to be addressed in a decision-making process.

product liability The liability of manufacturers, sellers, lessors, and others for injuries caused by defective products.

productivity The amount of output produced per unit of input.

profession A calling that requires specialized knowledge and often long and intensive academic preparation.

professional code of ethics A statement of the principles and core values that are essential to the work of a particular occupational group.

professional malpractice Breach of the duty of care by a professional.

project safety engineer An individual on a safety-critical system project who has explicit responsibility for the system's safety.

quality assurance (QA) Methods within the software development cycle designed to guarantee reliable operation of the product.

quality management Business practices that focus on defining, measuring, and refining the quality of the development process and the products developed during its various stages.

qui tam A provision of the False Claims Act that allows a private citizen to file a suit in the name of the U.S. government, charging fraud by government contractors and other entities who receive or use government funds.

Radio Frequency Identification (RFID) chip A microchip that listens for a radio query and responds by transmitting its own unique ID code.

reasonable assurance A concept in computer security that recognizes that managers must use their judgment to ensure that the cost of control does not exceed the system's benefits or the risks involved.

reasonable person standard A legal standard that defines how an objective, careful, and conscientious person would have acted in the same circumstances.

reasonable professional standard A legal standard that defendants who have particular expertise or competence are measured against.

redundancy The use of multiple interchangeable components designed to perform a single function—in order to cope with failures and errors.

reliability The probability of a component or system performing without failure over its product life.

résumé inflation Falsely claiming competence in a skill, usually because that skill is in high demand.

reverse engineering The process of taking something apart in order to understand it, build a copy of it, or improve it.

risk The probability of an undesirable event occurring times the magnitude of the event's consequences if it does happen.

risk assessment The process of assessing security-related risks from both internal and external threats to an organization's computers and networks.

rootkit A set of programs that enables its user to gain administrator level access to a computer without the end user's consent or knowledge.

safety-critical system A system whose failure may cause injury or death.

script kiddie A technically inept hacker. *See also* lamer.

security audit A process that evaluates whether an organization has a well-considered security policy in place and if it is being followed.

security policy A written statement that defines an organization's security requirements, as well as the controls and sanctions needed to meet those requirements.

sexting Sending sexual messages, nude or seminude photos, or sexually explicit videos over a cell phone.

slander An oral defamatory statement.

smart card A form of debit or credit card that contains a memory chip that is updated with encrypted data every time the card is used.

social audit A process whereby an organization reviews how well it is meeting its ethical and social responsibility goals, and communicates its new goals for the upcoming year.

social network advertising Advertising using social networks to inform, promote, and communicate the benefits of products and services.

social networking Web site A Web site that creates an online community of Internet users that enables members to break down time, distance, and cultural barriers and interact with others by sharing opinions, insights, information, interests, and experiences.

social shopping Web site A Web site that brings shoppers and sellers together in a social networking environment in which members can share information and make recommendations while shopping online.

software defect Any error that, if not removed, could cause a software system to fail to meet its users' needs.

software development methodology A standard, proven work process that enables systems analysts, programmers, project managers, and others to make controlled and orderly progress in developing high-quality software.

software piracy The act of illegally making copies of software or enabling others to access software to which they are not entitled.

software quality The degree to which a software product meets the needs of its users.

spear-phishing A variation of phishing in which the phisher sends fraudulent e-mails to a certain organization's employees. The phony e-mails are designed to look like they came from high-level executives within the organization.

spyware Keystroke-logging software downloaded to users' computers without the knowledge or consent of the user.

stakeholder Someone who stands to gain or lose depending on how a situation is resolved.

standard A definition that has been approved by a recognized standards organization or accepted as a de facto standard by a particular industry.

static testing The use of special software programs called static analyzers to look for suspicious patterns in programs that might indicate a defect.

store-and-forward telemedicine Acquiring data, sound, images, and video from a patient and then transmitting everything to a medical specialist for later evaluation.

strict liability A type of product liability in which a defendant is held responsible for injuring another person, regardless of negligence or intent.

submarine patent A patented process or invention that is hidden within a standard and which is not made public until after the standard is broadly adopted.

sunset provision A provision that terminates or repeals a law or portions of it after a specific date, unless further legislative action is taken to extend the law.

system testing A form of software testing in which various subsystems are combined to test the entire system as a complete entity.

telemedicine The use of modern telecommunications and information technologies to provide medical care to people who live far away from healthcare providers.

telework A work arrangement in which an employee works away from the office—at home, at a client's office, in a hotel—literally anywhere; also known as telecommuting.

Title III of the Omnibus Crime Control and Safe Streets Act A component of a 1968 law (amended in 1986) that regulates the interception of wire and oral communications; also known as the Wiretap Act.

trade secret Information, generally unknown to the public, that a company has taken strong measures to keep confidential. It represents something of economic value that has required effort or cost to develop and that has some degree of uniqueness or novelty.

trademark A logo, package design, phrase, sound, or word that enables a consumer to differentiate one company's products from another's.

trap and trace A device that records electronic impulses to identify the originating number for incoming calls.

Trojan horse A program in which malicious code is hidden inside a seemingly harmless program.

trustworthy computing A method of computing that delivers secure, private, and reliable computing experiences based on sound business practices.

U.S. Foreign Corrupt Practices Act (FCPA) A federal law that makes it a crime to bribe a foreign official, a foreign political party official, or a candidate for foreign political office.

USA PATRIOT Act A law passed in 2001 that gave sweeping new powers both to domestic law enforcement and to intelligence agencies, including increasing the ability of law enforcement agencies to search telephone, e-mail, medical, financial, and other records, and easing restrictions on foreign intelligence gathering in the United States.

user acceptance testing Independent testing performed by trained end users to ensure that a system operates as expected.

utilitarian approach An approach to ethical decision making that states that you should choose the action or policy that has the best overall consequences for all people who are directly or indirectly affected.

vice   A moral habit that inclines people to do what is generally unacceptable to society.

viral marketing   An approach to advertising that encourages individuals to pass along a marketing message to others, thus creating the potential for exponential growth in the message's exposure and influence.

virtual private network (VPN)   A technology that uses the Internet to relay communications, maintaining privacy through security procedures and tunneling protocols, which encrypt data at the sending end and decrypt it at the receiving end.

virtue   A moral habit that inclines people to do what is generally acceptable to society.

virtue ethics approach   An approach to ethical decision making that focuses on how you should behave and think about relationships if you are concerned with your daily life in a community.

virus   A piece of programming code, usually disguised as something else, that causes a computer to behave in an unexpected and usually undesirable manner.

virus signature   A specific sequence of bytes that indicates to antivirus software that a specific virus is present.

warranty   An assurance to buyers or lessees that a product meets certain standards of quality.

whistle-blowing   An effort to attract public attention to a negligent, illegal, unethical, abusive, or dangerous act by a company or some other organization.

white-box testing   A form of dynamic testing that treats the software unit as a device that has expected input and output behaviors and whose internal workings are known. White-box testing involves testing all possible logic paths through the software unit with thorough knowledge of its logic.

worm   A harmful program that resides in the active memory of the computer and duplicates itself.

zero-day attack   An attack that takes place before the security community or software developer knows about the vulnerability or has been able to repair it.

INDEX

D

E

H

I

M

M. A. Mortenson Company, 243
MAA (Machine Accountants Association), 48
Mac OS X Leopard, 205
Machine Accountants Association (MAA), 48
Mack, Ronald, 199
Mackey, John, 188–189
MAF (Master Address File) database, 362
Magedson, Ed, 189–190
Maiffret, Marc, 108
malicious insiders, 83–84, 95
malpractice, 53–54
Mangan, Joseph, 364
MapleStory, 316
MapQuest, 362
Mariner I space probe, 248
Marshalls, 352
Massachusetts Institute of Technology, 316
Master Address File (MAF) database, 362
MasterCard, 134, 352
Matasar, Richard, 34, 35
material breach of contract, 43
Maxtor, 28
McGraw-Hill, 228–229
Media Grid, 316
MeF system, 115
Meier, Megan, 310, 321–323
Meitai Plastics and Electronics, 332
Melissa worm, 76
Metro (Washington, D.C.), 248
Micron Technology, 206–207
Microsoft
    black hats hired by, 108
    certification offered by, 49
    contingent workers, 338
    cross-licensing agreements, 205
    Cygnus's patent infringement case against, 205
    demographic filtering, 138
    EICC Code of Conduct, 354
    in FOSI, 172
    free e-mail service offered by, 79
    green computing, 354
    hacker conferences hosted, 108
    H1-B employees, 340
    Internet Explorer, 205, 206
    outsourcing, 343
    partnering by, 62
    patent licensing, 206
    socially responsible activities, 7
    statement of values, 7
    suit against OnlineNIC, 220
    suppliers, 332
    trustworthy computing initiative, 87–88
    University of California's suit against, 206
    Vista operating system, 205, 239–240
    in W3C, 139

Mill, John Stuart, 381
Miller, Marvin, 169
Miller v. California, 169
Mir 3, 316
misrepresentation, 42
    intentional, 243–244
MMC Technology, 28
mobile phones, 279–280
mobile technology in healthcare industry, 284
modularization and IT professional services, 38
monitoring services for identity theft prevention, 136
Monsanto, 215–216
Moore, Demi, 170
moral code, 3
moral relativism, 376
morality, 3, 373–387
    deontology, 378–380
    egoism vs. altruism, 376–378
    feminism, 385–386
    Greek philosophers, 374–375, 383–385
    pluralism, 386–387
    question of goodness, 374–375
    relativism, 375–378
    utilitarianism, 380–383
morals, 4–5
Moss, Jeff, 108
Motorola
    chips used by Boeing, 364
    green computing, 353
    outsourcing, 343
Mousavi, Mir Houssein, 302
MY GRATE, 62
MySpace, 304, 305, 312, 313, 321–323
    in FOSI, 172

N

NASA, security report card, 96
NASDAQ, 12
National Association for the Advancement of Colored People (NAACP) v. Alabama, 175
National Association of Criminal Defense Lawyers, 9
National Association of Manufacturers, 9
National Center for Victims of Crime, 312
National Council of Examiners for Engineering and Surveying (NCEES), 52–53
National Highway Traffic Safety Administration (NHTSA), 259
National Infrastructure Protection Center, 85–86
National Institute of Standards and Technology (NIST), 92
National Labor Committee, 332
National Oncology Institute (Panama City, Panama), 248
National School Lunch Program, 277
National Science Foundation, 105
National Security Agency (NSA), 123, 347

Q

V

W

X

451